CW01572443

HANDBOOK ON THE GEOGRAPHIES OF CORRUPTION

RESEARCH HANDBOOKS IN GEOGRAPHY

Series Editor: Susan J. Smith, *Honorary Professor of Social and Economic Geography* and *The Mistress of Girton College, University of Cambridge, UK*

This important new *Handbook* series will offer high quality, original reference works that cover a range of subjects within the evolving and dynamic field of geography, emphasising in particular the critical edge and transformative role of human geography.

Under the general editorship of Susan J. Smith, these *Handbooks* will be edited by leading scholars in their respective fields. Comprising specially commissioned contributions from distinguished academics, the *Handbooks* offer a wide-ranging examination of current issues. Each contains a unique blend of innovative thinking, substantive analysis and balanced synthesis of contemporary research.

Titles in the series include:

Handbook on Geographies of Technology
Edited by Barney Warf

Handbook on the Geographies of Money and Finance
Edited by Ron Martin and Jane Pollard

Handbook on the Geographies of Regions and Territories
Edited by Anssi Paasi, John Harrison and Martin Jones

Handbook on the Geographies of Power
Edited by Mat Coleman and John Agnew

Handbook on the Geographies of Corruption
Edited by Barney Warf

Handbook on the Geographies of Corruption

Edited by

Barney Warf

Department of Geography, University of Kansas, USA

RESEARCH HANDBOOKS IN GEOGRAPHY

Cheltenham, UK • Northampton, MA, USA

© Barney Warf 2018

All rights reserved. No part of this publication may be reproduced, stored in a retrieval system or transmitted in any form or by any means, electronic, mechanical or photocopying, recording, or otherwise without the prior permission of the publisher.

Published by
Edward Elgar Publishing Limited
The Lypiatts
15 Lansdown Road
Cheltenham
Glos GL50 2JA
UK

Edward Elgar Publishing, Inc.
William Pratt House
9 Dewey Court
Northampton
Massachusetts 01060
USA

A catalogue record for this book
is available from the British Library

Library of Congress Control Number: 2018939986

This book is available electronically in the **Elgar**online
Social and Political Science subject collection
DOI 10.4337/9781786434753

ISBN 978 1 78643 474 6 (cased)
ISBN 978 1 78643 475 3 (eBook)

Typeset by Servis Filmsetting Ltd, Stockport, Cheshire
Printed and bound by CPI Group (UK) Ltd, Croydon CR0 4YY

Contents

Contributors

Agnes Batory, Central European University

Secil Bayraktar, Ozyegin University

Cleo Calimbahin, De La Salle University-Manila

Sufyan Dabbous, University of Pennsylvania

Filippo De Danieli, Padova, Italy

Eugen Dimant, University of Pennsylvania

Nasr G. Elbahnasawy, Kent State University and Mansoura University

Dominik H. Enste, Institut der deutschen Wirtschaft Köln

Mesut Eren, Marmara University

Asma Guizani, University of Sousse-Tunisia

Christina Heldman, Institut der deutschen Wirtschaft Köln

Alfredo Jiménez, Kedge Business School

Feisal Khan, Hobart and William Smith Colleges

Johannes Leitner, University of Applied Sciences Vienna

John M. Luiz, University of Sussex

Marcus Marktanner, Kennesaw State University

Hannes Meissner, University of Applied Sciences Vienna

Karl Z. Meyer, University of Cape Town

Marcus Mietzner, Australian National University

Stephen Morris, Middle Tennessee State University

Mohammad Nurunnabi, Prince Sultan University

Vanesa Pesqué-Cela, University of London

Günther G. Schulze, University of Freiburg

Kelly Senters, University of Illinois at Urbana-Champaign

Asma Sghaier, Université du Sud Toulon Var

Helena Olofsdotter Stensöta, University of Göteborg

Lena Wängnerud, University of Göteborg

Barney Warf, University of Kansas

Maureen Wilson, Kennesaw State University

Matthew S. Winters, University of Illinois at Urbana-Champaign

Nikita Zakharov, University of Freiburg

1. Introduction to the *Handbook on the Geographies of Corruption*
Barney Warf

Corruption consists of an insidious set of practices through which government officials use their offices for private gain. Corruption exists in virtually all countries across the planet and has numerous debilitating economic, political, social, and environmental costs. Manifested in various forms, including bribery, extortion, nepotism, and graft, corruption lowers economic growth, makes public institutions less efficient, hampers foreign investment, undermines trust in the state, and reduces the quality of life for billions of people. Most countries in the world are corrupt, and its presence makes life more inconvenient, less safe, and lowers standards of living.

Given its extent and severity, it comes as no surprise that there is a robust literature that examines the history, causes, effects, and implications of corruption. Excellent overviews may be found in Jain (2001, 2012) and Rose-Ackerman (2006). Among the social sciences, economists have offered extensive analyses of corruption's impacts (see Aidt 2003; Bardhan 2006), including the effects on growth (Mauro 1995; Mo 2001), income inequality (Gupta et al. 2002), and foreign investment (Wei 2000; Habib and Zurawicki 2002). Political scientists have written insightfully about corruption and bribery (Bose 2004), the role of an uncensored media (Brunetti and Weder 2003), and anti-corruption strategies (Klitgaard 1998; Brinkerhoff 2000), often using game theory. Sociologists have studied corruption's influence on government (Rose-Ackerman and Palifka 2016), networks and actors (Jancsics 2014), and its relations to democracy (Moreno 2002).

However, within this body of work, issues of geography and spatiality have been marginalized. Just as there is a history and sociology of corruption, so too is there a geography. Corruption occurs unevenly throughout the globe, and reflects diverse national economic, political, and cultural circumstances. For this reason, this Handbook attempts to address this void. While corruption is most obvious in less developed countries, particularly those with totalitarian governments and centralized systems of power, it also exists in wealthier ones, including the United States (Teachout 2014). The causes and consequences of corruption vary among states, as have strategies to contain it.

THE NATURE OF CORRUPTION

Corruption may be defined in different ways, occurs in varying levels of severity, and takes various forms in time and space, depending on local political cultures and institutional frameworks. Broadly, it refers to the misuse of public office for private gains, the abuse of power, and dishonest and fraudulent conduct, typically through bribery, graft, or extortion. In this sense, corruption should be differentiated from individual misbehavior or law-breaking. Its primary practitioners are politicians, legislators, bureaucrats, civil service employees, and military officials. Technically, corruption is by definition illegal, although in different cultural contexts its legality, and hence acceptability, range considerably. Not all illegal acts involve corruption because most crimes do not involve the misuse of public offices. At times, public policies that serve an elite few may be corrupt but not illegal, as in the case of post-retirement offers of employment (e.g., lobbyists) to high ranking civil servants. Corruption is thus inherently both political and economic in nature.

Corruption takes a variety of forms, all of which involve the transfer of assets from public to private hands. It ranges from individual acts such as accepting bribes to petty theft among small groups to grand larceny on an organized, institutional scale, in which elites engineer an entire government to serve their own purposes. It may involve the diversion of monies to private purposes, including foreign bank accounts, leading officials to enrich themselves at public expense. Perhaps the most common form is bribery, the improper use of gifts and money to obtain favors. In many countries, "fees" or "tea money" are euphemisms for facilitating transactions, moving contracts through a bureaucracy, or not reporting crimes. In many cases, bribery is seen simply as the "cost of doing business," that is, a normalized part of economic functioning. In Russia, a terminological distinction exists between *mzdoimstvo*, taking remuneration to do what a public official is supposed to do anyway, and *likhoimstvo*, taking remuneration for what the official is not supposed to do (Bardhan 1997).

Other forms of corruption involve graft, embezzlement, and theft, which may be accompanied by blackmail or extortion. In hiring, nepotism and political favoritism may occur. Other cases include the sale of military commissions and siphoning of monies from inflated payrolls, overinvoicing, and selling of licenses and permits, allocations of government contracts, and mining and land concessions. Customs inspectors are notorious agents of corruption in many cases. Additionally, corrupt judiciaries flout the rule of law, refusing to penalize corrupt officials and minimizing penalties. The purchase of legislative votes constitutes yet another form. All of these examples share in common the use of public office to promote private gains, the abuse of state power, and exercise of public bureaucratic or financial authority for purposes other than they were intended.

THE CAUSES AND CONSEQUENCES OF CORRUPTION

Corruption is a form of rent-seeking behavior that occurs when the benefits of the misuse of public power exceed the expected costs (i.e., the probability of being apprehended and the penalties that might follow). It requires that an official have discretionary power over the allocation of state resources; the greater the power, the stronger is the temptation to abuse it. Corruption thus involves avoiding or changing state regulations through the use of political authority; as Jain (2001, p. 78) argues, "therefore, we would expect to find larger discretionary powers, and hence more corruption, in regulated and controlled economies as opposed to market economies." The prevalence and severity of corruption are therefore a function of the expected gains it yields, how much those gains exceed legitimate forms of income (e.g., salaries), local cultural and political mores, the probability of being caught, and the punishments associated with exposure.

Klitgaard (1998) famously proposed that corruption occurs when the gains (monetary or political) exceed the expected costs, that is, the penalties multiplied by the probability of being caught. In this logic, corruption is a function of the discretion and power of corrupt officials as well as the degree of accountability and transparency they face. Greater authoritative oversight and penalties should, at least in theory, reduce the magnitude of corruption.

Corruption flourishes in states where important decisions are made out of view of the public, lack an independent media, have a judiciary highly amenable to the influence of power-holders, restrict civil liberties, and where legal channels are not equally available to everyone. In such environments, nepotism, favoritism, and political connections take the place of market-based criteria or meritocracies (Fisman 2001). Essentially, the practice requires lack of effective oversight and enforcement of existing laws, leading to inadequate accountability; corruption thus tends to be most common in highly centralized political systems. Bardhan (2006) usefully differentiates between political corruption, in which extra-legal means are used to gain power, and administrative or bureaucratic corruption, in which control over state resources is used for private gains. In countries where there is not much difference between political parties and the state, however, such as Vietnam or China, this distinction is small.

Another common cause of corruption is underpaid public employees, notably police officers and customs officials. Goel and Nelson (1989) found that in the United States, higher salaries were associated with reduced bribe-taking, although van Rijckeghent and Weder (2001) discovered empirically that large salary increases are needed to reduce corruption.

The ties between globalization and corruption are complex and geographically variable. Lalountas et al. (2011), using cross-sectional data for 127 countries, found that globalization (in the forms of foreign direct investment and import penetration) mitigated corruption in relatively developed countries but had no impact on corruption in poorer ones. Corrupt practices such as smuggling or black market money exchanges flourish when government policies are overly restrictive, unduly complicated, irrational, rigid, or unrealistic (such as setting official exchange rates too high). Corrupt countries tend to have porous borders through which drugs, weapons, or slaves may be moved easily.

Corruption owes much to insufficient transparency, as it is typically conducted in secrecy, including backroom deals wherein local elites manipulate the state for their own ends. When bribery is mutually beneficial for public officials and those bribing them, neither has an incentive to report the payments made. For this reason, it is commonly argued that democratic societies tend to have lower levels of corruption because they create mechanisms for accountability and the enforcement of laws (Moreno 2002). Indeed, many of the most notoriously corrupt governments today are profoundly anti-democratic. Countries with a relatively unfettered media, in which the press can publicize public sector misbehavior, have lower levels of corruption (Brunetti and Weder 2003).

The geographic variability of corruption also reflects national and cultural differentials in social norms. As Bardhan (1997, p. 1330) puts it, "What is regarded in one culture as corrupt may be considered a part of routine transaction in another." Interpersonal interactions may be monetized, such as with the payment of *baksheesh* in many Middle Eastern and South Asian countries. Gift exchanges are often held to be a regular part of doing business in many developing countries. But such explanations border on the tautological: a country is deemed corrupt because its culture is corrupt. Gender roles also play a role: Swamy et al. (2001) demonstrate that corruption tends to be mitigated when women hold larger shares of political office and high-level administrative positions (see also Goetz 2000).

Widespread, endemic corruption has a variety of corrosive social and economic effects. Notably, it engenders despair and resignation, saps the morale of the public, and leads to widespread cynicism and distrust of the government. Corruption undermines the credibility of the state, erodes public confidence, contributing to what Habermas (1973) famously called a legitimation crisis, in which the legitimacy of the state becomes problematic for social reproduction.

Ironically, at times, when the machinery of government moves too slowly, corrupt practices such as bribery may actually improve the functioning of the state. For example, in many countries "speed money" is used to cut through bureaucratic red tape and reduce queues for government contracts (Bose 2004). This phenomenon has generated a debate between those who argue that corruption "greases the wheels" of the state and those who maintain it "throws sand in the gears." In addition, corruption increases the size of the underground economy and the black market, including smuggling and the drug trade.

Corruption undermines the efficiency and effectiveness of government policies. Endemic corruption may push an administrative system to the point of collapse. It may lead, for example, to shortages of medical and military supplies and delays in the delivery of funds. Bribes may lead to the shoddy construction of buildings and infrastructure using substandard materials such as concrete diluted with too much sand, which increases the risk from earthquake damage, resulting in unnecessary fatalities, as recent quakes in Turkey and China revealed. Foreign aid, such as famine relief, may end up being sold on the black market, never reaching those it was intended to help. Nepotistic hiring practices lead to overstaffing of government offices and unqualified personnel. Corruption also inhibits effective natural resource management: in a case study of Indian forest management, Robbins (2000) concludes that corruption selectively stressed some elements of the ecosystem and not others.

A raft of economic studies concludes that corruption tends to alarm investors, misallocates talent and capital, and retards economic growth (Mauro 1995; Bardhan 1997; Aidt 2003; Rose-Ackerman 2006). Bribes and similar payments distort the allocation of public resources, leading to a schism between the intended and the privately appropriated benefits of capital, as when public funds for construction are used to build luxury homes for government officials. Corruption and economic inequality are thus closely related (Gupta et al. 2002). Corruption also reduces legitimate rent-seeking investments, including the introduction of new products and technologies, for which the demand is price-inelastic. Innovators generally lack extensive political connections and are more susceptible to demands for bribes than are most firms. High levels of corruption are thus associated with reduced foreign direct investment (Wei 2000; Habib and Zurawicki 2002). In a sense, corruption raises the transactions costs of economic activities, reducing efficiency and lowering productivity.

Corruption raises the barriers to entry for non-privileged groups, notably those lacking in political connections and funds for bribes and kickbacks. Thus Mo (2001) found that a 1 percent increase in the level of corruption reduces economic growth by 0.72 percent. By diminishing the quality of governance, corruption acts as a sort of tax (Jain 2001), reducing efficiency, raising production and transportation costs, lowering standards of living, and enhancing inequality.

OUTLINE OF THIS VOLUME

This Handbook consists of two major parts. Part I offers conceptual overviews of corruption, while Part II consists of a series of case studies of different countries. In Part I, chapters address issues such as the causes of corruption, the role of human capital, the frequently overlooked topic of gender, the global geographies of corruption, its consequences, and the role – potential as well as real – of electronic government in its mediation. Part II offers a series of national case studies that collectively highlight how and why corruption plays out unevenly in different contexts. These include two studies of Latin American countries, three drawn from Europe and Russia, two from the Middle East, one from Africa, and six from Asia. More detailed chapter summaries follow.

In Chapter 2, Sufyan Dabbous and Eugen Dimant offer an extended literature review of the causes and consequences of corruption, focusing on recent empirical studies. They note that recent works have addressed issues such as the contagion effects that facilitate the diffusion of corruption, wherein one state influences corruption in its neighbors, the roles of economic development and educational levels, and touch on the increasingly important role of e-government. They also examine the relation between immigration and corruption, a topic that rarely receives attention, as well as the Internet. Their summary of the effects of corruption includes analyses of the brain drain from developing countries, fiscal deficits, human capital, and the shadow (or underground) economy.

The third chapter, by Asma Sghaier and Asma Guizani, focuses on the interrelations among corruption, human capital, and economic growth in developing countries. Whereas the literature on corruption is dominated by Western scholars, they offer an Islamic interpretation. They also summarize the different causes of corruption, as well as the impacts, focusing on the creation of human capital. They use a Cobb-Douglas production function to examine corruption's relations to economic growth and institutional variables empirically. They show that corruption and the growth of human capital are simultaneously determinant: corruption impedes the creation of skills by discouraging the young from pursuing their studies, which in turn facilitates corruption.

Chapter 4, authored by Helena Olafsdotter Stensöta and Lena Wängnerud, delves into the much-overlooked issue of gender and corruption. Are men, by virtue of male entitlement, more prone to corrupt behaviors than women? Their chapter explores the institutional linkages between gender and corruption and its effects. Women are more likely to suffer the effects of corruption than are men, and the evidence indicates that having more women in high political offices tends to reduce corruption levels. Male-dominated networks and patriarchal regimes in most countries are ripe for corruption. In contrast, women frequently provide the "raw material," that is, an assemblage of characteristics, that lead them to favor policies that facilitate social reproduction. Finally, they turn to the gendered nature of accountability mechanisms, including social movements.

World regional geographies of corruption are the subject of Chapter 5, by Barney Warf. He examines global patterns using Transparency International's Corruption Perception Index, and proceeds to summarize corruption in Europe, the Middle East/North Africa, Asia, sub-Saharan Africa, and Latin America, using maps of each. The point of this survey is to demonstrate that there is no "one-size-fits-all" model of corruption; rather, it must be understood within widely varying economic, political, and cultural contexts.

Dominik Enste and Christina Heldman, the authors of Chapter 6, write about the consequence of corruption, which are numerous, nefarious, and debilitating. They trace corruption's effects on private investment, foreign direct investment and capital inflows, foreign trade, government expenditures and services, gross domestic product, inequality, and the shadow or underground economy. These effects are not straightforward and the evidence is not always consistent, and poorly functioning economies may generate corruption as much as corruption retards economic growth. However, there is convincing proof that corruption rewards a few officials but makes everyone else poorer.

In Chapter 7, Nasr Elbahnasawy turns to electronic government, or e-government, the use of the Internet by states around the world. E-government has occasionally been celebrated as a panacea to control corruption by improving transparency and making it easier for citizens to report public malfeasance. The reality is more complicated. Certainly countries with low levels of e-government readiness tend to be more corrupt, but this relation may reflect other forces such as poverty, illiteracy, and lack of a free media as much as anything. Elbahnasawy also explores the nexus between e-government and corruption, in which the former expands available information about public agents and may limit their discretionary powers. The empirical evidence about this issue is mixed, with some studies concluding e-government is a useful tool to limit corruption but only in specific cultural and institutional environments.

The second part of the book – national case studies – opens with Chapter 8, by Stephen Morris, which concerns Mexico. As with many countries, corruption there is deeply entrenched. Under the PRI (Institutional Revolutionary Party), corruption may have contributed to political stability by allowing government officials to enjoy the spoils. More recently, despite a gradual shift away from one-party PRI authoritarianism, corruption in Mexico has increased. The decline in the PRI's influence enhanced the power of state officials, with an associated splurge in local corruption. The shift involves new, and more corrupt, relations between firms and the state. Even democratization does not immunize a country from corruption, and Mexico's efforts to stem corruption have been half-hearted at best. One of the most serious consequences has been a horrific increase in deaths related to drug trafficking and associated gang activity. Another victim has been public confidence in the Mexican state.

Kelly Senters and Matthew Winters write about the ongoing corruption scandals in Brazil in Chapter 9, which have felled two presidents to date. Beyond the recurrent scandals lie deeply embedded forces that recreate corruption on a daily basis. Bribes, misallocations of public funds, non-competitive procurement processes, fake receipts, and the like testify to corruption's pervasiveness at all levels of the Brazil government bureaucracy. They also write about Brazil's attempts to curb corruption through institutional changes, such as random auditing of municipalities. They conclude by examining the growing public outrage over this matter and voter backlash against corrupt officials.

In the context of Eastern Europe, Agnes Batory discusses in Chapter 10 whether joining the European Union (EU) has helped or hindered the growth of corruption in post-communist states. Charges of misusing EU subsidies have been leveled in countries such as Croatia and the Czech Republic. She outlines the legacy of corruption during the dark days of Soviet occupation, such as clientelistic political structures under the *nomenclatura* system, which persisted in the form of shadowy networks that hamper democratic reforms in the region. She also discusses new forms that have arisen in the post-accession era such as transnational criminal groups and the opportunities for graft embedded in the transition to privatization. Adherence to the EU's legal rules was often more evident on paper than in practice. Corruption varies throughout the region: in addition to the east-west divide that distinguishes the region from low-corruption countries in Western Europe, there is also a north-south divide that places the continent's most corrupt states in the Balkans, where state capture by corrupt officials is evident.

Chapter 11, by Johannes Leitner and Hannes Meissner, focuses on Ukraine, where successive generations of politicians have pronounced corruption their top priority, to no avail. The Soviet legacy, a stuttering economy, and a rocky transition to a privatized economy set the stage for widespread corruption of various forms. Clientelism ensures that favors are handed down in return for loyalty, a practice that has become normalized over time in business networks. Notably, these practices vary considerably between the eastern and western parts of the country. The result has been widespread distrust of the state among Ukrainians and failed attempts to curb corruption.

Russian corruption is the focus of Chapter 12, by Günther Schulze and Nikita Zakharov. They put forth an intriguing argument: the state holds wages artificially low, and in return officials reap the benefits of kickbacks and embezzlement, which cements their political loyalties. Unlike democracies, corruption in autocratic states may help to promote stability. They hold that this state of affairs is the path-dependent result of a long historical trajectory that finds its origins during the Mongol occupation in the thirteenth through the fifteenth centuries. This so-called *kormlenie* system of systemic corruption persisted after the Bolshevik revolution, all throughout the seven decades of the USSR, and into the current kleptocracy under Putin. They conclude by looking at strategies to rein in corruption, including higher salaries and legislative initiatives.

In Turkey, the fight against corruption has been long, arduous, and largely unsuccessful, according to Alfredo Jiménez, Secil Bayraktar, and Mesut Eren in Chapter 13. Rapid economic growth in a neoliberalizing state has done little to curtail corruption, where bribery is endemic. They embed this phenomenon within Turkish culture, including the high levels of "power distance" or access to power among varying groups, collectivism, or the tendency to privilege members of in-groups at the expense of outsiders, the country's low level of interpersonal trust, and a paternalistic set of values that legitimizes the patronage networks set up by authority figures. Although Turkey has taken steps to reduce corruption, partly due to attempts to join the EU, it has little to show for its efforts.

Wasta – the Arabic term for nepotism or using personal connections to obtain government jobs and services – is the focus of Chapter 14, by Marcus Marktanner and Maureen Wilson. Not all *wasta* is bad, and some forms are culturally acceptable: it may refer to the use of legitimate intermediaries, but bad *wasta*, that is, corruption, is condemned by the Quran. A legacy of tribalism, the phenomenon is ubiquitous in the Arab world. Marktanner and Wilson use a game-theory approach to understand why some people use *wasta* and others do not, concluding that corrupt institutions promote the bad forms. Thus, as bribery becomes increasingly normalized, control of corruption becomes ever more difficult. They conclude by situating Arab corruption within an international dataset that reveals that while the prevalence of bribery there is not significantly greater than most of the world, attempts to control it have enjoyed less success.

South African corruption, studied in Chapter 15 by Karl Meyer and John Luiz, has roots in the colonial era. Under apartheid, the state resorted to money laundering and organized crime. The shift to black majority rule starting in 1994 did little to change its prevalence, allowing a small elite to capture the state as a means of extensive rent-seeking. The corruption of former President Jacob Zuma, who faces 783 charges of corruption, fraud and racketeering, is a case in point. An unstable political environment and intense income inequality have fueled corruption in different corners, a burden carried largely by the poor. As a result, distrust of the state is rampant. They then examine private corruption, such as in the country's construction industry, and conclude by turning to corporate responses.

"Drugs and corruption in former Soviet Central Asia," the title of Chapter 16 by Filippo De Danieli, offers a fascinating glimpse into the transnational crime networks that operate in Turkmenistan, Uzbekistan, Kazakhstan, Kyrgyzstan, Tajikistan, and Afghanistan. Borders mean little to nothing. The collapse of the Soviet Union, and the prolonged wars in Afghanistan, led to a river of opium spewing out of Afghanistan into the five former Soviet republics. Afghan warlords and Islamic groups alike partook of the bounty. Cities that once prospered due to the old Silk Road, such as Osh, found a new lease on life. De Danieli charts the decentralized organized crime groups that have also taken advantage of the situation. Not surprisingly, opium-derived revenues have tainted government officials and professionals throughout the region, whose countries are some of the world's most corrupt.

Pakistan, with almost 200 million people, is depressingly corrupt, as Feisal Khan lays out in Chapter 17. He notes that the trauma that accompanied partition in 1947 essentially has continued to the present, with administrative collapse and weak state institutions. After detailing the severity of the country's corruption, Khan charts its impacts on government revenues, such as taxes and the minuscule share of the budget dedicated to social and welfare programs. Remittances from Pakistanis working abroad have also contributed to the crisis. Corruption has also flourished despite – or perhaps because of – tens of billions of dollars of US aid. Indeed, remittances and foreign aid act as a version of the famed "resource curse." Efforts to control the plague are ineffective: military dictatorships turn a blind eye to the problem, and Pakistan's Anti-Corruption

Agency is a joke. The result is a deeply dysfunctional state plagued by social unrest, creating fertile ground for terrorist networks.

Mohammad Nurunnabi addresses the sad state of corruption in the Bangladeshi financial sector in Chapter 18. Corrupt and unstable financial institutions impede economic growth. After situating it within the broader context of a country with poor governance and widespread bribery, he proceeds to a case study of the Hallmark Group, which received hundreds of millions of dollars in illegal loans, which it then redistributed through fictitious companies, resulting in high levels of non-performing loans. He concludes with recommendations to address this problem.

China is not only the world's most populous country, it is also one of the most corrupt. In Chapter 19, Vanesa Pesqué-Cela charts the growth of Chinese corruption since its incorporation into the world economy began in the late twentieth century, noting its shift from individual behavior to an organized collective practice. In a country with highly centralized power structure and severe censorship, corruption is one of the major threats to the Communist Party, a hideously corrupt organization if there ever was one. From there, corruption has radiated out into the judiciary, law enforcement, and finance. Widespread bribery, kickbacks, the buying and selling of promotions, embezzlement, illegal taxes, the stripping of state-owned enterprise, nepotism, and cronyism testify to the capture of the Chinese state by corrupt parties. China's failure to develop a Weberian bureaucracy, that is, a meritocracy with clear lines of responsibility and accountability, has played a central role in this disaster. The reliance of local governments on discretionary transfers of funds from the national state and absence of an independent judiciary compounded the problem. As a result, the anti-corruption efforts of Xi Jinping, including the arrest of 1.3 million officials since 2014, have had minimal impacts. Tellingly, only 3 percent of those were criminally prosecuted. However, China also exhibits a "corruption paradox," in which high levels of malfeasance have not led to political instability or lowered its high rates of economic growth.

Cleo Calimbahin documents corruption in the Philippines in Chapter 20. Centuries of Spanish and American colonialism led to the formation of a weak state rife with systemic corruption. The long dictatorship of Ferdinand Marcos was a case study in kleptocracy. Clan-based oligarchs use a clientelistic power structure and crony capitalism that offer impunity to corrupt officials and led to an unholy marriage with organized crime. Corrupt elites and dynastic ruling families engage in widespread election fraud. More recently, Filipino corruption has taken a horrifically violent turn: President Duterte's campaign against illegal drugs has led to the deaths of 7,000 people, many murdered extra-judicially.

Lastly, in Chapter 21 Marcus Mietzner turns to Indonesia, the world's fourth-most populous country and largest Muslim nation. From the Dutch colonial era to the tyranny of Suharto, corruption became deeply interwoven with the Indonesian state. Aristocratic families came to view the state as a personal cash machine. Mietzner focuses on the important issue as to why democratization there has not led to a concomitant decline in corruption, and lays blame at the expensive and inefficient system of funding political campaigns. Rather than swirling around the presidency, corruption became multi-tiered and decentralized. Corruption has not been uncontested however; the Corruption Eradication Commission has arrested hundreds of politicians, mayors, judges, and others, although the actual prison time served is typically minimal and some convicted officials were re-elected. However, he emphasizes throughout that corruption persists in Indonesia because the likely gains exceed the costs.

These case studies serve to illustrate the diversity of corruption around the world. With varying historical legacies, economic and class structures, cultures, and political systems, it is not surprising that both the causes and effects of corruption differ greatly among countries. Corruption in Mexico is wrapped up in the near-monopoly of the PRI and the horrific violence of the drug trade; in Eastern Europe, it reflects the residues of the Soviet system; in the Arab world, in cultural attitudes that accept some nominally corrupt behaviors as normal; in South Africa, the legacy of apartheid continues to infect the government; in Pakistan, endemic corruption led to a failed state; in China, the lack of an effective government bureaucracy and a corrupt Communist Party are responsible; and in Indonesia, corruption arises in part from a competitive but expensive political system. There is, in short, a rich geography to corruption, and its spatiality is an inescapable part of understanding how and why it occurs as well as attempts to combat it. Such comments should give one pause in adopting broad theorizations that ignore local and national contexts. Rather, corruption can only be understood by viewing it through the path-dependent trajectories that give rise to the unique circumstances of each social formation. For this reason, strategies to reduce corruption must likewise take into account specific constellations of factors; one country's successful strategy may fail in another.

REFERENCES

Aidt, T. 2003. Economic analysis of corruption: a survey. *Economic Journal* **113**(491), 632–52.
Bardhan, P. 1997. Corruption and development: a review of issues. *Journal of Economic Literature* **35**, 1320–46.
Bardhan, P. 2006. The economist's approach to the problem of corruption. *World Development* **34**(2), 341–8.

Bose, G. 2004. Bureaucratic delays and bribe-taking. *Journal of Economic Behavior and Organization* **54**(3), 313–20.

Brinkerhoff, D. 2000. Assessing political will for anti-corruption efforts: an analytic framework. *Public Administration and Development* **20**(2), 239–52.

Brunetti, A. and B. Weder. 2003. A free press is bad news for corruption. *Journal of Public Economics* **87**(7–8), 1801–24.

Fisman, R. 2001. Estimating the value of political connections. *American Economic Review* **91**(4), 1095–102.

Goel, R. and D. Nelson. 1989. On the economic incentives for taking bribes. *Public Choice* **61**, 269–75.

Goetz, A. 2000. Political cleaners: women as the new anti-corruption force? *Development and Change* **38**(1), 87–105.

Gupta, S., H. Davoodi, and R. Alonso-Terme. 2002. Does corruption affect income inequality and poverty? *Economics of Governance* **3**(1), 23–45.

Habermas, J. 1973. *Legitimation Crisis*. Boston, MA: Beacon Press.

Habib, M. and L. Zurawicki. 2002. Corruption and foreign direct investment. *Journal of International Business Studies* **33**(2), 291–307.

Jain, A. 2001. Corruption: a review. *Journal of Economic Surveys* **15**(1), 71–121.

Jain, A. (ed.). 2012. *Economics of Corruption*. New York: Springer.

Jancsics, D. 2014. Interdisciplinary perspectives on corruption. *Sociology Compass* **8**(4), 358–72.

Klitgaard, R. 1998. *Controlling Corruption*. Berkeley, CA: University of California Press.

Lalountas, D., G. Manolas, and I. Vavouras. 2011. Corruption, globalization and development: how are these three phenomena related? *Journal of Policy Modeling* **33**(4), 636–48.

Mauro, P. 1995. Corruption and growth. *Quarterly Journal of Economics* **110**(3), 681–712.

Mo, P.-H. 2001. Corruption and economic growth. *Journal of Comparative Economics* **29**, 66–79.

Moreno, A. 2002. Corruption and democracy: a cultural assessment. *Comparative Sociology* **1**(3–4), 495–507.

Robbins, P. 2000. The rotten institution: corruption in natural resource management. *Political Geography* **19**(4), 423–43.

Rose-Ackerman, S. (ed.). 2006. *International Handbook on the Economics of Corruption*. Cheltenham, UK and Northampton, MA, USA: Edgar Elgar.

Rose-Ackerman, S. and B. Palifka (eds). 2016. *Corruption and Government: Causes, Consequences, and Reform*. Cambridge: Cambridge University Press.

Swamy, R., S. Knack, and O. Azfar. 2001. Gender and corruption. *Journal of Development Economics* **64**(1), 25–55.

Teachout, Z. 2014. *Corruption in America: From Benjamin Franklin's Snuff Box to Citizens United*. Cambridge, MA: Harvard University Press.

van Rijckeghent, C. and B. Weder. 2001. Bureaucratic corruption and the rate of temptation: do wages in the civil service affect corruption, and by how much? *Journal of Development Economics* **65**(2), 307–31.

Wei, S. 2000. How taxing is corruption on international investors? *Review of Economics and Statistics* **82**(1), 1–11.

PART I

THEMES FOR UNDERSTANDING CORRUPTION

2. Causes and effects of corruption: new developments in empirical research
Sufyan Dabbous and Eugen Dimant

Corruption has long been an enshrined facet of modern and even ancient human civilization. Corrupt behavior has afflicted the complex and ancient societies of Egypt, Greece, and Rome much to the same degree that corruption has plagued the modern day societies that now exist in their place. Different economic systems (capitalism, communism, and feudalism), as well as distinct religious institutions (Muslim, Christian, Buddhist, and Hindu) have all seen varying degrees of corruption. The reality is that corruption as a social phenomenon is a deeply engrained part of human society with motivations that are multifaceted and the result of interactions at the micro-, meso-, and macro levels (Bicchieri and Ganegonda 2016; Dimant and Schulte 2016). The effects of corruption can be wide ranging, inducing poverty and brain drain, and hindering economic growth. Yet for all of corruption's purported ingratiation and influence on human societies, it has only recently become a focus of academic study. Up until the 1980s, corruption was predominantly a topic of political, sociological, historical, and criminal law research and has only recently come to the fore as a topic of interest within economics. Recent interest and empirical undertakings have very much been fueled by the increased and improved quality, size, and availability of data since the late 1990s. The availability of these large datasets have enabled researchers to conduct multi-country analyses of corruption over varying time periods to better understand the factors that cause corruption as well as the effects that corruption can have on societies.

Since the late 1990s, a number of significant, expansive, and frequently cited surveys have been published. Notably, studies by Rose-Ackerman (1999), Tanzi (1998), Jain (2001), Aidt (2003), Lambsdorff (2006), Seldadyo and de Haan (2005), and Treisman (2000) have all sought to address different assumptions and findings regarding the causes and socioeconomic consequences of corruption.

It is the aim of this chapter to follow these seminal works by providing a survey of relevant literature to provide an understanding of the causes and consequences of corruption. However, while the aforementioned studies have focused on the entirety of available literature, this study differentiates itself by focusing on new developments in our understandings of both causes and effects of corruption. New developments are classified as either determinants or consequences of corruption for which academic/empirical studies have only been published post-2006. It is important to note that this chapter builds upon and is to an extent a modified version of the work of Dimant and Tosato (2018), which included a section on recent developments in our understandings of corruption, as well as a section surveying more established and older studies relevant to our understanding of the causes and effects of corruption.

By focusing on the causal and consequential relationships put forth over the last decade, it is the aim of this chapter to provide a comprehensive summary and survey of literature surrounding the most recent developments in our understandings of corruption. While surveys similar to this exist, few, if any, focus solely on and provide in-depth discussions of empirical studies in more recent fields regarding both causes and effects of corruption, such as brain drain as a mal-effect or the implications of e-government on corruption levels. Moreover, this chapter seeks not only to convey the findings of different studies, but also to discuss the implications of the different methodologies (datasets, time periods, and regions) employed by various studies.

In focusing on these various methodologies, we observe that there are two general tendencies that researchers take when tackling the causes and effects of corruption. The first is, as mentioned, to conduct analyses through recently available datasets. The second is to opt for laboratory-based experiments. Overwhelmingly, however, we note that researchers and economists have opted to employ analyses relying on newly available datasets, attempting to understand what we have called "new" developments. In fact, in many ways the availability of large datasets has enabled further empirical research that has brought to light many of the new causes and effects discussed hereafter. In laboratory-based experiments researchers are often limited to answering a specific question rather than addressing larger trends. The recent prevalence of multi-country and multi-year datasets appears promising as they lead to wide-scale observations. The measures and indexes that different researchers employ differ greatly and is a topic that we both focus on and discuss.

We structure our chapter as follows. In the next section we review literature relevant to new developments in our understandings of the causes of corruption. The following section emphasizes the social and economic consequences of corruption on society. In the last section, we give some concluding remarks. Tables 2.1 and 2.2 summarize the statistical findings of the relationship between the causes and effects of corruption respectively.

NEW DEVELOPMENTS IN THE DETERMINANTS OF CORRUPTION

Contagion Effects

Many theorists argue that the propagation of corruption is contagious and that the level of corruption in a given country is largely dependent on the level of corruption in neighboring nations. Akin to the spread of a disease, theorists contend that corruption in one state inevitably influences corruption in neighboring states. This, of course, poses many issues and suggests that often corruption may be beyond the control of a dominant governing body or society. An empirical study focused on the United States added merit to this theory and found that a 10 percent increase in the levels of corruption in neighboring states increases corruption in a state by 4–11 percent, seemingly confirming the contagious nature of corruption (Goel and Nelson 2007). While this study did not focus on multiple countries, opting to focus on American states, they were able to benefit from the breadth of data made available by US institutions such as the Census Bureau, Department of Justice, and Bureau for Economic Analysis. Moreover, Goel and Nelson opted to focus on data spanning all 50 states and averaged over the period from 1995 to 2004. A 2009 study added strength to the notion that corruption is contagious by examining multiple countries, finding that corruption can be viewed as a regional phenomenon and that any attempts at decreasing corruption in one nation will lead to a decrease in neighboring countries (Becker et al. 2009). Their study relies primarily on Transparency International's Corruption Perception Index (CPI), a commonly used measure of corruption by numerous academics of corruption, focusing on data for the years 2000–05. In corroborating their results, however, they employ supplementary corruption indexes, including the (Graft) Kaufman Index, as well as the index provided by the International Country Risk Guide (ICRG), finding consistent results. Interestingly, they note that corruption is contagious at a local geographical level and that it does not disseminate over vast distances, regardless of cultural (dis)similarities. A 2013 study looks specifically at 20 European nations and is critical of the standard use of CPI and ICRG indexes to measure corruption. Instead, Lee and Guven rely on and argue that the European Social Survey, which measures cultural values and behaviors, is a stronger measure of country-wide corruption. They find corruption to be contagious within countries, stating that one's likelihood of taking part in a corrupt act is significantly greater if exposed to a corrupt act in the previous five years (Lee and Guven 2013).

While the studies of Becker et al. (2009) and Goel and Nelson (2007) offer robust evidence for the contagious nature of corruption, an interesting means through which research in this field may develop is to rely more on corruption measures that gauge cultural values and behavior akin to those employed by Lee and Guven.

Economic Prosperity

Intuitively, it would seem that richer countries would have lower levels of corruption given the more advanced institutions of governance that wealth can buy and enable. Essentially, this relationship may lead many to conclude that less economically prosperous nations are fated to have higher levels of corruption. Less developed countries typically have weaker legal institutions, inefficient governments, lower levels of education, and sub-optimal government systems, all of which are associated with higher levels of corruption. This relationship was given some empirical backing in a paper that, using data from 62 countries over the period from 1990 to 1998, found richer countries have lower levels of corruption (Serra 2006). The study relies on the Graft index to measure corruption, developed by Kaufman et al. in 1998, and corroborates its findings by replicating its results with the use of Transparency International's CPI. Moreover, the study finds that strong democratic institutions, enabled by greater wealth, only lead to lower levels of corruption when those institutions have been continuously held for decades. A later study investigating the relationship between income inequality and corruption using data from 1980 to 2004 for 50 US states found that both in the short run and in the long run there is bidirectional causality between income inequality and corruption (Apergis et al. 2010). The study measured corruption by the number of government officials convicted in a state for crimes related to corruption in a specific year, as made available by the US Justice Department. A 2008 study instrumented the measure of income and corruption with biogeographical considerations to assert that while decreased corruption generally leads to higher levels of income in the short term, a long-term analysis shows that increased income is the driving cause of lower levels of corruption (Paldam and Gundlach 2008).

The studies of Serra (2006), Apergis et al. (2010), and Padlam and Gundlach (2008) offer robust findings. However, research in this field would benefit from further exploration of the directionality of the relationship between economic prosperity and corruption, similar to that carried out by Apergis et al. (2010) in order to definitively determine the direction of the relationship between the two.

Education

Theoretically, higher levels of education should reduce the levels of corruption in a country. Individuals with higher levels of education generally tend to be more committed to civil liberties and less tolerant of government oppression and of corrupt behavior (Truex 2011). Truex makes these findings by looking into the correlation between education levels and corruption in Nepal. Corruption was measured through a survey that asked residents of Nepal's capital about their views of various types of corrupt acts. Most respondents agreed that large-scale bribery, for instance, was inadmissible, while there was disagreement on whether petty corruption was similarly unacceptable. Education was found to be the primary determinant of these varied responses, indicating that increased education can reduce corruption in developed countries. Further, higher levels of education tend to lead to higher awareness of international standards, and, thus, in theory should reduce a person's tolerance for corruption. Initial empirical studies found that higher levels of education are correlated with lower levels of corruption, such as Glaeser and Saks (2006), where the study focused on all 50 US states and corruption was measured by the number of federal corruption convictions in each state. A 2015 study examining 53 African countries from 1996 to 2010 found that not only does education work to reduce corruption, but that higher levels of education (i.e., education through to the tertiary level) have a stronger effect in reducing corruption. As such, encouraging education through the tertiary level has corruption mitigating effects (Asongu and Nwachukwu 2015). Interestingly, however, not all studies have found there to be a negative correlation between education and corruption. A 2012 study of 20 sub-Saharan African countries found there to be a positive correlation between education and corruption, that is, increased education leads to increased corruption (Kaffenberger 2012). This is explained by the fact that often education systems are themselves corrupt and that in countries and cultures where schoolchildren must pay bribes for good grades individuals become desensitized and more accepting of corrupt behaviors.

Interestingly, studies of the relationship between corruption and education have been limited to nations and particular geographic regions. Truex focuses on Nepal, Glaeser and Sak on the United States, Asongu and Nwachukwu on Africa, and Kaffenberger on sub-Saharan Africa. Kaffenberger's findings prove to be less definite given their reliance on a single survey by AfroBarometer. Further research in this field may opt to consider multi-country, multi-continent relationships between education and corruption.

E-government

Perhaps among the most recent of the new developments, access to the internet and the existence of an e-government should, in theory, reduce corruption levels. E-government allows the transactions for permits or civil applications to be done online, thus increasing efficiency, transparency, and accountability, all of which inherently reduce corruption. A study found evidence to support the link between e-government and reduced corruption using a panel of 149 countries in two time observations, 1996 and 2006 (Andersen 2009). Given that e-governance is a recent phenomenon, Andersen looks at two time observations in order to determine whether the adoption of e-government is correlated with reduced corruption. Employing the Corruption Control Index (CCI) developed by Kaufmann (2007), which measures the "extent to which public power is exercised for private gain, including both petty and grand forms of corruption," Andersen finds that increased e-governance led to lower levels of corruption in the decade ending in 2006. This indicates that further technological advancements may enable increased transparency and reduced corruption. A later empirical study, which relied on Transparency International's CPI and a dataset of 160 countries from 1995 to 2009, also reached the same conclusion and determined a unidirectional causality between e-government and reduced corruption (Elbahnasawy 2014). Other studies produce similar results and suggest that e-government can work to reduce corruption. A 2012 study, which employed the CPI to measure corruption and the E-government Development Index (EDI) to measure the level of e-government adoption, found that as the use of information and communications technology (ICT)-related e-government increases, corruption decreases and that the effect is greater in developing versus developed nations (Mistry and Jalal 2012). Likewise, a 2013 study which measured e-government as based on the readily available global e-government reports and corruption as based on the Worldwide Governance Indicators report found that increased e-government levels led to decreased levels of corruption (Krishnan et al. 2013). All of the mentioned studies offer robust findings while using a range of corruption measures.

Moving forward, easier access to time series data will allow researchers to understand the mid- and long-term impact of e-government.

Immigration

As isolationist policy seems to regain popularity, the definitive effects of immigration and their effects on corruption can determine a country's immigration policy. As Dimant et al. (2015) point out, immigration from a highly corrupt country could have an effect on the levels of corruption. First, large-scale emigration due to strong push factors does not only apply to a handful of honest citizens, but rather to the entire population, corrupt or not corrupt. In addition, if corruption is part of their cultural beliefs this will emigrate with them. Further, it might take some time for individuals to fully adapt to their host country, and in this transitory period they may be more prone to corrupt behavior due to greater need (Dimant et al. 2015). Recent empirical evidence based on data from 207 countries for the period 1984 to 2008 indicates that while immigration in general has no significant impact on corruption levels, immigration from highly corrupt countries indeed increases corruption in the destination country in the short run, with the effect vanishing in the medium run (Dimant et al. 2015).

The Internet

It has been argued that the advent and increased use of the internet can lead to decreased levels of corruption, as the internet enables the expedient dissemination of information regarding corrupt practices. A more aware and informed population is better able to report and counter corrupt actions. An empirical study from 2011 supported this theory and found that increased awareness of corruption via the internet, measured through corruption-related internet searches via Google and Yahoo using "corruption," "bribery," and "country name" as keywords, correlated with decreased incidences of corruption in a number of countries (Goel and Nelson 2011). In this study, corruption is measured through the use of the following three indexes: Transparency International's CPI, the World Banks's corruption perceptions index, and the World Business Environment survey. Similarly, a 2010 study found that the increased use and availability of the internet was correlated with lower levels of corruption (Andersen et al. 2011). Uniquely, Andersen et al.'s study used lightning density as a means to measure the levels of internet diffusion, as lightning interrupts internet capabilities and lessens the rate of internet diffusion. In the same vein as Glaeser and Saks's 2006 study, Andersen et al. relied on the number of federal corruption convictions as made available by the Justice Department's *Report to Congress on the Activities and Operations of the Public Integrity Section* to measure corruption. A study from 2010 examines the causal directionality between the internet and corruption, and examines 70 countries over the 1998–2005 period (Lio et al. 2011). The authors found that the relationship between corruption and internet adoption is bidirectional and while the effects of internet adoption in reducing

*Table 2.1   Causes of corruption: summary of statistical results from the
literature*

Contagion Effects
Goel and Nelson 2007 (+); Becker et al. 2009 (+)*; Lee and Guven 2013 (+)*

Economic Prosperity
Serra 2006 (+)*; Apergis et al. 2010 (+); Paldam and Gundlach 2008 (+/–)*

Education
Glaeser and Saks 2006 (–); Truex 2011 (–); Asongu and Nwachukwu 2015 (–)*;
Kaffenberger 2012 (–)*

E-government
Andersen 2009 (–)*; Elbahnasawy 2014 (–)*;Mistry and Jalal 2012 (–)*; Krishnan
et al. 2013 (–)*

Immigration
Dimant et al. 2015 (+)*

Internet
Andersen et al. 2011 (–); Goel and Nelson 2011 (–)*; Lio et al. 2011 (–)*

Note: A plus (minus) sign indicates that the relationship between corruption and its
respective cause was found to be positive (negative). An asterisk (*) denotes that a cited
study considers multiple countries, lack of one notes that the study only assessed one nation.

corruption are significant, they are not substantial, concluding that the
internet's capacity for reducing corruption is "yet to be fully realized."

With regards to internet diffusion, Andersen et al. rely uniquely and perhaps less convincingly on measures of lightning density to quantify the levels
of internet usage/diffusion. Future studies may opt to measure internet
diffusion through a more reliable index. Moreover, given Lio et al.'s conclusion that the internet's capacity for corruption reduction is yet to be fully
realized, future studies may look at disparate geographic areas to determine
where the internet has had a greater impact on decreasing corruption.

NEW DEVELOPMENTS IN THE EFFECTS OF CORRUPTION

Brain Drain

Higher levels of corruption could increase a country's brain drain.
Corruption is associated with a number of unfavorable outcomes which
might act as push factors to potential migrants. It has been argued

that returns on education would be particularly affected (high levels of unemployment, lack of social advancement, slower economic growth, and so on), thus those particularly sensitive to such a push factor (highly skilled individuals) would be more likely to emigrate due to this (Dimant et al. 2013). Few empirical studies were done prior to 2006; however, this theory has received strong support in more recent years. Empirical evidence using data from 111 countries between 1985 and 2000, and relying on the ICRG's index measuring corruption, found that corruption was particularly significant in fueling skilled emigration (Dimant et al. 2013). Further empirical evidence was provided from a paper that found that high levels of corruption affect emigration levels of skilled labor far more than unskilled labor using data from 20 Organisation for Economic Co-operation and Development (OECD) destination countries between 1980 and 2010 and 115 origin countries between 1995 and 2010 (Cooray and Schneider 2014). Even more empirical evidence was provided from a paper that, using data from 230 countries, found corruption to be a significant push factor in emigration (Poprawe 2015). A 2016 study that examines the effects of corruption on the medical landscape of 50 African states found that corruption had a direct impact on the brain drain, manifested in the form of the emigration of those trained as physicians (Okey 2016). Using Transparency International's CPI to measure corruption, a 2010 study by Bhargava et al. sought to ascertain the level of physician emigration, finding that the most corrupt countries experience higher levels of physician emigration and that income levels (especially when considered in terms of returns on education) are a primary channel through which corruption promotes physician emigration; this leads them to conclude that in order to preserve and improve the state of the African medical landscape, governments must do more to combat corruption. Using the ICRG index to measure corruption, a 2013 study further confirms the idea that corruption pushes skilled natives to emigrate to countries where they can find employment based on meritocratic criteria (Ariu and Squicciarini 2013).

While all mentioned studies offer robust evidence for the correlation between corruption and emigration, future studies should focus on examining policy measures designed to reduce brain drain and even bring back lost human capital.

Fiscal Deficit

It has been argued that as corruption reduces public income (lower levels of growth, higher levels of inequality) and increases public expenditure (more inefficient spending), fiscal deficits will increase. Depken and Lafountain's (2006) study supports this with empirical evidence; US states with higher levels of corruption were found to have lower bond ratings, thus inducing taxpayers to pay more to borrow and increasing the likelihood of fiscal deficit (Depken and Lafountain 2006). The study examined US states over the period from 1995 to 2000. A later paper provided further evidence that showed that corruption leads to deviations from the optimal public expenditure structure, reducing growth and public income (De la Croix and Delavallade 2009). This study examined 63 countries from 1996 to 2004. It relied on financial statistics from the International Monetary Fund and employed the aforementioned CCI to measure corruption. More evidence is provided in a later study that, using data from Italian public works between 2000 and 2005, shows that public contracts execution is more inefficient in areas with higher corruption and, as a result, increases government expenditures (Castro et al. 2014). Giving closer examination to the nature of public spending which in turn leads to a fiscal deficit, a 2006 study found that higher levels of corruption lead to greater distortion in the structure of public spending in reducing the portion of social spending (i.e., spending on education, health, and social protection), while increasing the portion spent on public services and order, fuel/energy, culture, and defense (Delavallade 2006). The study examined 64 countries from 1996 to 2001.

All of the mentioned studies offer robust evidence for the relationship between corruption and fiscal deficits. None of the studies offer a definitive multi-country survey (the greatest number of surveyed nations is 64). As more data become available, researchers ought to conduct more encompassing analyses.

Human Capital

Corruption ought to have a negative impact on human capital. Higher levels of corruption are associated with lower levels of education, health, socioeconomic development, *and* lower levels of human capital. Initial empirical evidence supports this. Using a sample of 63 countries, Akçay found a statistically significant negative relationship between corruption indices and levels of human development (Akçay 2006). A later study that was based in the Philippines and focused on the effects of corruption on health also found that corruption negatively affected the country's health levels (Azfar and Gurgur 2008). A 2010 study that employed a cross-sectional dataset from 120 countries found that quality of government, as measured through the use of CPI, and the World Bank's Rule Law of Indicator and Government Effectiveness measure is positively correlated with population health. That is, as quality of government improves, so does population health, indicating that corruption has a negative effect on health (Holmberg and Rothstein 2011). Interestingly, a 2010 study flips the direction of causality and claims, by studying data for 68 countries, that human development has a direct effect in reducing the level of corruption as opposed to higher levels of corruption impeding human development. In attempting to establish this particular direction of causality, the study lags the human development dataset by a year, but still concedes that the direction cannot at the present time be definitively ascertained (Sims et al. 2012).

Of the mentioned studies, all but that of Sims et al. (2012) offer robust evidence for the relationship between corruption and human capital. The directionality of this relationship – that Sims et al. sought to address – will be of increasing interest moving forward.

Shadow Economy

It is important to first note that the relationship between shadow economies and corruption is not as straightforward as the others we have examined insofar as they are often studied as both a cause and effect of corruption. In theory, corruption can be viewed as a measure of taxation that might make it an economically rational decision for entrepreneurs to go underground. However, as Dreher and Schneider (2010) point out, one should differentiate between high- and low-income countries. Countries with high incomes have a wealth of public goods (such as legal institutions) that may lead corruption to be a substitution to the shadow economy. In low-income countries, the two may be complements. An early study created a model that suggested that corruption and shadow markets are substitutes (Choi and Thum 2005). This received support in a subsequent paper that re-examined the relationship (Dreher et al. 2009). A later empirical study found that, using cross-sectional data for 98 countries, there is in fact no robust relationship when perception-based indices are used. However, when using an index based on a structural model, shadow economies and corruption are complements in countries with low incomes (Dreher and Schneider 2010). More evidence was found in a 2015 paper that focused on the effects of decentralization on both corruption and the shadow economy. This paper found a positive relationship between the two using data from a number of indices for corruption and the shadow economy from 145 countries (Dell'Anno and Teobaldelli 2015). A 2007 study that examined shadow economies in 145 countries from 1999 to 2006 found that the shadow economy reduces corruption in high-income countries, but actually increases corruption in low-income countries (Schneider 2007).

The relationship between corruption and the shadow economy is, as

*Table 2.2   Effects of corruption: summary of statistical results from the
          literature*

Brain Drain
Dimant et al. 2013 (+)*; Cooray and Schneider 2014 (+)*; Poprawe 2015 (+)*;
Okey 2016 (+)*; Ariu and Squicciarini 2013 (+)*

Fiscal Deficit
Depken and Lafountain 2006 (+); De la Croix and Delavallade 2009 (+)*;
Castro et al. 2014 (+); Delavallade 2006 (+)*

Human Capital
Akçay 2006 (–)*; Azfar and Gurgur 2008 (–); Holmberg and Rothstein 2011 (–)*;
Sims et al. 2012 (–)*

Shadow Economy
Dreher et al. 2009 (+)*; Dreher and Schneider 2010 (–/+)*; Dell'Anno and
Teobaldelli 2015 (+)*

Note: A plus (minus) sign indicates that the relationship between corruption and its
respective cause was found to be positive (negative). An asterisk (*) denotes that a cited
study considers multiple countries, lack of one notes that the study only assessed one nation.

mentioned, difficult given the complexity of the relationship's directionality. Moreover, studies have found at times contradictory evidence when
using different models and datasets. Specifically, researchers have noted
that the effect of the shadow economy varies between high- and low-
income nations.

CONCLUSION

In looking at and conducting a comprehensive review of literature related
to our understanding of corruption over the past decade, our aim has
been to shed light upon the more recent developments affecting our
understandings of both the antecedent causes and effects of corruption
in human social and economic development. In increasing knowledge of
these recent developments we hope to stimulate further research that could
inhibit some of the mal-effects that corruption can have on societal and
economic development. We are hopeful that in providing an overview of
the existing literature in the areas of these new developments we will not
only increase general awareness regarding these new understandings of
corruption, but also motivate and encourage increased interest and study
within the relevant scholarly communities.

Our survey highlights that the increased availability and access to large datasets has enabled vast advancements in research capabilities and, in turn, our understandings of both the effects and causes of corruption. In looking at the discussed causes/effects it is difficult to see researchers proclaiming their various findings without access to the datasets that have been made more and more readily available since the mid-1990s. Corruption indices such as Transparency International's Corruption Perceptions Index, as well as various governmental, economic, and development organizations provide the vital information that make the surveyed literature possible. Interestingly, the majority of the literature surveyed supports the notion that corruption is a hindrance to socioeconomic development and, as such, a facet of human society that governments, organizations, and societies themselves should seek to eradicate.

Moving forward, there are still many challenges and obstacles in our quest to fully understand the complex relationships between society and corruption. While datasets have enabled sweeping cross-cultural, multi-country, and multi-year studies, the challenge remains to deal with and understand the behavior and implications of corruption that cannot be measured through country-wide statistics or surveys. While the vast majority of the literature reviewed has looked at empirical data, there are some researchers who argue that there is more to gain through experimental studies that look at human behavior at a far more detailed micro level.

Finally, we still struggle to understand the multifaceted nature of corruption and its interactions at the micro, meso, and macro levels, and the debate versus experimental and empirical studies will continue to take place. Undoubtedly, however, the quality of empirical research on corruption is still advancing and has to settle important issues (including the right way to measure corruption) before being able to both settle the debate of conflicting empirical findings and answer the plethora of unanswered questions. As we saw in the discussed studies, different researchers opt for different methods of measuring corruption. One promising approach is the use of more objective micro data instead of using subjective perception-based data. A number of recent seminal studies have paved the way for a more reliable attempt to measure corruption (Gorodnichenko and Peter 2007; Ferraz and Finan 2011; Chatterjee and Ray 2012). Capitalizing on micro data to overcome the shortcomings of cross-country macro data is likely the future of empirical work on corruption. However, such micro data are usually hard to obtain, thus reducing the area of application of such an approach (cf. Heywood and Rose 2014). It will be the respective government's responsibility to be more transparent and provide the much needed data to facilitate future research.

REFERENCES

Aidt, T. 2003. Economic analysis of corruption: a survey. *Economic Journal* **113**(491), F632–F652.

Akçay, S. 2006. Corruption and human development. *Cato Journal* **26**, 29–48.

Andersen, T. 2009. E-government as an anti-corruption strategy. *Information Economics and Policy* **21**(3), 201–10.

Andersen, T., Bentzen, J., Dalgaard, C., and Selaya, P. 2011. Does the Internet reduce corruption? Evidence from US states and across countries. *World Bank Economic Review* **25**(3), 387–417.

Apergis, N., Dincer, O., and Payne, J. 2010. The relationship between corruption and income inequality in US states: evidence from a panel cointegration and error correction model. *Public Choice* **145**(1–2), 125–35.

Ariu, A. and Squicciarini, M.P. 2013. The balance of brains – corruption and migration. *EMBO Reports* **14**(6), 502–4.

Asongu, S. and Nwachukwu, J. 2015. The incremental effect of education on corruption: evidence of synergy from lifelong learning. AGDI Working Paper WP/15/036. https://www.econstor.eu/bitstream/10419/123685/1/agdi-wp15-036.pdf (accessed July 28, 2017).

Azfar, O. and Gurgur, T. 2008. Does corruption affect health outcomes in the Philippines? *Economics of Governance* **9**(3), 197–244.

Becker, S., Egger, P., and Seidel, T. 2009. Common political culture: evidence on regional corruption contagion. *European Journal of Political Economy* **25**, 300–10.

Bicchieri, C. and Ganegonda, D. 2016. Determinants of corruption: a socio-psychological analysis. In P. Nichols and D. Robertson (eds), *Thinking about Bribery, Neuroscience, Moral Cognition and the Psychology of Bribery*, pp.179–205. Cambridge: Cambridge University Press.

Castro, M., Guccio, C., and Rizzo, I. 2014. An assessment of the waste effects of corruption on infrastructure provision. *International Tax and Public Finance* **21**(4), 813–43.

Chatterjee, I. and Ray, R. 2012. Does the evidence on corruption depend on how it is measured? Results from a cross-country study on microdata sets. *Applied Economics* **44**(25), 3215–27.

Choi, J.P. and Thum, M. 2005. Corruption and the shadow economy. *International Economic Review* **46**(3), 817–36.

Cooray, A. and Schneider, F. 2014. Does corruption promote emigration? An empirical examination. IZA Discussion Paper Series No. 809.

De la Croix, D. and Delavallade, C. 2009. Growth, public investment and corruption with failing institutions. *Economics of Governance* **10**(3), 187–219.

Delavallade, C. 2006. Corruption and distribution of public spending in developing countries. *Journal of Economics and Finance* **30**(2), 222–39.

Dell'Anno, R. and Teobaldelli, D. 2015. Keeping both corruption and the shadow economy in check: the role of decentralization. *International Tax and Public Finance* **22**(1), 1–40.

Depken II, C. and Lafountain, C. 2006. Fiscal consequences of public corruption: empirical evidence from state bond ratings. *Public Choice* **126**(1–2), 75–85.

Dimant, E. and Schulte, T. 2016. The nature of corruption: an interdisciplinary perspective. *German Law Journal* **17**(1), 54–72.

Dimant, E. and Tosato, G. 2018. Causes and effects of corruption: what has past decade's empirical research taught us? A survey. *Journal of Economic Surveys* **32**(2), 335–56.

Dimant, E., Krieger, T., and Meierrieks, D. 2013. The effect of corruption on migration, 1985–2000. *Applied Economics Letters* **20**(13), 1270–4.

Dimant, E., Krieger, T., and Redlin, M. 2015. A crook is a crook . . . but is he still a crook abroad? On the effect of immigration on destination-country corruption. *German Economic Review* **16**(4), 464–89.

Dreher, A. and Schneider, F. 2010. Corruption and the shadow economy: an empirical analysis. *Public Choice* **144**(1–2), 215–38.

Dreher, A., Kotsogiannis, C., and McCorriston, S. 2009. How do institutions affect corruption and the shadow economy? *International Tax and Public Finance* **16**(6), 773–96.

Elbahnasawy, N.G., 2014. E-government, internet adoption, and corruption: an empirical investigation. *World Development* **57**, 114–26.

Ferraz, C. and Finan, F. 2011. Electoral accountability and corruption: evidence from the audits of local governments. *American Economic Review* **101**, 1274–311.

Glaeser, E. and Saks, R. 2006. Corruption in America. *Journal of Public Economics* **90**(6), 1053–72.

Goel, R. and Nelson, M. 2007. Are corrupt acts contagious? Evidence from the United States. *Journal of Policy Modelling* **29**, 839–50.

Goel, R. and Nelson, M. 2011. Measures of corruption and determinants of US corruption. *Economics of Governance* **12**(2), 155–76.

Goel, R.K., Nelson, M.A., and Naretta, M.A. 2012. The internet as an indicator of corruption awareness. *European Journal of Political Economy* **28**(1), 64–75.

Gorodnichenko, Y. and Peter, K. 2007. Public sector pay and corruption: measuring bribery from micro data. *Journal of Public Economics* **91**, 963–91.

Heywood, P. and Rose, J. 2014. "Close but no cigar": the measurement of corruption. *Journal of Public Policy* **34**(3), 507–29.

Holmberg, S. and Rothstein, B. 2011. Dying of corruption. *Health Economics, Policy and Law* **6**(4), 529–47.

Jain, A. 2001. Corruption: a review. *Journal of Economic Surveys* **15**(1), 71–121.

Kaffenberger, M. 2012. The effect of educational attainment on corruption participation in sub-Saharan Africa. PhD Dissertation. Vanderbilt University.

Kaufmann, D. 2007. *Governance Matters VI: Aggregate and Individual Governance Indicators, 1996–2006*, Vol. 4280. Washington, DC: World Bank.

Kaufmann, D. 1998. Challenges in the next stage of corruption. In *New Perspectives in Combating Corruption*. Washington, DC: Transparency International and the World Bank.

Krishnan, S., Thompson, S., and Lim, V. 2013. Examining the relationships among e-government maturity, corruption, economic prosperity and environmental degradation: a cross-country analysis. *Information and Management* **50**(8), 638–49.

Lambsdorff, J. 2006. Causes and consequences of corruption: what do we know from a cross-section of countries. In S. Rose-Ackerman (ed.), *International Handbook on the Economics of Corruption*, pp. 3–51. Cheltenham, UK and Northampton, MA, USA: Edward Elgar.

Lee, W.-S. and Guven, C. 2013. Engaging in corruption: the influence of cultural values and contagion effects at the microlevel. *Journal of Economic Psychology* **39**, 287–300.

Lio, M.-C., Liu, M.-C., and Ou, Y.-P. 2011. Can the internet reduce corruption? A cross-country study based on dynamic panel data models. *Government Information Quarterly* **28**(1), 47–53.

Mistry, J. and Jalal, A. 2012. An empirical analysis of the relationship between e-government and corruption. *International Journal of Digital Accounting Research* **12**, 145–76.

Okey, M. 2016. Corruption and emigration of physicians from Africa. *Journal of Economic Development* **41**(2), 27–52.

Paldam, M. and Gundlach, E. 2008. Two views on institutions and development: the grand transition vs the primacy of institutions. *Kyklos* **61**(1), 65–100.

Poprawe, M. 2015. On the relationship between corruption and migration: empirical evidence from a gravity model of migration. *Public Choice* **163**(3–4), 337–54.

Rose-Ackerman, S. 1999. *Corruption and Government: Causes, Consequences and Reforms.* Cambridge: Cambridge University Press.

Schneider, F. 2007. Shadow economies and corruption all over the world: what do we really know? CESifo Working Paper Series No. 1806. https://papers.ssrn.com/sol3/papers.cfm?abstract_id=938369 (accessed July 28, 2017).

Seldadyo, H. and De Haan, J. 2005. The determinants of corruption. *The Economist*, 66.

Serra, D. 2006. Empirical determinants of corruption: a sensitivity analysis. *Public Choice* **126**(1–2), 225–56.

Sims, R., Gong, B., and Ruppel, C. 2012. A contingency theory of corruption: the effect of human development and national culture. *Social Science Journal* **49**(1), 90–7.

Tanzi, V. 1998. Corruption around the world: causes, consequences, scope, and cures. *IMF Staff Papers* **45**(4), 559–94.

Treisman, D. 2000. The causes of corruption: a cross-national study. *Journal of Public Economics* **76**(3), 399–457.

Truex, R. 2011. Corruption, attitudes, and education: survey evidence from Nepal. *World Development* **39**(7), 1133–42.

3.   Effects of corruption on human capital and economic growth in developing countries
Asma Sghaier and Asma Guizani

Corruption, defined as the abuse of a public office in order to obtain a private gain, has become a universal phenomenon affecting all countries and, in particular developing countries, to different degrees. Corruption can affect the countries that are democratic and non-democratic, the rich and poor ones. Over several decades, several authors have demonstrated that corruption retards economic growth (Leff 1964; Huntington 1968). Others have shown that corruption's effects on the management of public affairs generates effects such as the poor allocation of resources: public expenditures are less effective in countries that are most corrupt, with the fact that the corrupt officials will defend the investment projects to the creators of "bribes" and not the more productive (Shleifer and Vishny 1993). Corruption restricts the efficiency of both the private and public sectors and diverts flows of finance inappropriately. Finally, it introduces political instability and anarchy in the political process.

Concerning the question of the influence of corruption on economic growth, one of the themes least studied is whether corruption is detrimental to the accumulation of human capital, for example, by discouraging young people to pursue their graduate studies. In effect, in an environment of strong corruption, easy earnings encourage young people to stop their studies. Corruption also modifies the distribution of public expenditure in education and health, and decreases the quality of the services supplied (Ablo and Reinikka 1998).

Reducing this misfortune would achieve significant improvements. Pranab Bardhan (1996) stipulates that the ability to control corruption depends on the credibility of the government as well as the establishment of credible institutions. Good governance of the institutions of any state is an important part of any policy to fight against corruption.

The objective of our work is to illustrate the slowdown of economic growth that can result from corruption, especially as it forms an obstacle to human development by reducing the effectiveness of public spending. Several economists have explored the direct impacts of corruption on economic growth (e.g., in terms of investment, training of human capital,

31

law, and so on), but they have not examined the channels through which corruption can influence economic growth. We are interested in the inter-actions between corruption and human capital formation and on proving their effect on economic growth. We study the impact of corruption on the investment in human capital.

The chapter is structured as follows. First, we examine the concept of corruption, its causes and effects. A review of the literature on the impacts of corruption on the macroeconomic performance of the countries, as well as an empirical application illustrating the impact of the interaction between corruption and human capital on economic growth are then presented. Finally, we conclude with recommendations to the decision makers of economic policy for interventions that are more rational and effective in developing countries.

CORRUPTION: CONCEPTS, CAUSES, AND CONSEQUENCES

Corruption has repeatedly been shown to be a central problem of the economy of development. Its magnitude in developing countries is sufficient that its economic and social costs generate obstacles to the achievement of economic reforms, making them a main objective which is needed in the analysis of the economics of development. Because of the magnitude of corruption, it is difficult to find a single definition that satisfies the different practices in each country. Several definitions have been proposed.

According to the Islamic religion, corruption is a set of moral disrup-tions. The Koran addresses this issue by using the concept of *fasad*, an Arabic word that can be loosely translated as corruption. Discussion of corruption in the Koran is broader than the simple concept expressed as the misuse of power. Iqbal and Lewis (2002) have shown through the Islamic perspective that corruption is explicitly condemned by Sharia law. The term *fasad* brings together all human behaviors that disrupt the lives of individuals, create social stability, and are harmful to sustainable development. This theme is illustrated in Chapter 30, verse 41, which sets out that the *fasad* affects the earth and the sea following the human behaviors and actions: "Corruption appears on the Earth and in the sea because of what humans have done with their own hands, so that it leaves the taste of wrongdoing of their work" (Surah Arroum 30, V41). Allah recommends ignoring corruption: "and do good as God has made thee, and do not search for corruption on earth. Because in truth, God loves not the corrupters" (Surah al Kasas 28, V77).

Corruption is generally defined as the abuse of a public service with the intent to improve personal gain. Alesina and Weder (2002) presented corruption as the misuse of state property by a public servant to draw an individual profit. Mishra (2005) adopts the same definition of corruption, holding that corruption is sensitive to the bribes that individuals face. This phenomenon can exist in the form of an exchange, a favor, or access to a public service in return for a monetary or reciprocal favor (Ganuza and Celentani 2002).

Public opinion varies among countries with regard to certain practices which may be defined as corruption but which do not necessarily constitute certain offenses according to the national criminal laws. As the form of corruption is not the same in different countries, this makes the existence of a common definition difficult.

Most theoretical studies on the causes of corruption point to profound institutional weaknesses. Tanzi and Davoodi (1998), Mauro (1996), Gray and Kaufmann (1998), Sekrafi and Sghaier (2016) and the World Bank (2002) classified the causes of corruption into four categories: institutional, political, social, and economic. Corruption is stronger in countries where the public sector is well developed and where we find significant regulations, taxes, and trade restrictions. Institutional environments with an opacity of decision-making processes are likely sources of corruption and inequality. Corruption flourishes in countries where public institutions such as the judiciary, civil society, and non-governmental organizations are marginalized. Corruption is also a function of how political power is exerted, including over the institutions that have an influence on corruption. Indeed, much evidence indicates that civil society has a capacity to combat corruption effectively in developed countries. However, studies of economies in transition suggest that their policies and laws tend to benefit private commercial interests rather than individual freedoms (Hellman et al. 2000). The social environment can also be conducive to corruption: corruption is blatant in cases where the members of the society, especially when the majority is illiterate, are unaware of their rights and which public services are free of charge to the public. Corruption is commonly viewed as an economic crime. It tends to be highest in economies with low levels of growth. Low wages and salaries of public officials, including police and customs officials, are common drivers.

If the causes of corruption are many and vary from one country to another, the consequences are also diverse. Mauro (1997) and Sekrafi and Sghaier (2016) suggested that corruption lowers the economic growth of countries. The primary manifestations of this include: the under-utilization of stakeholders in society, notably women (Murphy et al. 1991); low levels of domestic and foreign investment (Mauro 1997); the emergence of an underground or black market economy that distorts the development of firms (Johnson et al. 1998); the poor distribution of expenditures and public investments; and the deterioration of the physical infrastructure. Tanzi and Davoodi (1997) estimated that corruption promotes investments that are unproductive. Public expenditures shaped by corruption negatively affect the economy. This decreases the productivity of public investments by lowering the quality of infrastructure and public services and by increasing the costs of goods and services. Circumvention of the laws and the police by some companies also limits production and investment in the private sector. Huntington (1968), Leff (1964) and Liu (1985) reported a different rationale, that is, that bribes can play a positive role in the development of firms.

Finally, and despite these perverse effects, especially for the case of developing countries, some believe that corruption is not really a problem since it produces economic benefits and may reduce the inefficiency of public bureaucracies. Several international institutions are concerned with putting in place policies and anti-corruption strategies in order to eliminate corruption or at least to reduce their costs.

The relationship between the quality of public institutions and economic growth has been the subject of a renewed interest with the development of analyses of endogenous growth and the construction of databases on institutions. At the core of recent theories of endogenous growth is human capital, which is one of the key factors driving economic growth. However, there are very few empirical studies concerning the relationship between corruption, public spending, and economic growth in developing countries.

By studying the causes and consequences of the performance of local public authorities in a climate of great corruption, Rotberg (2004) and Shera et al. (2014) showed that corruption is primarily harmful to economic growth through its impacts on investment. Mauro (1995) also studied the relationship between corruption and investment, specifically, investment as a share of gross domestic product (GDP). He showed that a high degree of corruption leads to a decrease in investment and concluded that corruption is unfavorable to growth and economic development.

Some economists believe that corruption decreases investment by affecting its composition and reducing the quality of the infrastructure of a country. These perverse effects affecting the process of the construction of infrastructures and projects are studied by Laffon and N'Guessan (1999) and by Laffon and N'Gbo (2000). Braguinsky (1996) reached the same result that indicates that the asymmetry of information between the state and corrupt parties is an important factor behind corruption.

Similarly, Tanzi and Davoodi (1997) argued that corruption is able to increase public investment while reducing its productivity. They demonstrated that a high level of corruption is associated with high costs of infrastructure maintenance, which has a negative impact on economic growth. Countries that are the most corrupt have lower levels of public expenditure on education (Mauro 1998). This result was confirmed by Gupta et al. (1998), who show that anti-corruption strategies reduce income inequalities and poverty.

More recently, Pellegrini and Gerlaugh (2004) and Mauro (2000, 2001) analysed empirically the direct influence of corruption on economic growth as well as the indirect effects. In general, these studies have shown that corruption negatively affects economic growth through its effects on investment and international trade. Pellegrini and Gerlaugh (2004) suggest that there is a statistically significant relationship between corruption and economic growth if all the determinants are controlled.

Martinez-Vasquez (2005) have shown that corruption, in addition to its direct influence on the growth of GDP per capita, can indirectly affect the accumulation of human capital. For this reason, any effort to reduce corruption has positive effects on the level of GDP per capita. Accordingly, and following this review of theoretical and empirical literature, we offer the work that follows specifying the nature of the relationship between corruption, human capital, and economic growth of developing countries.

THE MODEL

On the basis of the literature showing that human capital is one of the determinants of economic growth, we assess the impacts of corruption on this variable. Like the work of Mankiw et al. (1992), Knight et al. (1993), Ghura and Hadjimichael (1996), and Demetriades and Law (2004), we use the Cobb-Douglas production function:

$$\ln (y_{it}) = \alpha_i + \beta_1 \ln k_{it} + \beta_2 \ln h_{it} + \beta_3 \ln FL_{it} + \beta_4 INS_{it} + \varepsilon_{it} \quad (3.1)$$

where $i = 1, \ldots, 25$ and $t = 1994, \ldots, 2015$, y is the real GDP per capita in constant US dollars for the year 2000, k is the physical capital, FL is the rate of growth of the labor force, h is human capital, INS is the quality of the institutions, and ε_{it} is the error term.

Since the quality of the institutions, in addition to their direct impacts, has an indirect impact via human capital, in order to examine the effect of the interaction between corruption and human capital on economic growth, equation (3.1) is expanded to include the interaction term as follows:

$$\ln (y_{it}) = \alpha_i + \beta_1 \ln k_{it} + \beta_2 \ln h_{it} + \beta_3 \ln FL_{it} + \beta_4 INS_{it} + \beta_5$$
$$(h_{it} * INS_{it}) + \varepsilon_{it} \tag{3.2}$$

Equations (3.1) and (3.2) represent the empirical models that will be used in this work.

The data from our study include a sample of 25 developing countries that are divided among Africa, Asia, and Latin America. These countries are classified by the World Bank, which provides data on the institutional characteristics of these countries. The observation period extends from 1994 to 2015, or 550 observations.

The stock of physical capital is calculated using the Perpetual Inventory Method devised by Van Pottelsberghe (1997). The index of the level of instruction in education (Education Index) is used as a proxy for human capital, calculated from the rate of literacy and the enrolment rate in primary, secondary, and higher education.

The institutional variables used are corruption, which has a score ranging from 0 to 6, and political stability, which ranges between 0 and 12. In general, the higher values indicate lower levels of corruption and a low instability of the government. The indicator of corruption (CORR) is a measure of the abuse of the public service for personal gain. Lower scores indicate that the senior officials are very corruptible and that corruption exists through the entirety of the national administration. The indicator of political stability (GS) reflects the political violence and instability in the country.

RESULTS AND INTERPRETATIONS

The majority of previous studies, such as Hall and Jones (1999), Acemoglu et al. (2001), Rodrik et al. (2004), and Easterly and Levine (2003), used heterogeneous samples of developed and developing countries in order to analyse corruption in terms of a composite index of the quality of public institutions. The results of these works are not convincing because they do not take into account the institutional differences among countries and subsequently do not indicate the relationship that may exist between the institutional environment and economic growth. For this reason, we used a homogeneous sample composed of 25 developing countries during the period from 1994 to 2015, including several that have not been included in previous studies, such as Syria, Jordan, Iran, El Salvador, and Peru. In addition, we used institutional indices that are specific to each country in order to uncover the impact of institutional failures on economic growth. We used a growth model that includes human capital as well as institutional variables.

After having carried out tests of stationarity for all the series included in our model (see Appendix), we found that five of the six series are stationary. We took into account the individual effects that are specific to each country to capture differences both economically and institutionally. This individual heterogeneity is verified by the results found in the descriptive statistics that show there was a difference between the minimum value and the maximum value of the economic variables and institutional categories used in each country. The results of the estimation of these two regressions are presented in Table 3.1. The dependent variable is ln GDP per capita.

For the macroeconomic variables, the results are similar to those expected: in effect, the labor force and investment in physical capital and human capital are statistically significant. These confirm the results of Levine and Renelt (1992) in their study of the main determinants of growth, where they found that the physical investment was positively and significantly correlated with the rate of economic growth. These results have been reaffirmed in particular by the study of Easterly et al. (1997), which stipulated that investment in education and human capital leads to the acquisition of skills and encourages technological advances.

The empirical evidence provided by Barro (1991, 1997), Benhabib and Spiegel (1994), and various other researchers suggests that the level of education is an important determinant of future growth. In addition, technology, labor markets, human capital, and natural resources also explain growth and economic development. Nevertheless, these are only the direct causes and they have limited effects in the absence of favorable institutions.

One of the main characteristics of developing countries is political instability, which is stronger when levels of corruption are high. For this reason, we have excluded from our sample countries (e.g., Lebanon, Palestine) whose political instability is serious in order to control for the role of the stability of the government in the establishment of an effective economy.

To estimate the contributions of institutional variables in explaining corruption, we add them to the analysis one by one. When we add the indicator of the stability of the government (GS), R^2 rises to 0.927, and with the introduction of the indicator of corruption (CORR), it rises to 0.921. These two indicators are statistically more significant at the 99 percent confidence level. Political stability is positively correlated with economic growth; in other words, a stable environment is more conducive to the sustainability of economic growth. The relation of growth to corruption is negative, as expected. The results for all 25 countries confirm the conclusions of Barro (1991) and Londregan and Poole (1996)

Table 3.1 Direct and indirect impact of the quality of institutions on economic growth

Variables	M1		M2		M3		M4		M5	
	Within	MCG	Within	MCG	Within	MCG	Within	Within	Within	MCG
Cst	6.34 (35.00)	7.30 (55.28)	6.18 (35.43)	6.94 (51.29)	6.32 (35.27)	7.32 (55.96)	6.11 (35.70)	6.83 (52.04)	6.24 (34.86)	7.19 (54.86)
Lnk	0.26 (6.05)***	0.02 (1.39)	0.21 (5.21)***	0.02 (1.26)	0.27 (6.40)***	0.03 (1.59)	0.21 (5.03)***	0.02 (1.29)	0.27 (6.30)***	0.03 (1.76)
Lnh	0.47 (4.02)***	0.49 (4.24)***	0.24 (2.05)**	0.24 (2.15)	0.36 (3.03)**	0.40 (3.43)	–	–	–	–
LnFL	−0.29 (−9.69)***	−0.33 (−11.07)	−0.19 (−6.02)***	−0.21 (6.57)	−0.28 (−9.18)***	−0.32 (−10.69)	−0.19 (−5.77)***	−0.20 (−6.22)	0.31 (−10.65)***	−0.35 (−12.27)
GS	–	–	0.02 (7.15)***	0.02 (7.99)	–	–	–	–	–	–
CORR	–	–	–	–	−0.03 (−3.43)***	(−0.02) (−2.82)	–	–	–	–
h*GS	–	–	–	–	–	–	0.03 (8.19)***	0.04 (9.08)	–	–
h*CORR	–	–	–	–	–	–	–	–	−0.05 (−3.88)***	−0.04 (−3.27)
No.d'ob	550	550	550	550	550	550	550	550	550	550
R^2	0.919	0.255	0.927	0.330	0.921	0.264	0.926	0.330	0.919	0.246
t-Haus	–	41.91	–	34.74	–	48.11	–	30.51	–	46.30
P-values	–	(0.00)	–	(0.00)	–	(0.00)	–	(0.00)	–	(0.00)
t-DWH	–	–	2.74	–	0.04	–	2.07	–	3.06	–
P-values	–	–	(0.12)	–	(0.83)	–	(0.24)	–	(0.14)	–

Note: *** significant at 1%, ** significant 5%, and * significant at 10%.

Source: Calculations by the author.

regarding political stability and corruption, and that of Mauro (1995, 1996) for economic growth.

Good public institutions and a better quality of governance are important not only in and of themselves, but they also help developing countries to improve their economic performance. Similarly, stability seems to be an important prerequisite for developing countries. Several researchers have studied the relation between political stability and economic growth in developing countries. Venieris and Gupta (1986) and Devereux and Wen (1996) showed that political instability, including the number of changes of governments and sometimes the number of changes of high level offices, reduces foreign and domestic investment. Similarly, Edwards and Tabellini (1991) and Alesina et al. (1996) insisted on strengthening policies that promote stability because instability led to a poor management of public affairs and negatively affects economic growth. This pattern is confirmed by Murphy et al. (1991), Shleifer and Vishny (1993), and Mauro (1995), who all showed that corruption reduces economic growth.

We demonstrate that institutional variables, that is, political stability and corruption, in addition to their direct effects on economic growth have indirect effects through the accumulation of human capital. This observation is confirmed by results showing that the significance of human capital decreased to 5 percent when we added the variables political stability and corruption. Given that the quality of institutions may have an indirect effect on economic growth through the accumulation of human capital, we deployed two interaction terms. The first measures the interaction between human capital and political stability, which is positive and statistically significant at the 99 percent confidence level. This result indicates that in a stable environment, military expenditures decrease the benefits of public expenditures in education. Political stability is based on investments in physical and human capital and encourages citizens to complete their graduate studies, creating productive human capital. In other words, political stability sustained by public expenditures in education is productive and increases economic growth.

The relationship between political stability and growth can be explained by the accumulation and the efficiency of the factors of production (FOSU 1992; Dixit and Pindick 1994). Stability ensures the protection of property rights, which increases their return to investments. FOSU (1992) proves that the same thing applies to the accumulation of human capital, since stability ensures people can use their skills. Guillaumont et al. (1999) call into question these results in the case of African countries and show that political instability, defined as a combination of coups and civil wars, directly affects growth.

Political instability can also decrease the accumulation of human capital. Thus, political instability in Latin America has been one of the main reasons why the region's countries have relatively low levels of human capital.

The second term added in our model was the interaction between human capital and corruption, which is negatively correlated with economic growth and is statistically significant at the 99 percent confidence level. This result is consistent with that of studies indicating that corruption alters the structure of public expenditures. Specifically, a level of corruption deemed high has a tendency to reduce the productivity of human capital (Devarajan et al. 1996).

Our empirical test enriches the economic analyses showing that corruption reduces economic growth. It follows that the allocation of public expenditures depends on the quality of economic policies, which is confirmed by the work of Anderson and Tverdora (2003) and Lambsdorff (2003). Guetat (2006) similarly showed that the indirect impact of corruption on the long-term economic growth of the Middle East/North Africa region is transmitted via its effects on investment and human capital.

This result leads to a reflection on the nature of the socio-economic environment and on the need and efficiency of public contributions to growth. Inefficient public spending reduces economic performance. For this reason, good governance is imperative in developing countries, including the restructuring of the public sector, improved budgeting and financial management, and more efficient tax administration (Ciocchini et al. 2003). The prevalence of inefficiency reflects that agencies providing public services must be held responsible for their actions to certify the effectiveness and inclusiveness of these services.

The negative impact of the interaction between human capital and corruption is explained in two ways. First, the expenditure must be financed by taxes, which alter economic decisions and can generate losses of well-being. Second, any public expenditure may divert activity toward functions that are not productive, which acts as a second tax. Devarajan et al. (1996) concluded similarly when they examined public expenditures, notably in countries where expenditures on equipment are excessively high and expenditures on education and health are insufficient.

In conclusion, developing countries suffer from certain institutional failures that include the ineffectiveness of the taxation system, lack of management skills, insufficient technical skills, inefficient financial markets, and the low credibility of the states (Laffont and N'Guessan 1999; Laffont and N'Gbo 2000). We have stressed the impacts of human capital on growth in recalling the problems that are specific to education in developing countries, which are closely linked to institutional failures.

CONCLUSION

Throughout this work, we have advanced different definitions of corruption that summarize the debate over its impacts. We have emphasized the effect of corruption on economic growth as it is conceived in the economic literature and tried to verify its impacts empirically. The objective of this study is to analyse the interactions between political stability, corruption, and human capital and their impacts on growth. The results obtained show that policies that promote stability have a direct effect as well as a positive indirect effect through the accumulation of human capital, while corruption acts negatively on the accumulation of human capital.

It is clear from these results that developing countries characterized by institutional deficiencies and shortcomings related to high levels of corruption suffer from low rates of economic growth. In effect, malfunctioning public institutions discourage education and result in a negative impact on economic growth. Although these countries may possess natural and human resources, a high level of corruption means that the political institutions are less democratic and effective. For this reason, any policy in the fight against corruption has significant effects on economic growth and on public expenditures on education.

REFERENCES

Ablo, E. and Reinikka, R. 1998. Do budgets really matter? Evidence from public spending on education and health in Uganda. World Bank Policy Research Working Paper, Washington, DC.

Acemoglu, D., Johnson, S., and Robinson, J. 2001. Reversal of fortune: geography and institutions in the making of the modern world income distribution. NBER Working Paper No. 8460, National Bureau of Economic Research, Washington, DC.

Alesina, A. and Weder, B. 2002. Do corrupt governments receive less foreign aid? *American Economic Review* **92**(4), 1126–37.

Alesina, A., Ozler, S., Roubini, N., and Swagel, P. 1996. Political instability and economic growth. *Journal of Economic Growth* **1**, 193–215.

Anderson, V. and Tverdora, Y. 2003. Corruption, political allegiances and attitudes toward government in contemporary democracies. *American Journal of Political Science* **47**(1), 91–109.

Bardhan, P. 1996. The economics of corruption in less developed countries: a review of issues. Center for International and Development Economics Research Working Papers C96-064, University of California at Berkeley.

Barro, R. 1991. A cross-country study of growth, saving, and government. In B.D.

Bernheim and J.B. Shoven (eds), *National Saving and Economic Performance*, pp. 271–304. Washington, DC: National Bureau of Economic Research.

Barro, R. 1997. Myopia and inconsistency in the neoclassical growth model. NBER Working Paper 6317, National Bureau of Economic Research, Washington, DC.

Benhabib, J. and Spiegel, M. 1994. The role of human capital in economic development: evidence from aggregate cross-country data. *Journal of Monetary Economics* **34**(2), 143–73.

Braguinsky, S. 1996. Corruption and Schumpeterian growth in different economic environments. *Western Economic Association International* **14**(3), 14–25.

Ciocchini, F., Durbin, E., and David, T. 2003. Does corruption increase emerging market bond spreads? *Journal of Economies and Business* **55**, 503–28.

Demetriades, P. and Law, S. 2004. Finance, institutions and economic growth. Working Paper in Economics 4/5, Department of Economics, University of Leicester.

Devarajan, S., Swaroop, V., and Zou, H.-F. 1996. The composition of public expenditure and economic growth. *Journal of Monetary Economics* **37**, 313–44.

Devereux, M. and Wen, J. 1996. Political uncertainly, capital taxation and growth. Mimeo, University of British Colombia.

Dixit, A. and Pindick, R. 1994. *Investment under Uncertainty*. Princeton, NJ: Princeton University Press.

Easterly, W. and Levine, R. 2003. Tropics, germs, and crops: how endowments conferences influence economic development. *Journal of Monetary Economics* **50**(1), 3–39.

Easterly, W., Norman, L., and Montiel, P. 1997. Has Latin America's post-reform growth been disappointing? *Journal of International Economics* **43**(3–4), 287–311.

Edwards, S. and Tabellini, G. 1991. The political economy of fiscal policy and inflation in developing countries: an empirical analysis. Policy Research Working Paper Series 703, World Bank, Washington, DC.

Fosu, A. 1992. Political instability and economic growth: evidence from Sub-Saharan Africa. *Economic Development and Cultural Change* **40**(4), 829–41.

Ganuza, J. and Celentani, M. 2002. Corruption and competition in procurement. *European Economic Review* **46**(7), 1273–303.

Ghura, D. and Hadjimichael, M. 1996. Growth in sub-Saharan Africa. *International Fund Monetary Staff Papers* **43**, 605–34.

Gray, C. and Kaufmann, D. 1998. Corruption and development. *Finance and Development* **35**(1), 7–10.

Guetat, I. 2006. The effects of corruption on growth performance of MENA countries. *Journal of Economics & Finance* **30**(2), 208–21.

Guillaumont, P., Guillaumont, J.. and Brun, J. 1999. How instability lowers African growth. *Journal of African Economies* **8**(1), 87–107.

Gupta S., Davoodi, H., and Alonso-Terme, R. 1998. Does corruption affect income inequality and poverty? IMF Working Paper 98-76.

Hall, R. and Jones, V. 1999. Why do some countries produce so much more output per worker than others? *Quarterly Journal of Economics* **114**(1), 83–116.

Hellman, J., Jones, G., and Kaufmann, D. 2000. Seize the state, seize the day: state capture, corruption, and influence in transition. World Bank Policy Research Working Paper No. 2444, Washington, DC.

Huntington, S. 1968. *Political Order in Changing Societies*. New Haven, CT: Yale University Press.

Iqbal, Z. and Lewis, M. 2002. Governance and corruption: can Islamic societies and the West learn from each other? *American Journal of Islamic Social Sciences* **19**(2), 1–33.

Johnson, S., Kaufmann, D., and Zoido-Lobatón, P. 1998. Regulatory discretion and the unofficial economy. *American Economic Review* **88**(2), 387–92.

Knight, N. et al. 1993. What we know about the socio-economic impacts of Canadian megaprojects: an annotated bibliography of post-project studies. Centre for Human Settlements, University of British Columbia, Vancouver.

Laffont, J. and N'Guessan, T. 1999. Competition and corruption in an agency relationship. *Journal of Development Economics* **60**, 271–95.

Laffont, J. and N'Gbo, A. 2000. Cross-subsidies and network expansion in developing countries. *European Economic Review* **44**, 797–805.

Lambsdorff, J. 2003. How corruption affects persistent capital flows. *Economics of Governance* **4**(3), 229–44.

Leff, N. 1964. Economic development through bureaucratic corruption. *American Behavioral Scientist* **8**(3), 8–15.

Levine, R. and Renelt, D. 1992. A sensitivity analysis of cross-country growth regressions. *American Economic Review* **82**(4), 942–63.

Londregan, J. and Poole, K. 1996. Does high income promote democracy? *World Politics* **49**, 1–30.

He, F. 1996. Three aspects of corruption. *Contemporary Economic Policy* **14**(3), 26–9.

Liu, F.T. 1985. An equilibrium queuing model of bribery. *Journal of Political Economy* **93**(4), 760–81.

Mankiw, N., Romer, D. and Weil, D. 1992. A contribution to the empirics of economic growth. *Quarterly Journal of Economics* **107**(2), 407–37.

Martinez-Vasquez, J. (2005). Corruption, investment and growth in developing countries. Defense Resources Management Institute Working Paper Series. http://hdl.handle.net/10945/32540 (accessed May 2018).

Mauro, P. 1995. Corruption and growth. *Quarterly Journal of Economics* **110**, 681–712.

Mauro, P. 1996. The effects of corruption on growth, investment, and government expenditure. International Monetary Fund Working Paper, September, pp. 1–28. https://ssrn.com/abstract=882994 (accessed May 21, 2018).

Mauro, P. 1997. *The Effects of Corruption on Growth, Investment and Government Expenditure: A Cross Country Analysis.* Washington, DC: Institute for International Economics.

Mauro, P., 1998. Corruption and composition of government expenditure. *Journal of Public Economics* **69**, 263–79.

Mauro, P. 2000. Stock returns and output growth in emerging and advanced economies. IMF Working Papers 00/89, International Monetary Fund, Washington, DC.

Mauro, P. 2001. Corruption and growth. *Journal of Comparative Economics* **29**(1), 66–79.

Mishra, A. 2005. Indian venture capitalists' (VCs) investment evaluation criteria. *Quarterly Journal of Economics* **106**(2), 503–30.

Murphy K., Shleifer, A. and Vishny, R. 1991. The allocation of talent: implications for growth. *Quarterly Journal of Economics,* **106**(2), 503–30.

Pellegrini, L. and Gerlaugh, R. 2004. Corruption's effect on growth and its transmission channels. *Kyklos* **57**(3), 429–57.

Rodrik, D., Subramanian, A., and Trebbi, F. 2004. Institutions rule: the primacy of institutions over geography and integration in economic development. *Journal of Economic Growth* **9**(2), 131–65.

Rotberg, I. 2004. The failure and collapse of nation-states: breakdown, prevention, and repair. In I. Rotberg (ed.), *When States Fail: Causes and Consequences,* pp. 1–51. Princeton, NJ: Princeton University Press.

Sekrafi, H. and Sghaier, A. 2016. Exploring the relationship between tourism development, energy consumption and carbon emissions: a case study of Tunisia. *Management Strategies Journal* **32**(2), 34–43.

Sekrafi, H. and Sghaier, A. 2018. The relation between corruption, energy consumption and CO_2 emissions: the case of Tunisia. *PSU Research Review* **2**(1), 81–95.

Shera, A., Dosti, B., and Grabova, P. 2014. Corruption's impact on economic growth: an empirical analysis. *Journal of Economic Development, Management, IT, Finance and Marketing* **6**(2), 57–77.

Shleifer, A. and Vishny, R. 1993. Corruption. *Quarterly Journal of Economics* **108**(3), 599–617.

Tanzi V. and Davoodi, H. 1997. Corruption, public investment and growth. IMF Working Paper 97/139, Washington, DC.

Tanzi, V. and Davoodi, H.R. 1998. *Roads to Nowhere: How Corruption in Public Investment Hurts Growth.* Washington, DC: International Monetary Fund.

Van Pottelsberghe, B. 1997. Issues in assessing the effect of interindustry R&D spillovers. *Economic Systems Research* **9**(4), 331–56.

Venieris, Y. and Gupta, D. 1986. Income distribution and socio-political instability as determinants of savings: a cross-sectional model. *Journal of Political Economy* **94**, 873–83.

World Bank. 2002. Literature survey on corruption 2000–2005. http://www1.worldbank.org/publicsector/anticorrupt/ACLitSurvey.pdf (accessed May 21, 2018).

APPENDIX

Table 3A.1   Results of IPS stationarity tests (2003)

Variable	Stationarity in level		Stationarity in first difference	
	With constant	With constant & trend	With constant	With constant & trend
Ln GDP/	1.59504	2.39797	−1.51708	−2.43042
capita	(0.9446)	(0.9918)	(0.0646)	(0.0075)
Ln K	−7.21893*	−8.69787*	−33.7277	−25.3933
	(0.0000)	(0.0000)	(0.0000)	(0.0000)
Ln H	−1.94162*	−2.65163*	−15.8463	−12.9244
	(0.0023)	(0.0005)	(0.0000)	(0.0000)
Ln FL	−3.23567*	−3.46397*	−4.80842	−3.21568
	(0.0004)	(0.0003)	(0.0000)	(0.0007)
H*GS	−2.30210*	−3.93705*	−13.5973	−9.61145
	(0.0016)	(0.0000)	(0.0000)	(0.0000)
H*CORR	−2.22197*	−3.19542*	−11.3326	−7.81612
	(0.0031)	(0.0007)	(0.0000)	(0.0000)

Note: If the realization of the statistic of Im, Pesaran and Shin is lower than the threshold of the normal centered reduced law (for a non-symmetric test at 5%, this threshold is equal to −1.645), we reject the null hypothesis of unit root for all series.

Source: Estimate made by the author with Eviews 6. With * means stationary level series.

4. Gender and corruption: institutions and mechanisms of accountability
Helena Olofsdotter Stensöta and
Lena Wängnerud

Corruption, often defined as "the misuse of public office for private gain" (Rose-Ackerman 1999), reduces economic growth (Mauro 1995) and general trust (Morris and Klesner 2010) as well as various dimensions of human well-being such as health, access to clean water, and education (Ciccone et al. 2014; Gupta et al. 2000). Further, corruption negatively affects various subjective dimensions of human well-being such as life satisfaction and happiness (Holmberg and Rothstein 2012; Tavits 2008; Treisman 2007).

Gender is related to corruption in several ways (Stensöta and Wängnerud 2017). First, women are negatively affected by the consequences of corruption to a larger extent than men (Hossain and Nyamu-Musembi 2010). This is mainly because they are generally responsible for everyday tasks of life, which makes them more dependent on public service delivery that might be negatively affected by corruption. As several reports from the United Nations (UN) have established, women, and especially poor women, are vulnerable to pressures of different kinds in order to access social services (Goetz 2008; Hossain and Nyamu-Musembi 2010). Second, corruption can hinder women's access to positions in the public sphere. Corruption seems to be more frequent when power is located in informal networks, which often are dominated by men, and some studies focusing on the importance of networks have found these to exclude women from appointments in elected assemblies (Bjarnegård 2013). Third, women in power seem to be less corrupt than men, as higher proportions of female representation in government correlate with lower national levels of corruption. Moreover, similar patterns – showing women as less corrupt than men – are found when individual-level data are examined, including in the area of business, examining managers' involvement in bribery (Swamy et al. 2001).

The research focusing on how women in power affect the level of corruption was sparked some 20 years ago by research groups linked to the World Bank, through the report by Dollar et al. (2001) titled "Are women really the fairer sex? Corruption and women in government." The authors demonstrated that higher proportions of female representation in national parliaments were associated with lower national levels of corruption. Corruption was measured through data from the International Country Risk Guide, and control variables that could have been responsible for a spurious relationship, such as civil liberties (Gastil's civil liberties index), were included in the analysis. When interpreting the findings, the authors drew on literature holding that women have higher standards of ethical behavior and are more concerned with the common good. Hence, they concluded that "if women are less likely than men to behave opportunistically, then bringing more women into government may have significant benefits for society in general" (p. 427).

In another study presented the same year, also from a group linked to the World Bank, Swamy et al. (2001) provided further evidence on the link between gender and corruption and presented new ideas on why the relationship appears. They used World Values Survey data showing that women, at the individual level, were more eager to condemn bribe-taking than men, and they suggested the possibility that women may be more inclined than men to follow laws because they consider laws as protection, for example, against violence. In addition, they argued that girls may be brought up to endorse higher levels of self-control than boys, which might prevent them from engaging in criminal acts such as corruption. Swamy and colleagues also explored relationships between women in government and corruption and argued that women in power may curb corruption by not engaging in it themselves, or by taking more explicit actions against it, for example, by launching policies against corruption or by recruiting staff who are less corrupt.

However, the proposition of a link between gender and corruption was questioned in a study by Sung (2003, 2012), who instead suggested that the factors driving the relationship were to be found at the systemic level, and more precisely in the democratic foundations of societies, and not within a particular group of actors, such as women. Using more specific measurements of the institutional features that characterize liberal democracy than were used in previous research, including the Rule of Law Index and Freedom of Press ratings, Sung provided a tough test for the impact of "fair women" versus "fair system" and concluded that the relationship between higher proportions of women in government and lower levels of corruption was spurious.

Since then, research on gender and corruption has in various ways deepened the understanding of the complexity of relationships between gender, corruption, and various system-level factors. Far from having "solved" the puzzle of the exact circumstances through which the relationship between gender and corruption unfolds, a growing field has paid attention to precisely this complexity, demonstrating that there is not one single mechanism driving the relationship, but instead a number of ways that the relationship between gender and corruption can play out.

This chapter presents a review of contemporary research in the area of gender and corruption and suggests a framework that orders some main dimensions that are dominant. On the one hand, research has elaborated on individual-level mechanisms for the relationship between gender and corruption. Research in this area has used a variety of methods, including experimental, both within a single context (Alatas et al. 2009; Barnes and Beaulieu 2014; Rivas 2013) and through comparative experiments in different contexts (Alhassan-Alolo 2007). On the other hand, cross-country comparisons have produced knowledge about macro-level features affecting the relationship, but these studies seldom go into detail on the micro-foundations for the detected tendencies. In this area, research has used large-sample studies, both cross-country and/or over time (Esarey and Chirillo 2013; Schwindt-Bayer and Esarey 2017; Stensöta 2016), and more in-depth case studies (Bjarnegård 2013). In this research, a picture has emerged in which the relationship between gender and corruption is non-linear and the mechanisms involved are far from "automatic" but dependent on the context as well as the specific actors involved – for example, whether we talk about women as elected politicians or as appointed bureaucrats (Stensöta et al. 2015a).

We contend that the field is in need of a more systematic theoretical foundation accounting for the specific circumstances that affect the relationship between gender and corruption. As we read the literature, this foundation can fruitfully be built upon an inquiry into *how* and *when* gender curbs corruption. The use of these terms – *how* and *when* – indicates that we regard institutions as moderating the relationship between gender and corruption.

The second building block that we regard as promising for the field is a focus on accountability mechanisms. Accountability mechanisms have become a frequent theme in studies of good government (Bovens 2010; Feraz and Finan 2011; Hellwig 2012; Sattler et al. 2007). It is much less common, however, to discuss accountability mechanisms in gendered terms. We see two main ways that accountability mechanisms can be gendered. First, gender can strengthen or weaken already existing and acknowledged accountability mechanisms. This is the case, for example, when women react more strongly than men to the controlling power of the free press in democracies. Second, gender itself can work as an accountability mechanism. This is the case, for example, when women embody symbolic power to enhance good government, or when discrimination and exclusion are recognized as important accountability mechanisms for the improvement of government.

On a more general level, combining an institutional perspective with a focus on accountability mechanisms emphasizes accountability not as a "pure" principal-agent relationship, but how this relationship is induced by norms that may affect women and men differently. The most common way to comprehend accountability mechanisms is to consider how a principal, often captured as the public, holds an agent, often captured as their elected representatives, to account. Rooted in rational choice theory, the norms that may affect this relationship are seldom of primary interest for scholars; rather, the courses of action tend to be seen as driven by universal strategic reasoning. The way accountability mechanisms have been discussed within the field of gender and corruption challenges this dominant principal-agent picture and emphasizes norms.

The chapter is organized as follows. We first present research on how the definition of corruption may be expanded, before illustrating the range of the problem using common definitions; how the relationship between gender as female political representation in national assemblies and control of corruption varies, by a figure that illustrates both patterns, and the large variation that is to be explained. We proceed by presenting the main theoretical and empirical contributions to the field, through which the two theoretical perspectives outlined above – institutional theory and focus on accountability mechanisms – become clear. We conclude by pointing out the direction that we would like to see research on how gender plays into the dynamics of corruption explored in further detail.

EXPANDED DEFINITIONS OF CORRUPTION

An important and innovative research branch is conducted by feminist scholars and non-governmental organizations examining how the definition of corruption can be expanded from money-based forms to also include non-delivery of public services and types of "kickbacks" other than bribes. The United Nations Development Programme (Hossain and Nyamu-Musembi 2010) has highlighted that corruption often occurs in the form of illicit commissions at the point of procurement, which reduce the overall amount of public resources available for distribution and affect their equitable distribution among different population segments. Because women in most countries are the primary users of basic public services such as health, education, water, and sanitation, they are disproportionately affected by corruption in service delivery. When grassroots women in corrupt communities are asked about their perceptions of corruption, they tend to emphasize such non-delivery of goods and services. Moreover, when grassroots women are asked about tools to curb corruption, they bring forward monitoring of service delivery as one of the most important aspects (UNDP 2012).

In a study of the implementation of welfare reforms in Mexico, Hevia de la Jara (2007) documented cases of recipients, most of whom are women, being asked to do extra work for the city, that is, cleaning and sweeping streets in order to avoid losing benefits. Goetz (2007) also pointed to sexual danger for women in patronage networks, and sexual abuse might be seen as a kind of kickback.

Towns (2015) picked up this thread and developed the understanding of sexual corruption in the context of diplomacy. According to Towns, sexual corruption can be defined as "transactional relationships that involve the trade of sex for services, benefits or goods tied to public office" (p. 51). Towns argues that sexual corruption is not new, but we are not used to thinking about it as corruption. We are familiar with the phenomenon of how sex is traded against personal gain, as "stories about the relationship between sex and public power abound in popular media and popular culture" (p. 51), but this is not explicitly called corruption. For example, the term "honey trap" is used derogatorily when a person uses sex, the appeal of sex, or enticing sex to force another person into a course of action. Further, the likewise derogatory term "casting couch" refers to a person trading sex for entry into a public professional organization. Both of these examples refer to situations in which a person uses sex to gain something, similar to the way a person can use money as currency to gain something. Therefore, Towns argued, we can use the term "sexual corruption," when sex is the trade used for a corrupt act. Diplomacy is especially interesting, since this institution has not been subjected to the same pressure of transparency as other state institutions. This context has served as a breeding ground for various forms of corruption. The intersection of gender/class and the comparatively unconstrained power of diplomats (immunity) is illustrated by examples of low-status female applicants providing sex to male officials in exchange for a visa.

Recently, Lindberg and Stensöta (2018) elaborated on this line of reasoning and argued that feminist materialist theories are useful for the study of gender and corruption because they help us to expand on additional types of corrupt "currencies," and especially the inclusion of sex as currency, as in the term "sexual corruption." Second, because the notion of asymmetrical power relations, and notions of power as exploitative, can help us theorize on why women lose more than men on corrupt arrangements. From a feminist materialist perspective, power and gender is intertwined, and this makes exploitative power part of any analysis on gender (and corruption). The authors provide a summary picture of different types of sexual corruption and discuss to what extent these transactions respectively expand the study of corruption to new forms and mechanisms of corruption.

GENDER AND CORRUPTION: THE VARIATION TO BE EXPLAINED

Taking a birds-eye view of the problem of gender and corruption, we present a graph plotting the relationship between the proportion of seats held by women in national parliaments and one of the common measures of corruption, control of corruption, in order to provide an introductory picture of the problem discussed.

Figure 4.1 first shows that there is a distinction between democratic and authoritarian regimes. While there is no relationship between the proportion held by women in national parliaments and the level of corruption in authoritarian regimes (open circles), there is such a relationship in democracies (solid circles) where a higher share of female representatives in parliament correlate with a higher control of corruption. Hence, we can conclude that regime matters for how the relationship between women and corruption plays out at the government level, however, as clearly shown in Figure 4.1, there is considerable variation in between these two regime types that needs to be explained.

HOW INSTITUTIONS MODERATE THE RELATIONSHIP

Previous research on gender and corruption underpin the notion of considerable variation in gender and corruption at the meso level. Previous studies have analysed a range of institutions such as areas of government within countries (input side, such as parliament versus output side, such as bureaucracy); welfare state regimes (how encompassing and women-friendly they are); and the role of male-dominated informal networks within political organizations. In the following, we present some of this research, which demonstrates both the variation found and how this variation can be productively theorized on using institutional theory. We start, however, by briefly describing the institutional perspective.

Briefly put, meso-level analysis is situated between larger macro structures of society, such as socio-economic patterns, and micro-level analysis, focusing on individuals' attitudes and behavior. The general problem of macro-structural theory is how to account for change without having a

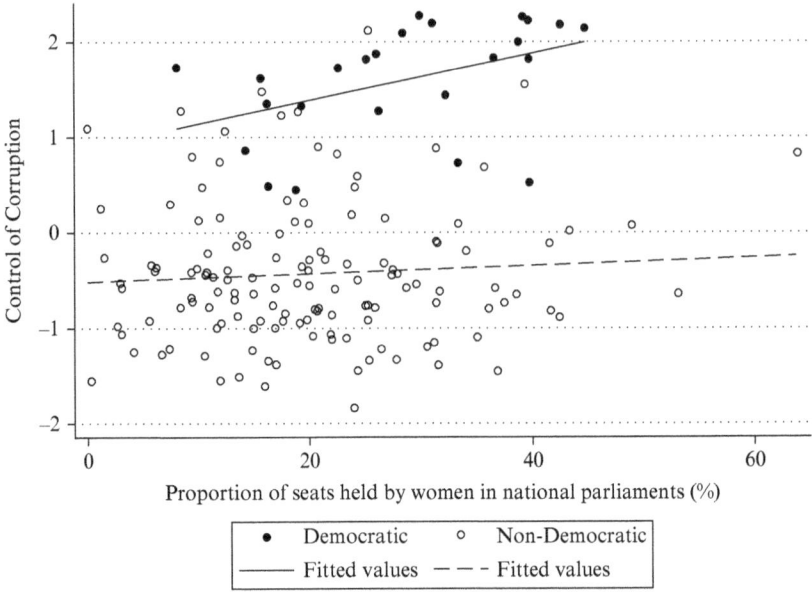

Figure 4.1   The relationship between the proportion of women in national parliaments and control of corruption

Note: Data in Figure 4.1 reflect the situation in 2014. Control of corruption is a measure produced by the World Bank where higher values mean lower levels of corruption. Information on the proportion of seats held by women in national parliaments comes from the Inter-parliamentary Union. *The Economist* has produced an index that runs from 0 to 10 where values 8–10 are considered "full democracy" and 0–8 are considered "non-democratic" states.

Source: Jan Teorell, Stefan Dahlberg, Sören Holmberg, Bo Rothstein, Anna Khomenko and Richard Svensson, 2017. The Quality of Government Standard Dataset, version Jan17. The Quality of Government Institute, University of Gothenburg. http://www.qogdata.pol. gu.se/dataarchive/qog_std_ts_jan17.dta (accessed May 24, 2018).

theoretical idea about the role of actors in promoting such change (see discussion by, e.g., Elster 1985). Micro-level approaches have no difficulty accounting for individual agency; however, they are confronted with other difficulties, for example, accounting for variations in individuals' comprehension of goals or the importance they give to norms of different kinds (Kahneman 2003). Both of these problems direct attention to the level between the macro and micro, hence the meso level. According to Douglas North's (1990, p. 4) widely held definition, institutions are "the framework within which human interaction takes place. They are perfectly analogous to the rules of the game in a competitive team sport." Hence, institutional theory offers analytical tools to capture how actors' reasoning and behavior are constrained by the formal and informal norms provided by surrounding institutions. The institutional theoretical perspective that was given the epithet "new" in the late 1980s pays specific attention to informal rules, often understood as specifying "taken for granted" notions of appropriate behavior (March and Olsen 1989).

Recently, the institutional perspective has been endorsed within feminist theory, establishing the feminist institutional perspective. Research in this area demonstrates, in a range of studies, how the institutional theoretical framework can be used to describe and understand prevailing male dominance (Kenny 2007, 2013; Krook and Mackay 2011; Larcinese and Sircar 2017; Mackay 2014; Mackay et al. 2010; Thomson 2017). Echoing the rationale for emphasizing meso-level institutions, feminist theory has a long history of focusing on structural patterns as important for patriarchy (MacKinnon 1989). In contrast to these approaches, feminist institutionalism and other contemporary approaches do not see structural theory as providing answers as to why there is variation in gender equality across time and/or across space. On the other hand, gender scholars seldom perform pure individual-level analysis, as the basic aim in this field is to recognize how actors as gendered beings do not have unconstrained free choice, but are affected by gendered societal and cultural norms. Hence, from this background, anchoring male dominance and constrained choices for females in surrounding institutional milieus seems like a compelling theoretical proposition.

Institutional Constraints Affecting Access to Power

In studies of gender and corruption, a first line of reasoning from an institutional perspective discusses the institutional constraints women face when running for office. Using the notions of male-dominated informal networks and homosocial capital, Bjarnegård (2013) has shown how entry into the political realm, in some contexts, is less a matter of merit or promoting the most qualified candidates and more the product of clientelistic exchange, social networks, and linkages. This approach of discussing corruption in terms of power, holding that it can lock women out, was initially put forward by Goetz (2007; see also Stockemer 2011).

In a recent study, Sundström and Wängnerud (2016) explored the relationship between corruption and women's political participation at the subnational level in Europe and found a correlation between higher levels of corruption and lower proportions of women in local councils. Using the informal networks hypothesis as a theoretical framework, Sundström and Wängnerud suggested that corruption indicates the presence of "shadowy arrangements" that benefit the already privileged and pose a direct obstacle to women when male-dominated networks influence political parties' candidate selection. The authors further suggested that there is a more diffuse, indirect, signal effect derived from citizens' experiences with a broad range of government authorities; the presence of corruption may be a signal of "no equal treatment" that makes women, who otherwise would have stepped forward, unwilling to stand as political candidates (see also Kenny 2013; Rothstein and Uslaner 2005).

The Importance of Regime

Institutional logics can also take place at the regime level. In fact, until recently, most research on gender and corruption was conducted in democracies, however, now, some research in authoritarian states is also available. The study by Bjarnegård (2013) mentioned above is one example. In a recent study, Nistotskaya and Stensöta (2018) addressed the importance of regime as institution by exploring in depth how the relationship between female representation in elected assemblies and bureaucracy affects child well-being at the regional level within Russia. They expand the analysis of gender and corruption to include how female representation in legislatures and bureaucracies at the regional level in Russia affects the policies regarding child mortality. Drawing from previous research, child well-being is considered part of women's interest, and operationalized as the absence of child mortality. The expected relationship from democracies would be that increased female representation in the legislature decreases child mortality. However, when Russian regions are compared, the reversed relationship is revealed, so that increased female representation is related to higher child mortality. This finding, that authoritarian regimes may use female representation to mimic democracy and possibly cover up bad policy outcomes, is a further novel mechanism introduced into the field of gender and corruption.

Further, some studies address regime as patriarchal regime. For example, the work of Alatas et al. (2009) directs attention to what they argue are patriarchal features of society. They conducted experiments in Australia, India, Indonesia, and Singapore and argued that these countries differ in regard to patriarchal structures, and that unequal gender structures in the developing countries can be expected to suppress gender differences in relation to corruption, whereas the more equal gender structures of Australia can allow them to emerge: "In relatively more patriarchal societies where women do not play as active a role in the public domain, women's views on social issues may be influenced to a greater extent by men's views" (Alatas et al. 2009, p. 678). The findings indeed indicated that gender differences emerged more clearly in Australia than in the other examined countries.

Suppressing or Enforcing?

A more comprehensive institutional theory on gender and corruption, however, needs to pay attention not only to the constraints put on women by institutions but also to how institutions may provide logics of appropriateness that may either constrain or enforce actors' strategies that impact how the relationship between gender and corruption unfolds. Here, the institutions are expected to affect norms of behavior.

Esarey and Chirillo (2013) established that the relationship between proportions of women in government and levels of corruption is evident in democratic but not in authoritarian regimes. Their interest was oriented towards the institutional arrangements that "change the incentives to appropriate public policy for private advantage" (p. 367). Their dependent variable was gender gaps in tolerance of bribes, derived from World Value Surveys. To assess institutionalized democracy and autocracy, they mainly used the Polity Score (Polity IV Project revised combined Polity Score) but also checked the robustness with other related indexes. They argued that

> When vote-buying, favoritism in government contracting, nepotism, bribery, personal loyalty over obedience to law, and other such behaviors are viewed as "corruption," when there are incentives to expose these corrupt behaviors, and when corruption is stigmatized and punished, we expect a gender gap: women will express more disapproval and be more reluctant to participate. (Esarey and Chirillo 2013, p. 367)

Since democracies entail a strong norm against corruption, the gender gap appears in these regimes but not in authoritarian states. In order to explain this pattern, the authors relied on research showing that women, on average, are more risk-averse than men.

In a previous study, we analysed the relationship between gender and corruption from an institutional perspective, distinguishing between the input and output spheres of government within countries (Stensöta et al. 2015a). We compared how the relationship between corruption and female presence in administrative institutions, on the one hand, and female presence in the electoral arena, on the other hand, plays out. Empirically, our analysis showed that while the relationship between a higher proportion of women in national parliaments and lower levels of corruption is strong and holds for a number of controls, the relationship between the proportion of women in administrative institutions and levels of corruption is much weaker. Drawing on institutional theory, we argued that the bureaucratic/administrative institution and the electoral arena provide different "logics of appropriate behavior and action." The bureaucratic logic can be described as enhancing impartial handling of cases by suppressing other preferences and orientations, thereby strengthening actors' abilities to see behind any personal characteristics of the client and realize impartiality (Rothstein and Teorell 2008). Thus, it can be hypothesized that the bureaucracy suppresses gender differences. Quite a contrary logic can be expected to thrive in national parliaments. Inspired by the work of Manin (2007), we reasoned that political candidates are motivated to "stand out" in the electoral arena and thus women can be prompted to use gendered attributes such as being a "clean" outsider in electoral races (Kostadinova and Mikulska 2015). The institutional perspective provided a way to anchor such varying logics of appropriate behavior affecting actors' attitudes and behavior in the surrounding institutions. Standing-out strategies can make use of group-specific experiences and/ or stereotypes of, for example, gender. Hence, the legislature provides hypothetically a logic that *enforces* individual or group qualities of those seeking entry, such as gender.

What do Institutions Suppress/Enforce? The Notion of Raw Material

If we reason that institutional logics suppress or enforce individual properties, this assumes that such properties are attached to actors *before* they enter the institutions. Actors are then comprehended not as "empty boxes" but as characterized by particular features. The notion of "raw material" (Stensöta 2018; Stensöta et al. 2015a) captures properties connected to gender in this latter way. Gender as raw material is not a biological or essentialist property of the individual, but composed from the socially constructed position that an individual or group has in a particular society. As previous feminist theory has outlined, macro structures in society affect the relationships between women and men: "we all find ourselves passively grouped according to structural relations" (Young 2002, p. 422). According to Young, these structural relations with importance to gender can be captured along three axes: the sexual division of labor that, in most modern societies, directs women to have more experience of reproduction than men; the normative heterosexuality; and the gendered hierarchies of power, which address the importance of power asymmetries. Asymmetries along these axes give rise to gender-specific power differences and experiences, and possibly preferences.

The notion of "raw material" may give the impression that its substance is "automatic" and part of the socialization of individuals, and, hence, that it does not require deliberate choice of action. The proposition here, however, is that it is based in choice. Many norms that we follow daily, for example, not exceeding the speed limit, waiting for the green light when crossing the street, or engaging in environmentally friendly garbage disposal, may represent social norms that we are more or less socialized into following; nevertheless, we make a choice when we decide to actually follow them and not to transgress.

In a further study, we took the first steps of combining theoretical reasoning on institutions with that of accountability mechanisms, by exploring gender gaps in citizen attitudes towards tolerance of corruption, measured as the preparedness to punish corrupt political parties, across Europe. We anchored differences in gender gaps in countries' welfare state regimes, and especially in the way the welfare state provides policies that improve self-determination for women citizens (Stensöta et al., 2015b). Hence, we argued that gendered welfare state regimes affect the relationship between gender and corruption, so that gender gaps are larger when the welfare state provides women with public services that make it easier for them to combine care and career. The assumption was that women's willingness to protect the state from "wrongdoing" should be seen as a refusal to "vote for" or "show up as a decision not to vote for" corrupt political parties. Our empirical analysis demonstrated that there is a significant gender gap in terms of how willing citizens are to punish corrupt parties, and that this gap is largest in the encompassing welfare states of the Nordic countries, especially in Sweden. In countries such as Spain and Italy, where actual levels of corruption are higher, the gender gap was much smaller. Inspiration for anchoring gendered attitudes and behavior, along the above-mentioned lines, was taken from Helga Maria Hernes's (1987) book on "women-friendly states," in which she argues that women have a special interest in having a state that encourages (economic) self-determination and offers high quality child-care facilities in combination with opportunities for paid work.

Going back to the theoretical model proposed here, the notion that institutions enforce or suppress "raw material" qualities might explain why the impact of institutions on women and men differs. However, we can also discuss how raw material qualities can vary between regime types, such as authoritarian, democratic, or welfare state regime. This can help to account for why the same type of institution may have different mediating power, depending on context. Hence, we can address outcomes due to both variation in institutions and variation in the raw material. To sum up, raw material may, however, also be divided along societal axes evoking asymmetries, and raw material qualities may also vary among countries.

The above examples demonstrate how studies have explored how meso-level variation in gender and corruption can be fruitfully theorized on, using institutional theory. We conclude that previous research has established (i) that variation on the meso level is significant, (ii) that institutions imbued with male-oriented norms may constrain female actors, but also (iii) that institutions more broadly provide logics of appropriateness that may affect the relationship between gender and corruption in multiple ways that either reinforce existing structures of power or trigger developments that change "the rules of the game."

MECHANISMS OF ACCOUNTABILITY

The second theoretical approach that we see is that research in the area of gender and corruption is increasingly using concerns about mechanisms of accountability. In her work on representation, Hanna Pitkin (1967) merged ideas from theorists who are interested in the conditions of political authorization and theorists who are interested in conditions of political accountability. Pitkin sees both these conditions as part of formal representation. Following Pitkin, accountability may refer to the ability of voters to *punish* their representatives for failing to act in accordance with their wishes (e.g., voting an elected official out of office) and also in a broader sense to the *responsiveness* of representatives to their constituents.

Accountability is an increasingly used concept in studies of good governance, where some argue that accountability should be seen as a virtue (Bovens 2010), while others argue that it should be discussed as a mechanism of control. Following the latter line of thought, there are a number of accountability mechanisms specified in previous research. In the input sphere of government, election is the basic mechanism of accountability as voters may punish leaders that they think have performed poorly by not voting for them. In the output sphere, bureaucratic institutions of accountability include performance targets and performance reviews. Moreover, meritocratic selection may enhance accountability in bureaucratic institutions (Dahlström et al. 2012). "Ombudsman" arrangements or auditing agencies are, very directly, ways to enable one set of government actors to monitor and sanction another set (Gustavson 2015). In addition, civil society provides mechanisms of accountability, such as when the free press puts out information on elite behavior that enhances transparency and accountability, but also through arrangements for incorporating citizen participation and oversight in the policy-making and policy implementation processes in between elections. We see a number of ways that accountability can be related to gender and corruption.

First, the art of holding representatives to account can be seen as especially important for subordinated groups because they are less able to rely on their own resources for survival, as these are limited. As several reports from the UN have established, women represent a group that especially benefits from increased monitoring and control of public institutions. The Goetz (2008) report titled *Who Answers to Women? Gender and Accountability* sees accountability as a core element of democratic politics and key for enhancing female power. The report understands accountability relationships as ensuring that decision makers adhere to "publicly agreed standards norms and goals" (p. 2), which may happen through two processes: when "power holders give an account of what they did with public trust and national revenue" and when "corrective action is taken as a process of enforcement of remedy, for example voting politicians out of office or setting up a judicial inquiry" (p. 2). The report further identifies two main forms of accountability that may ameliorate the situation for women; "'voice'-based approaches that emphasize collective action, representation of interests, and the ability to demand change, and 'choice'-based approaches that promote changes in the supply of responsive public service or fair market practices" (p. 4; see also Hossain and Nyamu-Musembi 2010). These more policy-oriented documents list several possible mechanisms by which governance can be enhanced through increased accountability. "Voice"-based approaches seek to publicize accountability failures and to demand accountability processes such as judicial investigations or legislative inquiries into abuses of women's rights. Further, there are additional tools for enhancing accountability such as gender mainstreaming initiatives, gender budgeting, traditional oversight mechanisms (e.g., elections, judicial review, public audits), promotion and performance review systems within public governance structures, quasi-public review bodies (e.g., ombudspersons, human rights commissions and vigilance commissions), and market regulators (Hossain and Nyamu-Musembi 2010, p. 32).

Moreover, there are indications that existing mechanisms of accountability may be gendered, meaning that they have different impacts on women and men. The work of Schwindt-Bayer and Esarey (2017) provides an example of this when they argue that the link between women's representation in national parliaments and corruption is strongest when the risk of corruption being detected and punished by voters is high. Exploring the hypothesis that a gender difference in corrupt behavior is proportional to the strength of electoral accountability, through a time series cross-sectional dataset of 76 democratic-leaning countries, the authors find support for the proposition that accountability is the key moderating factor between higher proportions of women in power and lower levels of corruption.

A similar line of reasoning is used by Schwindt-Bayer and Tavits (2016), when they put forward a "theory of clarity of responsibility" to explain variation in corruption perceptions, in which accountability plays a major role. The authors argue that "when clarity of responsibility is high, voters can easily identify and directly attribute responsibility to elected officials for corruption and this can withhold votes from incumbent elites at election time" (p. 12).

Another example of how existing mechanisms of accountability can have a stronger impact on women than men draws on how female politicians may be closer to social movements than their male counterparts. Rodríguez (2003) found, in a study of Mexico, that women politicians tend to have a strong base in civil society. In most societies, social movements serve as watchdogs for abuse of public office (Grimes 2008a, 2008b) and thus women engaging in corrupt behavior may ruin their power base for future elections. In essence, the possibly stronger connections that female politicians have to social movements might mean that they are more reluctant to engage in corrupt behavior. This means that social movements, as a form of accountability mechanism, may have a stronger effect on women than on men.

A final way that accountability mechanisms can be important in relation to gender and corruption focuses explicitly on the role of gender and sees gender itself as a possible accountability mechanism. Hence, here, gender does not interact with already acknowledged accountability mechanisms but represents a "new" mechanism. This could be the case, for example, if citizens perceive female candidates as more honest than male candidates, or when females are socialized into more altruistic behavior, paying increased attention to the "common good," as has been discussed in previous research. A new accountability mechanism could also come forth, however, if women are looked upon as discriminated against and excluded from inner circles of power, and citizens conclude from this that they will not be part of corrupt networks. Even if this situation can apply to other groups as well, we argue that an accountability mechanism of discrimination is a hitherto neglected mechanism where gender itself plays a major role.

INTEGRATING INSTITUTIONAL THEORY AND ACCOUNTABILITY MECHANISMS

The above sections have outlined how institutional theory and concepts of accountability mechanisms have been used in previous research on gender and corruption. Hence, building on previous insights, this section presents in more detail a foundation for a theory on gender and corruption merging institutional theory with ideas on mechanisms of accountability.

Table 4.1   Integrating institutional theory and accountability mechanisms to enhance research on gender and corruption

		Institutions	
		Citizen-affecting level	Decision-making level
The role of gender for accountability	Strengthens gender differences in established accountability mechanisms	When welfare regimes provide services strengthening women's position in society	When there is public exposure, e.g., through media, of wrongdoing among elites
	Being an accountability mechanism in itself	When women are socialized into rule-following and/ or collective enhancing behavior	When women in office have symbolic power as clean outsiders

More specifically, we suggest to distinguish between citizen-affecting and decision-making institutions, and between gender differences in the strength of how established accountability mechanisms work, versus gender as a "new" accountability mechanism.

First, as has been described above, previous research has used institutional theory to describe how actors are constrained by informal norms that, for example, enhance male dominance. We contend that most previous theory has been directed to describe and assess how institutions affect actors that are active within them but have not paid much attention to how these actors are moved before they enter the institutions and/ or when they move outside of these institutions. In previous research conducted within the framework of feminist institutionalism, this "outside" of institutions has been labeled the institutional environment or gendered institutional environment (Lowndes and Wilson 2001; Mackay 2014). Some importance is thus given to the outside, but the institutional environment captured in previous research takes the form of a residual category more than that of a circumstance with particular theoretical expectations tied to it.

We argue that the institutional environment should be analytically scrutinized and we suggest that this can be done by distinguishing institutions on different levels of society. Hence, in order to understand more precisely how and when gender shows varying relationships to corruption, we need to pay attention to how institutions on some levels affect what happens in institutions on other levels. Thus, rather than discussing institutional environments in a broad and imprecise way, scholars should theorize in more detail which institutions can be distinguished on which levels and what norms are attached to them. For example, the literature discusses how regime (democratic versus authoritarian) mediates the relationship between gender and corruption. Further, the literature on welfare state regimes also specifies how particular norms can be attached to policy institutions of this type. These different propositions can be systematized by bringing them together under the overarching heading of citizen-affecting institutions, specifying one level on which institutions can be distinguished. Among citizen-affecting institutions, we include institutions that generally affect citizens in a country, such as regimes (democratic or authoritarian) or gendered welfare state regimes (being more or less women-friendly). The other level on which we suggest distinguishing institutions is the decision-making level. On this level, we include legislatures, such as national parliaments, but also cabinets and the administration.

It is important to distinguish between these two levels of institutions as the decision-making institutions do not operate in a vacuum but in different country-specific contexts, in which overarching institutions and the norms attached to them can be specified. Hence, this makes it possible to discuss why national parliaments, for example, can be gendered in different ways as they exist in different citizen-affecting institutional milieus. Further, this model gives a certain amount of leeway in discussing changes within institutions. One possible way to account for changes in decision-making institutions is to account for how actors, such as female politicians, bring "newness" into existing institutions. Hence, actors can be seen as carriers of ideas that might change the "rules of the game," for example, the priority given to certain areas, such as effective and encompassing social services. In each case, however, the exact meaning of this "newness" needs to be specified through the anchoring in the institutional milieu on an overarching level.

Further, we turn to research on accountability for a more detailed discussion on the mechanisms that specify how and when the link between gender and corruption appears. Here, we gender the discussion on accountability mechanisms. More specifically, in previous sections we have argued that accountability mechanisms can be gendered in two main ways: first, the strength of their effect can be different for women compared to men, and second, gender can work as an accountability mechanism in itself. Hence, we argue that mechanisms of accountability that are already recognized in the literature can be gendered in the sense that they have a stronger effect on women than on men. For example, scholars note that the media tends to punish women politicians more harshly for wrongdoing and thus has a stronger effect of curbing corruption among women than men.

Further, social movements as watchdogs may also have gendered impacts. Building on the findings by Rodríguez (2003) and Grimes (2008a, 2008b) previously discussed, a more general theoretical expectation on the links between female representation and social movements can be formulated. On the one hand, women may be more dependent on social movements in order to reach and hold on to positions of power but, on the other hand, women might also actively seek to build alternative power bases by developing a political agenda that favors groups represented in social movements, such as women. Moreover, the existence of policies that diminish gender inequalities can also curb corruption among women more than among men, as women may feel more strongly that a state that provides services that help families and children in their everyday tasks should be protected. There are surely further ways that existing accountability mechanisms may be gendered, and we consider this an important area for future research.

Gender can work as an accountability mechanism in itself, however. The point we are making is that gender does not only mediate the strength of already acknowledged accountability mechanisms, but can itself be a mechanism of accountability. This can be the case, for example, if citizens perceive female candidates as more honest than male candidates, or when females are socialized into more altruistic behavior, paying increased attention to the "common good," but it can also be when women are looked upon as discriminated against and excluded from inner circles of power, and citizens conclude from this that they will not be part of corrupt networks.

We began by saying that corruption, or the act of using public power for private ends, can be considered a major destructive force for humans and human societies. In the quest to understand the mechanisms behind good government, which will most likely be an important issue for many years to come, we contend that the ways that gender plays into these dynamics should be further explored.

REFERENCES

Alatas, V., Cameron, L., Chaudhiri, A., Erkal, N., and Gangadharan, L. 2009. Gender, culture, and corruption: insights from an experimental analysis. *Southern Economic Journal* **75**(3), 663–80.

Alhassan-Alolo, N. 2007. Gender and corruption: testing the new consensus. *Public Administration* **27**(1), 227–37.

Barnes, T. and Beaulieu, E. 2014. Gender stereotypes and corruption: how candidates affect perceptions of election fraud. *Politics and Gender* **10**, 365–91.

Bjarnegård, E. 2013. *Gender, Informal Institutions and Political Recruitment: Explaining Male Dominance in Parliamentary Representation.* Basingstoke: Palgrave Macmillan.

Bovens, M. 2010. Two concepts of accountability: accountability as a virtue and as a mechanism. *West European Politics* **33**(5), 946–67.

Ciccone, D., Vian, T., Maurer, L., and Bradley, E. 2014. Linking governance mechanisms to health outcomes: a review of the literature in low and middle-income countries. *Social Science and Medicine* **117**, 86–95.

Dahlström, C., Lapuente, V., and Teorell, J. 2012. The merit of meritocratization. *Political Research Quarterly* **65**(3), 656–68.

Dollar, D., Fishman, R., and Gatti, R. 2001. Are women really the fairer sex? Corruption and women in government. *Journal of Economic Behavior and Organization* **26**(4), 423–9.

Elster, J. 1985. *Making Sense of Marx.* Cambridge: Cambridge University Press.

Esarey, J. and Chirillo, G. 2013. "Fairer sex" or purity myth? Corruption, gender and institutional context. *Politics and Gender* **9**(4), 361–89.

Feraz, C. and Finan, F. 2011. Electoral accountability and corruption: evidence from the audits of local governments. *American Economic Review* **101**(4), 1274–311.

Goetz, A. 2007. Political cleaners: women as the new anti-corruption force? *Development and Change* **38**(1), 87–105.

Goetz, A. 2008. *Who Answers to Women? Gender and Accountability. Progress of the World's Women 2008/2009.* United Nations Development Fund for Women (UNIFEM). http://www.unwomen.org/en/digital-library/publications/2008/1/progress-of-the-world-s-women-2008-2009-who-answers-to-women (accessed 24 May, 2018).

Grimes, M. 2008a. Contestation or complicity: civil society as antidote or accessory to political corruption. The Quality of Government Working Paper Series, 2008:8. http://qog.pol.gu.se/digitalAssets/1350/1350640_2008_8_grimes.pdf (accessed 24 May, 2018).

Grimes, M. 2008b. The conditions of successful civil society involvement in combatting corruption: a survey of case study evidence. The Quality of Government Working Paper Series. http://cmspres-vir-1.it.gu.se/digitalAssets/1350/1350673_2008_22_grimes.pdf (accessed 24 May, 2018).

Gupta, S., Davood, H.R., and Tiongson, E. 2000. Corruption and the provision of health care and education services. IMF Working Paper, Fiscal Affairs Department, Geneva.

Gustavson, M. 2015. Does good auditing generate quality of government? The Quality of Government Working Paper Series. http://qog.pol.gu.se/digitalAssets/1538/1538160_2015_15_gustavson.pdf (accessed 24 May, 2018).

Hellwig, T. 2012. Constructing accountability: party position taking and economic voting. *Comparative Political Studies* **45**(1), 91–118.

Hernes, H. 1987. *Welfare State and Woman Power: Essays in State Feminism.* Oslo: Oslo Norwegian Press.

Hevia de la Jara, F. 2007. Between individual and collective action: citizen participation and public oversight in Mexico's Oportunidades programme. *IDS Bulletin* **38**(6), 64–72.

Holmberg, S. and Rothstein, B. 2012. *Good Government: The Relevance of Political Science.* Cheltenham, UK and Northampton, MA, USA: Edward Edgar.

Hossain, N. and Nyamu-Musembi, C. 2010. *Corruption, Accountability and Gender: Understanding the Connections.* United Nations Development Fund for Women (UNIFEM). Primers in Gender and Democratic Governance. http://www.undp.org/content/dam/aplaws/publication/en/publications/womens-empowerment/corruption-accountability-and-gender-understanding-the-connection/Corruption-accountability-and-gender.pdf (accessed 24 May, 2018).

Kahneman, D. 2003. Maps of bounded rationality: psychology of behavioral economics. *American Economic Review* **93**(5), 1449–75.

Kenny, M. 2007. Gender institutions and power: a critical review. *Politics* **27**(2), 91–100.

Kenny, M. 2013. *Gender and Political Recruitment: Theorizing Institutional Change.* New York: Palgrave Macmillan.

Kostadinova, T. and Mikulska, A. 2015. The puzzling success of populist parties in promoting women's political representation. *Party Politics* **23**(4), 400–12.

Krook, M. and Mackay, F. 2011. *Gender, Politics and Institutions: Towards a Feminist Institutionalism*. Basingstoke: Palgrave Macmillan.

Larcinese, V. and Sircar, I. 2017. Crime and punishment the British way: accountability channels following the MPs' expenses scandal. *European Journal of Political Economy* **47**, 75–99.

Lindberg, H. and Stensöta, H. 2018. Corruption as exploitation: feminist exchange theories and the link between gender and corruption. In H. Stensöta and L. Wängnerud (eds), *Gender and Corruption: Political Corruption and Governance*, pp. 237–56. Cham: Palgrave Macmillan.

Lowndes, V. and Wilson, D. 2001. Social capital and local governance: exploring the institutional design variable. *Political Studies* **49**(4), 629–47.

Mackay, F. 2014. Nested newness, institutional innovation and the gendered limits of change. *Politics and Gender* **10**(4), 549–71.

Mackay, F., Kenny, M., and Chappell, L. 2010. New institutionalism through a gender lens: towards a feminist institutionalism? *International Political Science* **30**(5), 573–88.

MacKinnon, C. 1989. *Towards a Feminist Theory of the State*. Cambridge, MA: Harvard University Press.

Manin, B. 2007. *The Principles of Representative Government*. Cambridge: Cambridge University Press.

March, J. and Olsen, J. 1989. *Rediscovering Institutions: The Organizational Basis of Politics*. New York: Free Press.

Mauro, P. 1995. Corruption and growth. *Quarterly Journal of Economics* **110**(3), 681–712.

Morris, S. and Klesner, J. 2010. Corruption and trust: theoretical considerations and evidence from Mexico. *Comparative Political Studies* **43**(10), 1258–85.

Nistotskaya, M. and Stensöta, H. 2018. Is women's political representation beneficial to women's interests in autocracies? Theory and evidence from Post-Soviet Russia. In H. Stensöta and L. Wängnerud (eds), *Gender and Corruption: Political Corruption and Governance*, pp. 145–68. Cham: Palgrave Macmillan.

North, D. 1990. *Institutions, Institutional Change and Economic Performance*. Cambridge: Cambridge University Press.

Pitkin, H. 1967. *The Concept of Representation*. Berkeley, CA: University of California Press.

Rivas, F. 2013. An experiment on corruption and gender. *Bulletin of Economic Research* **65**(1), 10–42.

Rodríguez, V. 2003. *Women in Contemporary Mexican Politics*. Austin, TX: University of Texas Press.

Rose-Ackerman, S. 1999. *Corruption and Government: Causes, Consequences, and Reform*. Cambridge: Cambridge University Press.

Rothstein, B and Teorell, J. 2008. What is the quality of government? A theory of impartial government institutions. *Governance*. doi: 10.1111/j.1468-0491.2008.00391.x

Rothstein, B. and Uslaner, E. 2005. All for all: equality, corruption and social trust. *World Politics* **58**(1), 41–72.

Sattler, T., Freeman, J., and Brandt, P. 2007. Political accountability and the room to maneuver: a search for a causal chain. *Comparative Political Studies* **41**(9), 1212–13.

Schwindt-Bayer, L. and Esarey, J. 2017. Women's representation, accountability, and corruption in democracies. *British Journal of Political Science*. doi: 10.1017/S0007123416000478

Schwindt-Bayer, L. and Tavits, M. 2016. *Clarity of Responsibility, Accountability and Corruption*. Cambridge: Cambridge University Press.

Stensöta, H. 2018. Final thoughts: taking stock and reflections on ways forward. In H. Stensöta and L. Wängnerud (eds), *Gender and Corruption: Political Corruption and Governance*, pp. 275–88. Cham: Palgrave Macmillan.

Stensöta, H. and Wängnerud, L. 2017. Gender, politics, and corruption. In W. Thompson (ed.), *Oxford Research Encyclopedia of Politics*. Oxford: Oxford University Press. doi: 10.1093/acrefore/9780190228637.013.206

Stensöta, H., Wängnerud, L., and Svensson, R. 2015a. Gender and corruption: the mediating power of institutional logics. *Governance* **28**(4), 475–96.

Stensöta, H., Wängnerud, L., and Agerberg, M. 2015b. Why women in encompassing welfare states punish corrupt political parties. In H. Stensöta (ed.), *Elites, Institutions and the Quality of Government*, pp. 245–62. Basingstoke and New York: Palgrave Macmillan.

Stockemer, D. 2011. Women's parliamentary representation in Africa: the impact of democracy and corruption on the number of female deputies in national parliaments. *Political Studies* **59**(3), 693–712.

Sundström, A. and Wängnerud, L. 2016. Corruption as an obstacle to women's political representation: evidence from local councils in 18 European countries. *Party Politics* **22**(3), 354–69.

Sung, H.-E. 2003. Fairer sex or fairer system? Gender and corruption revisited. *Social Forces* **82**(2), 703–23.

Sung, H.-E. 2012. Women in government, public corruption, and liberal democracy: a panel analysis. *Crime, Law and Social Change* **58**, 195–219.

Swamy, A., Knack, S., Lee, Y., and Azfar, O. 2001. Gender and corruption. *Journal of Development Economics* **64**(1), 25–55.

Tavits, M. 2008. Representation, corruption, and subjective well-being. *Comparative Political Studies* **41**(12), 1607–30.

Thomson, J. 2017. Resisting gendered change: feminist institutionalism and critical actors. *International Political Science Review*. doi: 10.1177/0192512116677844

Towns, A. 2015. Prestige, immunity and diplomats: understanding sexual corruption. In C. Dahlström and L. Wängnerud (eds), *Elites, Institutions and the Quality of Government*, pp. 49–65. Basingstoke: Palgrave Macmillan.

Treisman, D. 2007. Causes of corruption from ten years of cross-national empirical research? *Annual Review of Political Science* **10**, 211–44.

UNDP (United Nations Development Programme). 2012. *Seeing Beyond the State: Grassroots Women's Perspectives on Corruption and Anti-corruption*. Report of the Huairou Commission. http://www.undp.org/content/undp/en/home/librarypage/democratic-governance/anti-corruption/Seeing-Beyond-the-State-Grassroots-Womens-Perspectives-on-Corruption-and-Anti-Corruption.html (accessed 24 May, 2018).

Young, I.M. 2002. Lived body vs gender: reflections on social structure and subjectivity. *Ratio* **15**(4). doi: 10.1111/1467-9329.00200

5. World regional geographies of corruption
Barney Warf

Corruption is extremely widespread throughout the world – almost all countries exhibit it to one degree or another – and has a variety of insidious social, political, economic, and environmental consequences. Unsurprisingly, the causes and consequences, as well as the type and severity, of corruption vary markedly across the planet.

The goal of this chapter is to explicate the uneven spatiality of corruption, that is, to understand how and why it varies so dramatically across the earth's surface. Because there are abundant and insightful discussions of the history, causes, and consequences of corruption, these issues are not addressed here. Rather, the chapter offers an overview of corruption in five major world regions: Europe, the Middle East and North Africa, Asia, Africa, and Latin America. It does not address corruption in Russia, North America, or Oceania. The goal is to illuminate the similarities and differences in the causes and consequences of corruption in different regional contexts. The conclusion summarizes the findings, criticizes anti-corruption campaigns, and points to steps that may alleviate corruption.

DATA

Empirical analyses of corruption are often hampered by the difficulty in measuring its severity as well as the conditions that may give rise or impede it. For this analysis, data on the perceived relative levels of government corruption in 2016 were obtained from Transparency International, a global non-governmental organization (NGO) dedicated to monitoring and combating public and private sector corruption. The group is involved in a variety of intertwined activities, including measuring corruption, exposing egregious cases, offering advice to companies to minimize extortion and bribery, and developing tools for combating it, such as integrity packs.

The most noteworthy and well-known product of this organization is its annual Corruption Perceptions Index (CPI) of government malfeasance, issued annually since 1995 as part of a global corruption report. The most widely used measure of corruption in the world, the CPI is a composite indicator based on surveys and interviews with public and private sector officials in each country and expert assessments by 13 sources, including the World Bank, Transparency International's Bribe Payers Survey, African Development Bank, the World Justice Project, the Bertelsmann Foundation, Freedom House, Political Risk Services International, and the Economist Intelligence Unit. A minimum of three of these units contributed to the assessment of corruption in each country; most were evaluated on the basis of seven to eight sources. Importantly, the index assesses the extent of corruption in a country, not its impacts. Statistical analyses of the inputs into the corruption index were conducted by the European Commission Joint Research Center in Italy. Scores were normalized on an ordinal scale of zero (most corrupt) to 100 (least corrupt).

Obviously, any quantitative measure of a phenomenon as elusive as corruption is problematic. The CPI, therefore, should not be regarded as a precise, absolute measure, but judged in relative or qualitative terms. As a broad indication of the level of severity of government malfeasance, it suffices, and is the best available measure. More nuanced approaches use several indices, including risk assessments. In this analysis, grouping countries into categories of severity of corruption helps to avoid giving the impression of an undue level of precision; rather, these groupings serve to paint a broad picture of the degree to which the public sector is compromised.

GLOBAL PATTERNS OF CORRUPTION

In many respects, the geography of corruption across the world mirrors global patterns of wealth. Figure 5.1 represents the global distribution of Transparency International's CPI score. Clearly, most countries in the world exhibit moderately high to severe levels of corruption. The worst offenders (scores under 20) are found primarily in Africa and the Middle East; notably, all the states in this group are predominantly Muslim, with the exceptions of Angola and North Korea, and, in the Western hemisphere, Haiti and Venezuela. The second-most corrupt group (scores 20–39) includes almost all of East, South, and Southeast Asia, with the exceptions of Japan, South Korea, Singapore, and Malaysia, as well as most of Africa and a broad swath of Spanish-speaking countries in Latin America. The third-most corrupt group (scores 40–59) includes wealthier countries in the developing world, including Brazil, South Africa, Turkey, Saudi Arabia, and several Balkan states. The fourth-most corrupt cluster (scores 60–79) includes the United States, Japan,

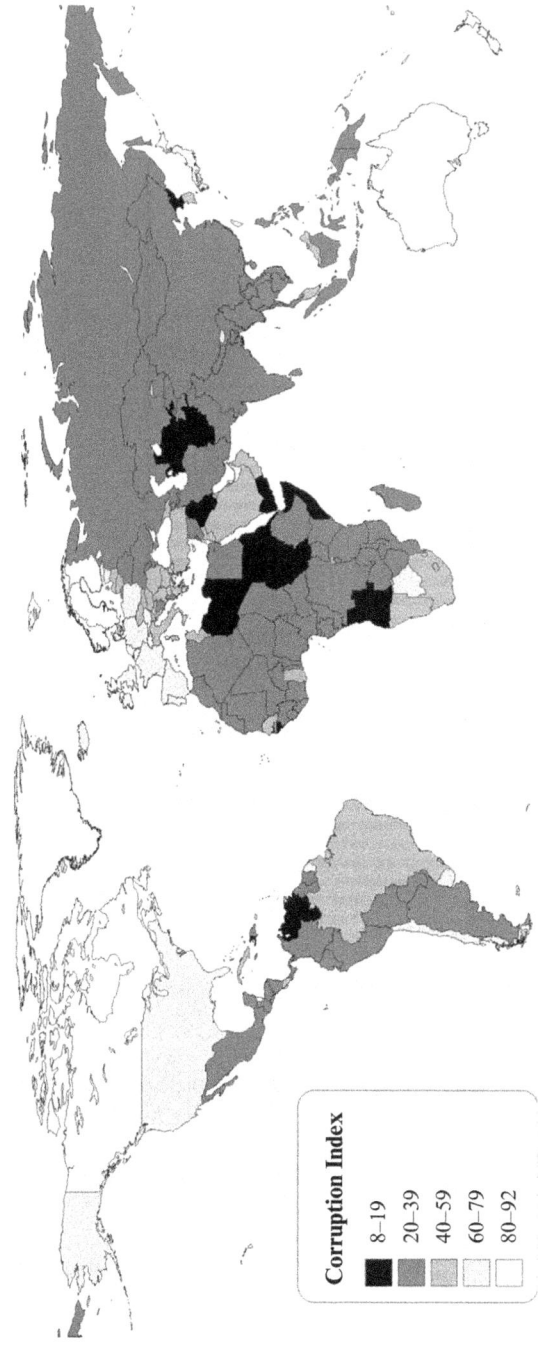

Source: Author, using data from https://www.transparency.org/news/feature/corruption_perceptions_index_2016 (accessed May 14, 2018).

Figure 5.1   Global distribution of Transparency International corruption scores, 2016

most of Europe, and, notably, Botswana, Africa's least corrupt country. Finally, the least corrupt (scores 80 and above), and smallest, group includes Canada, the Scandinavian states, Australia, New Zealand. Much of sub-Saharan Africa and war-torn states such as Iraq and Afghanistan are typically cited as supreme examples of corruption; in contrast, Northern and Western Europe are models of low levels of corruption, with vigorous policies in place to limit it. Wealthier countries, it is commonly argued, have a relatively unshackled media, high rates of literacy, and well-established institutions and regulatory structures to keep a lid on corruption (Bardhan 1997; Jain 2001; Brunetti and Weder 2003).

EUROPEAN CORRUPTION IN PERSPECTIVE

Compared to most of the world, corruption in Europe (here defined to exclude Russia and Turkey) is minimal. For example, Transparency International, a NGO that monitors corruption, rates many European countries as among the least corrupt on the planet. However, the relative lack of corruption in Europe, as in North America, Japan, and Australia and New Zealand may lead to the misperception that it is only a disease of the poor and hide its extent and influence among wealthier states. There is, of course, an extensive literature on corruption on the European continent. Della Porta and Mény (1997) present a series of case studies of different countries, mostly European, that detail the linkages between corruption and the lack of democracy in different national contexts. For example, the collapse of the Franco regime in Spain witnessed a notable decline in corruption, as also happened in East Germany when it merged with its western counterpart.

This body of work flourished following the collapse of communism in Eastern Europe in the 1990s. To take but one example, Sajo (1998) uses a legal perspective to interrogate corruption's linkages to clientelism in the region. Miller et al. (2001) examine how citizens in Eastern Europe coped with the explosion of corruption in this context, including dealing with public bureaucrats and calling for reforms. Holmes (2009) notes how corruption in the area has fostered organized crime, while Wallace and Latcheva (2006) point to corruption's role in encouraging the growth of the informal economy in Eastern and Central Europe. More recently, Kostadinova (2012) focuses on how networks of corruption infiltrate the state, with a focus on the Balkans.

More recently, European corruption studies have emphasized the impacts of joining the European Union (EU). Warner (2007) offers a compelling case refuting the notion that EU integration would reduce corruption, a promise held by those who argue that privatization, free trade, and regulatory harmonization would limit it; indeed, neoliberalism and corruption are perfectly compatible. Tänzler et al. (2012) offer a series of case studies of corruption, and anti-corruption policies, in several categories of European states, which they categorize as "disenchanted modernity," catch-up modernization, and semi-modern. The essays collected by Gounev and Ruggiero (2012) point to strong linkages between corruption in the public sphere and organized crime, which obviously vary among countries but have made several, such as Italy and Spain, into hotspots of such activity.

Corruption's role in electoral politics has also received considerable scrutiny. Polk et al. (2017) show that corruption is more common among political parties in countries with poor quality of governance, whereas Engler (2015) demonstrates that corruption has fueled the growth of new parties in Eastern Europe. Bågenholm (2013) demonstrates that European voters do toss out corrupt politicians when scandals emerge, albeit not routinely.

Certainly, many Europeans feel, understandably, that corruption in their countries is a serious issue. For example, Transparency International's (2016) Corruption Barometer notes that large shares of the population view it as "one of the three biggest problems facing their country." The proportion holding this view varies widely, ranging from a mere 2 percent in Germany to two-thirds in Spain and Moldova; the majority believe this in many Balkan states, Lithuania, and Portugal. Intriguingly, subjective perceptions of corruption do not map isomorphically on Transparency International's widely used corruption index: some countries seem to feel it is a worse problem than the statistics reveal, such as Spain and Portugal, whereas others seem to dismiss it in the face of evidence that indicates otherwise, such as Belarus, Poland, and Greece. These discrepancies point to the cultural contexts of corruption, in which its severity may be judged by varying norms as simply part of the cost of doing business with the state.

Figure 5.2 displays Europe's corruption scores in map form. The spatial distribution of corruption varies widely over the continent. Northern Europe fares best, with countries such as Germany, Switzerland, the Netherlands, Britain, and several Scandinavian states exhibiting high scores (low corruption). A secondary tier consists of France, Portugal, Ireland, Austria, Slovenia, Poland, Estonia, and Cyprus. The third group includes Italy, Spain and several countries in Eastern Europe and the Balkans. At the bottom of the scale, that is, the most corrupt countries, lie Ukraine, Moldova, Albania, Macedonia, and Serbia.

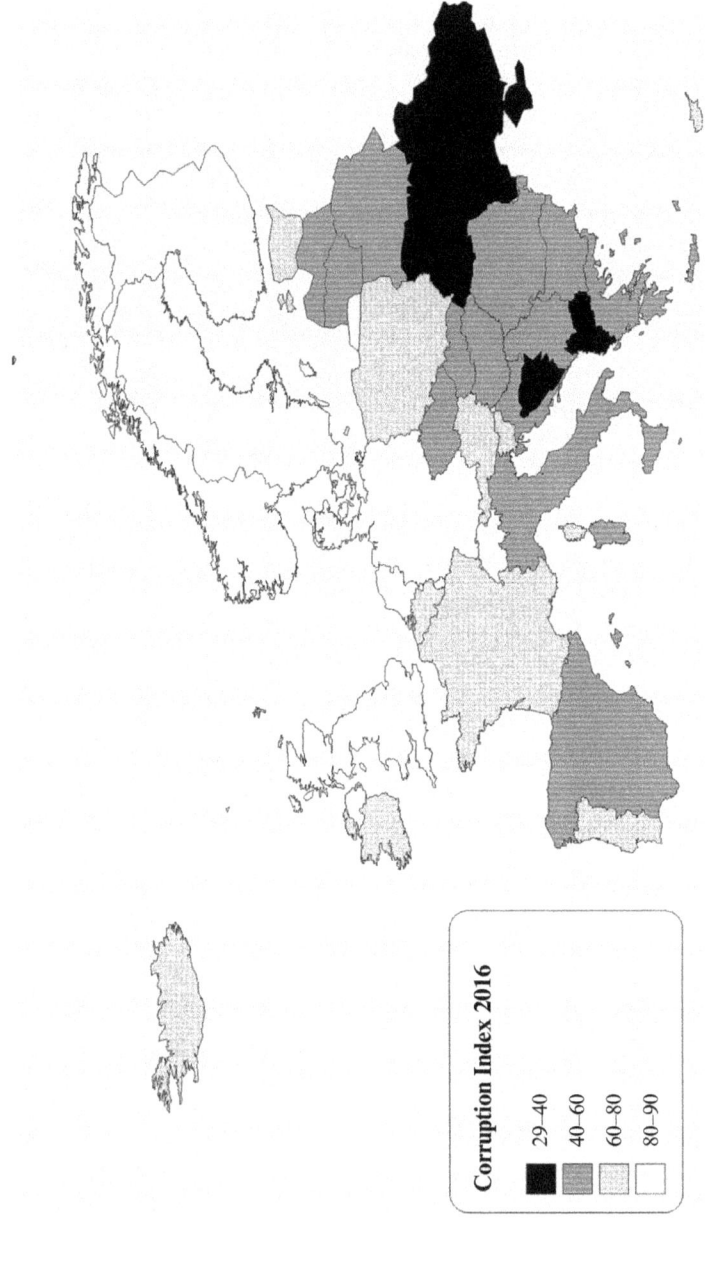

Source: Author, using data from https://www.transparency.org/news/feature/corruption_perceptions_index_2016 (accessed May 14, 2018).

Figure 5.2 Transparency International corruption scores across Europe

THE MIDDLE EAST AND NORTH AFRICA

Unfortunately, corruption is widespread and severe in the Middle East and North Africa (MENA) region. One need not descend into racist, Orientalist stereotypes to find evidence of this phenomenon. For example, Transparency International ranks three MENA countries (Sudan, Iraq, and Libya) among the ten most corrupt governments in the world, and the region's least corrupt state – the United Arab Emirates – is rated only "slightly corrupt." Gillespie (2006, p. 40) holds that

> Graft at the highest levels of government is commonly augmented by rampant petty bribery among low-level officials. Some observers have attributed this to a continuation of practices established by the earlier Ottoman bureaucracies in many Middle Eastern countries. Poorly paid bureaucrats are expected to supplement their incomes with unofficial tips. This petty corruption affects the average person, and especially businesspeople, the most as they attempt to operate in countries misgoverned by numerous and opaque rules, where permission is required for just about everything.

As Looney (2005, p. 6) asserts, "most MENA governments compensate for low popular support or poor legitimacy by granting opportunities for bribery to leading families or cliques."

The MENA region is particularly susceptible to corruption for several reasons. The dependence on petroleum exports is a major factor: oil-rich countries tend to have large public sectors, and the revenues offer large incentives for rent-seeking behavior. In this light the MENA region is a classic example of the "resource curse," the tendency of natural resource-based economies to generate rents appropriated by small, politically powerful elites that hold onto power through tight control of the military and the media (Haber and Menaldo 2011). Petroleum dependency argu-ably drowned the Arab Spring and strengthened the hand of despotic regimes in some MENA countries (Ross 2011), undermining democratic forces predisposed to curbing corruption. The deeply entrenched auto-cratic governments, patronage systems, lack of democracy, weak political opposition movements, low rates of literacy in some MENA countries, and heavily monitored press also contribute to the prevalence and severity of corruption. Among the many effects of this phenomenon in the area are lower rates of investment, productivity, and economic growth (Guetat 2006; Sayan 2009).

In Iraq, one of the world's most corrupt countries, pernicious corruption has generated a plethora of political and economic repercussions. Despite promises of the American-backed Provisional Authority put into place following the invasion of 2003 to implement a democratic meritocracy, corruption remains worse than it was under Saddam Hussein. The country's large oil reserves have facilitated a patronage system that has survived multiple governments (Le Billon 2005), an example of the "resource curse" notoriously known to hamper development in many countries rich in petroleum, diamonds, minerals, and other natural wealth (McFerson 2009). Whyte (2007, p. 177) claims that an enormous sum, likely more than $12 billion of Iraqi oil revenue, "has disappeared into the pockets of contractors and fixers in the form of bribery, over-charging, embezzlement, product substitution, bid rigging and false claims." Many companies involved in the food-for-oil program have been investigated for bribes and kickbacks (Gillespie 2006). Looney (2008) attributes the tidal wave of corruption there to the growth of the informal economy, the disintegration of trust among different religious groups and regions, and the rise of organized crime. Struggles among gangs over resources, notably oil smuggling, have undermined the government's flaccid attempts to implement the rule of law (Williams 2009). Corruption severely hampers the Iraqi government's efforts against the Islamic State (al-Ali 2014; Kirkpatrick 2014), including officers' theft of soldiers' salaries and sales of military equipment on the black market, including to insurgents.

Syrian corruption has a long-standing history stretching back decades (Sadowski 1987). One-party control by the Baathists provided the ideal context for flourishing back channels to circumvent official circles of power, leading to widespread nepotism and government expenditures that enriched a few at the expense of the Syrian masses (Volker 2004). Corruption became a vital tool for the Assad dynasty to retain its hold on the reins of power. Three families – the Assads, the Shalishes, and the Makhlufs – became an entrenched elite with monopoly control over many sectors, such as telecommunications (Borshchevskaya 2010). As a result, the economy languished for decades before the bloody civil war erupted in 2009. For example, the Makhluf family, related to the Assads, benefited from regulations giving them an unfair advantage in the cell phone market over their competitors (Leenders and Sfakianakis 2002). The World Bank ranks Syria among the world's worst countries for ease of doing business, and the country is plagued by inflation, unemployment, and a deteriorating infrastructure.

Another example of the unholy combination of authoritarianism and rampant corruption is Iran. The ruling theocracy there has steadily enriched itself while ruining the economy; much corruption is associated with the powerful sons of high-ranking clerics, who enrich themselves through their political connections (Alamdari 2005). President Ahmadinejad gave billions of dollars in no-bid contracts to firms associated with the elite Revolutionary Guards (Ross 2011). Pollack (2006, p. 2) observes tartly that

Iran's economy is in desperate need of reform. It is crippled by corruption, with an estimated 40 percent of Iranian GDP accounted for by the *Bonyads* – nominally charitable foundations established to administer the Shah's assets on behalf of the Iranian people, but in actuality massive corruption machines that bankroll the senior leadership. The problem of corruption has reduced liquidity, frightened off investment, boosted inflation, spurred widespread unemployment, diminished non-oil exports, impoverished the middle class, and created a very serious gap between rich and poor.

Smuggling across the borders with Turkey, Iraq, Pakistan, and Afghanistan is rampant. Occasional scandals periodically expose the corruption of the ruling clerics, although they tend to escape unpunished. During the 2005 elections, corruption emerged as a significant point of debate, but soon thereafter vanished from view. Mashali (2012) distinguishes between grand and petty corruption, and finds that in Iran the two are closely associated. In short, the structural conditions of corruption – oil revenues, oppressive ayatollahs, a censored media, and lack of effective opposition – are deeply intertwined with, and reproduced by, the everyday corrupt practices of petty officials and bureaucrats.

Turkish corruption flourished following the neoliberal reforms that began in the 1990s, leading to government malfeasance and the steady rise of organized crime. Aybar and Lapavitsas (2001, p. 303) hold that "Criminals wield considerable influence among politicians and the 'Istanbul bourgeoisie', indeed all sections of Turkish society." Finkel (2000) holds that corruption penetrates the Turkish press, inhibiting it from doing its watchdog duty. A particularly notorious example was the earthquakes of 1999 and 2003 that struck the Marmara and Düzce regions and claimed over 40,000 lives and destroyed 300,000 homes; Green (2005) notes that the catastrophic results were less the products of the quakes themselves and more the result of corrupt political decisions that led to shoddy construction of homes and other buildings. As might be expected, corruption has debilitating consequences for the country: Tekin-Coru (2006) found that corruption inhibits joint ventures in Turkey, as foreign investors are wary of sinking money in a political context that may yield lower than optimal returns. Corruption likely plays a role in hampering Turkey's efforts to join the EU (Doig 2010). Beyond the purely economic dimensions of Turkish corruption, it plays an important ideological role. Thus, Bedirhanoğlu (2007) argues that anti-corruption campaigns serve neoliberal interests by marginalizing developing countries, using ahistorical Western criteria to demonize long-standing practices that are regarded as corrupt in London or Washington, DC but not in Istanbul or Ankara. A large bribery scandal that erupted in 2013 destabilized the government of Recep Tayyip Erdogan and the ruling Justice and Development Party, leading to dozens of arrests, a cabinet shake-up, arrests of journalists, and accusations of a government cover-up (Arango 2013). Erdogan blamed the scandal on his political rival, the cleric Fethullah Gülen.

In Egypt, the MENA region's most populous nation, corruption has been a long-standing feature of national politics. Although it lacks the petroleum revenues of Iraq, Iran, and Persian Gulf states, Egypt nonetheless demonstrates many structural similarities in its political order, including an autocratic one-party system and an entrenched ruling elite. American-inspired neoliberal reforms under Mubarak widened opportunities for graft. As Grimaldi and Harrow (2011) note, "[t]he privatization and economic opening of recent years have created new opportunities for 'vertical corruption' at upper levels of government affecting state resources." The Egyptian police are particularly noteworthy for their high levels of bribe-taking (Arafa 2011). Not surprisingly, corruption takes a toll on the Egyptian economy, including reduced investment and higher inefficiencies (Ghalwash 2014). Rampant corruption also takes a toll on the everyday lives of Egyptians. For example, Soliman and Cable (2011) conclude that high-level corruption was responsible for the 2006 sinking of an Egyptian ferry in the Red Sea, which killed 1,034 people.

Sudan is governed by a militarized kleptocracy in which the patronage system leaves few funds for social services (Ismail 2011). With a Transparency International Score of 15 in 2014, it ranks as the most corrupt country in the MENA region and one of the most corrupt in the world. Another victim of the resource curse (Carmody 2009; Patey 2010), the government of Oman al-Bashir, indicted by the International Criminal Court, has long been a notorious abuser of human rights, including organized atrocities in Darfur. Cables leaked by Wikileaks indicate that al-Bashir has accumulated up to $9 billion in Lloyd's Bank in London (Hirsch 2010). Transparency International (Martini 2011) notes that the opaque government budgets, endemic embezzlement, and cronyism in Sudan have hampered small businesses; the majority of Sudanese report paying bribes to the police and government officials; the judiciary is dysfunctional; government contracts are awarded only to well-connected elites, notably Islamist parties; and high-level officials import goods without paying tariffs.

Other MENA states offer a depressing litany of similar examples. In Algeria, following the civil war that killed 200,000 people, mafia-like organized crime has undermined state institutions (Hadjadj 2007). The Khalifa affair, an enormous banking scandal, witnessed the collapse of a corrupt corporate conglomerate founded and run by Rafik Khalifa, a well-connected son of a politician. Libya, which has descended into anarchy, has a notoriously dysfunctional police force prone to bribery (Domoro and Agil 2012). While corruption certainly dates back to the Qaddafi regime, recent chaos there has undermined the rule of law and increased the incentives for extra-legal means of enhancing public sector incomes, while reducing the likelihood of being caught (Kamba and Sakdam 2012). In Saudi Arabia, government institutions often amount to little more than private fiefdoms for members of the House of Saud (Hertog 2011). Yemen, in which the uprising of 2011 against President Saleh led to another failed state now engulfed in civil war and Saudi bombings, has a government in which political patronage plays a central role (Alley 2010). Hill et al. (2013) note that the country's corruption is one reason it has suffered from serious capital flight, which erodes tax revenues for public services. The perception of stolen sovereign wealth fueled considerable anger against the Yemeni government, as in many Arab states. Yemen's case illustrates how corruption is often entangled in a mix of domestic and foreign factors, including international donors and aid agencies. Lebanon, traumatized by multiple cycles of violence between 1975 and 1990 that left over 150,000 people dead, has hardly offered a model of post-war recovery. Leenders (2012, p. 2) notes that "Many Lebanese look at their political leaders and state institutions primarily through the prism of omni-present corruption." For example, "The *baksheesh* for a building permit for a residential house can cost more than US $2,000" (Leenders and Sfakianakis 2002, p. 206). The litany of complaints in Lebanon includes embezzlement of funds intended for public health; widespread nepotism; and inflated military payrolls in which soldiers never receive their full salaries.

The MENA region's corruption scores are displayed cartographically in Figure 5.3.

UNEVEN LANDSCAPES OF ASIAN CORRUPTION

Corruption is widespread throughout Asia. According to Transparency International, North Korea has the dubious distinction of being the world's most corrupt state. A perfect storm of highly centralized power, lack of any political opposition, international isolation, and the rampant nepotism of the Communist Party created the ideal conditions for private gain at the expense of the impoverished population. The Kim dynasty has resorted to a variety of tools to retain authoritarian control (Byman and Lind 2010), effectively eliminating meritocracy and sowing the seeds for a political system highly susceptible to cronyism. Many North Koreans see corruption as the only way to get ahead, given the lack of formal opportunities (Haggard and Noland 2010). Refugees from the country report that

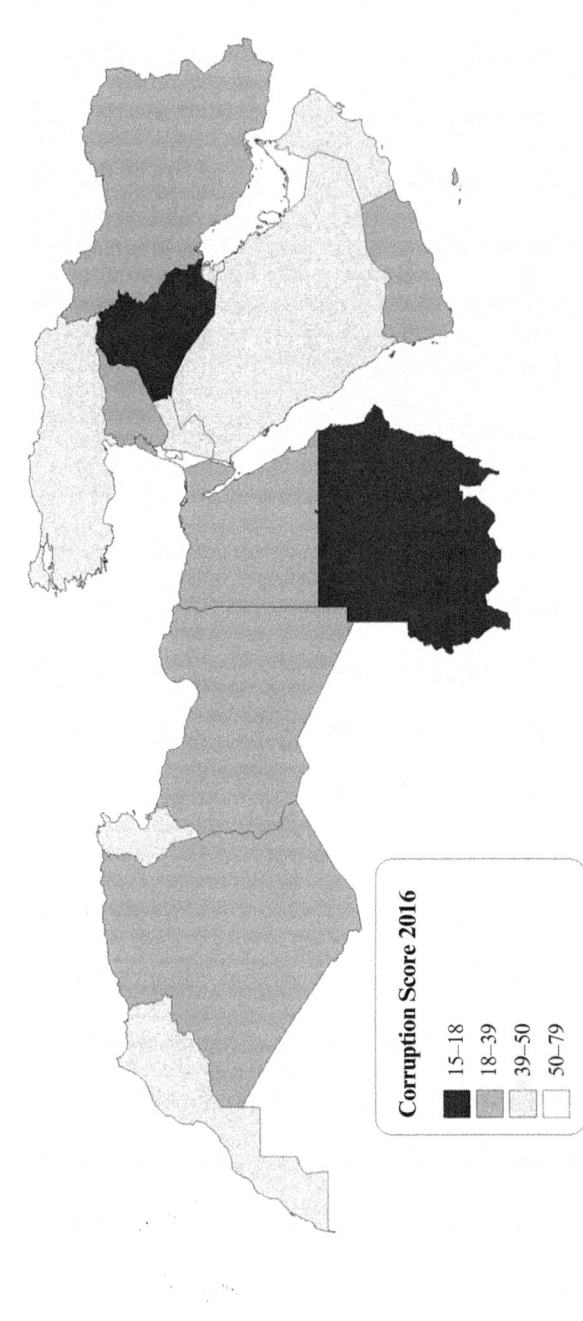

Source: Author, using data from https://www.transparency.org/news/feature/corruption_perceptions_index_2016 (accessed May 14, 2018).

Figure 5.3   Transparency International corruption scores across the Middle East and North Africa

bribery is found everywhere in North Korea, and is often the only means of obtaining travel documents, becoming a party member, acquiring a good job, and getting a child into a university (Chae et al. 2006).

China has a long history of corruption that continues today. Under the Ch'ing dynasty, magistrates were granted an allowance called *yang-lien yin* ("money to nourish honesty"). China's Transparency International score was 36, indicating widespread and severe corruption; however, Hong Kong earned a respectable 74. Whereas Hong Kong has noticeably reduced corruption from its maximum levels reached in the 1960s, China has witnessed an explosion of fraud and bribery (Manion 2004). A sizeable literature on corruption in China indicates that it is systemic and institutionalized, reaches the highest echelons of government, and has flourished over time, propelled in part by the country's enormous economic growth (Johnston and Hao 1995; He 2000; Wedeman 2012; Kwong 2015). Guo (2008) documents the steady rise in number of new cases. Chinese corruption takes numerous forms, including

> graft (*tanwu*), bribery (*xinghui*), and misappropriation of public property (*nuoyong gongkuan wu*), along with seeking illicit benefits for relatives and friends; neglecting official duties; nepotism and favouritism; shirking; retaliation; making false accusations; filing false reports; boasting and exaggerating; banqueting at public expense; running unauthorized businesses; profiteering; housing irregularities; living lavishly; engaging in improper sexual relations; forming cliques; gambling; whoring; excessive spending on marriages and funerals; engaging in superstitious activities; smuggling; selling state secrets; engaging in insider stock trading, engaging in real estate speculation and fraud; evading taxes; engaging in financial fraud; making illegal and irregular bank loans; and diverting and selling disaster relief goods. (Wedeman 2004, p. 897)

Ko and Weng (2012) argue that several factors have led to structural changes in Chinese corruption, including an emphasis on a merit-based civil service system and better monitoring of government expenditures, transforming the practice from an administrative issue to a transactional problem between the public and private spheres. This issue has been exacerbated by the widespread privatization of state-owned companies that has been underway in China for the last two decades. Moreover, like every other segment of Chinese society, corruption there has become globalized, as when bribes take the form of foreign trips, houses, and bank accounts (Li and Wu 2007).

Notably, corruption varies geographically within China. The incomplete devolution of power from Beijing to the local level has put many regional governments in a bind, having to act as agents of the state on the one hand and as engines of economic development on the other (Gong 2002), a process that places discretionary powers in the hands of local officials while it simultaneously offers significant rewards for malfeasance. This phenomenon helps to explain the illegal land grabs that have been common in China and have fueled much popular dissent. Provinces that are relatively less corrupt and better governed have succeeded more often in attracting foreign direct investment (Cole et al. 2009). Counties with active anti-corruption campaigns tend to have higher rates of gross domestic product (GDP) growth (Wu and Zhu 2011).

Chinese corruption extends well beyond the state to permeate business relations as well. *Guanxi* ("connections") is a famous and long-standing Chinese practice that refers to reliance upon interpersonal ties to facilitate actions, resolve problems, and improve access to resources, including friendships and extended family relations that may last a lifetime, blurring the boundaries between the personal and professional worlds (Gold et al. 2002). *Guanxi* and corruption tend to be deeply intertwined (Luo 2008). Using a series of case studies, Li (2011) notes that far from being a series of haphazard acts, *guanxi* instead forms a methodically organized informal institutional mechanism for facilitating corrupt exchanges. Because China has a weak legal system, business-to-government *guanxi* undermines corporate governance there (Braendle et al. 2005), leading to private gains at social cost (Fan 2002). Luo (2002) argues that *guanxi* is a supplement to established corporate law. Chinese corruption has led, among other things, to corrupt real estate practices (Zhu 2012), shoddy construction of schools, a thriving opium trade in the south, and mismanagement of public revenues. Frequently, business must pay bribes for procurements; parties in court cases bribe judges for favorable opinions; local government officials illegally confiscate land from peasants and benefit from the resale for development.

One might suspect that endemic corruption would act to the detriment of China's famously rapid economic growth. However, as Li and Wu (2007) argue, given the hidebound nature of the Chinese bureaucracy, corruption there may be "efficiency-enhancing" as opposed to predatory. Corruption here is seen as a form of investment that reaps future rewards, perhaps when the official being bribed retires. In this reading, China flourishes *because* it is corrupt, not despite it.

Episodic crusades to stop corruption, which included occasional executions of Communist Party officials, have failed to stem the tide (Ma 2008; Wedeman 2015). In 2001, Li Jizhou, Vice Minister of Public Security, was sentenced to death for accepting tens of millions of yuan in bribes. The latest attempt, launched by Xi Jinping in 2012 to catch "tigers and flies" (high- and low-level officials), included regional inspection teams in various provinces. For example, the vice mayor of Beijing and supervisor of the construction of buildings for the Olympic games was fired for taking bribes;

Politburo member Zhou Yongkang was implicated in a vast influence-peddling scheme; the Communist Party head of Shanghai was sentenced to 18 years in prison for loaning retirement funds to real estate speculators; the head of the China Food and Drug Administration was executed for kickbacks; the national chief statistician, Qiu Xiaohua, was fired after a pension fund scandal; the chief executive of China Petroleum was dismissed; and the official in charge of the national rail system was fired after a corruption investigation. In part these efforts stem from the threat that corruption poses to the legitimacy of the Communist Party and its autocratic rule. Heineman (2011) maintains that anti-corruption campaigns may deter low-level bribes and kickbacks, but do little to reduce that among high-ranking officials, and may even increase the size of bribes. He notes that the People's Bank of China estimated that "17,000 Communist Party members and state functionaries had illicitly obtained and then smuggled out of China an astonishing $124 billion from the mid-90s until 2008."

Tyrannical governments, kleptocratic elites, extensive censorship, the lack of political opposition, and deeply entrenched cultures of graft have combined to exacerbate Central Asian corruption (Gleason 1995). Oil revenues have transformed some of these into rentier states plagued by the famous "resource curse" found widely in Africa. In Turkmenistan, corrupt officials prosper from the blossoming Central Asian trade in opium and heroin (Peyrouse 2012). Regional patronage networks undermine the efficiency of the region's public administration (Collins 2006). However, Western perceptions of Central Asian bribery must take into account the important role of gift-giving as a medium of resource redistribution, such as at elaborate wedding celebrations (Werner 2000; Urinboyev and Svensson 2013). Nonetheless, graft, nepotism, and kickbacks can be debilitating: the *New York Times*, for example, documented in detail the cesspool of corruption in Afghanistan, leading to insecure borders, armed forces that sell weapons to the Taliban, widespread graft, and abuse of foreign aid (Rosenburg and Bowley 2012; Walsh 2014).

India, too, suffers from severe corruption (Guhan and Paul 1997). Quah (2008) notes that while anti-corruption campaigns there extended back to the period of British colonial rule, its Central Bureau of Investigation has been so ineffective that he deems eradicating corruption to be an "impossible dream." Sun and Johnston (2009) note that while democracy is often upheld as an antidote to corruption, India's severity in this regard is roughly equal to that in decidedly undemocratic China. Elections there are routinely plagued by accusations of corruption, such as stuffed ballot boxes. In a study of the National Rural Employment Guarantee Scheme (NREGS), which entitles every rural household to up to 100 days of paid employment per year, Niehaus and Sukhtankar (2013) discovered an almost 100 percent "marginal corruption rate," that is, almost none of the additional funds from a wage increase reached those to whom it was intended. Throughout India the "mafia raj" of politicians, real estate developers, and law enforcement officials acquire and develop land in illegal ways (Abdulraheem 2009). Even the mundane act of obtaining a driver's license involves extensive negotiations with corrupt officials (Bertrand et al. 2007). Anjaria (2011), however, in an ethnographic study holds that what is viewed as low-level corruption among street vendors' dealings with municipal officials may be discursively recast as an "ordinary space of negotiation," normalizing the practice as routine and accepted rather than unusual or criminal. Although Indian corporate governance is improving, weak enforcement has often undermined investor confidence (Chakrabarti et al. 2008).

Attempts to stem Indian corruption have generally failed when undermined by inadequate enforcement, lack of political will, and a cultural context that emphasizes tolerance and forgiveness (Tummala 2009). Whereas transparency laws (e.g., the Right to Information Act of 2005) can be almost as useful as anti-corruption strategies in aiding the poor (e.g., to obtain ration cards) (Peisakhin and Pinto 2010), Peisakhin (2012) asserts that transparency can be ineffectual when the gap between privileged bureaucrats and the poor is great. Thus, Shah (2009) concludes that Indian corruption is part of a broader moral economy, not simply the political economy, and that the rural poor often eschew the state as hopelessly corrupt. Heston and Kumar (2008) hold that India's robust economic growth has mitigated pressure on top government officials to reign in corruption, and that corruption hampers the country from realizing its potential rate of increase in economic wellbeing. Robbins (2000) investigated corruption in Indian forest management, Jeffrey (2002) studied popular protests against corrupt sugarcane marketing in India and its role in the reproduction of caste and material inequality, and Corbridge and Kumar (2002) examined jackfruit trees as a means of exploring how corrupt linkages transcended the state-civil society divide.

Similarly, Pakistan has experienced deeply entrenched and debilitating corruption that has rendered the government inordinately ineffective in providing services, maintaining law and order, or combating Islamist militants, to the point that it is sometimes regarded as a failed state (Khan 2007). Thus, corruption has increased water stress in the country (Kugelman and Hathaway 2009), rendered the police illegitimate in the public's eyes (Jackson et al. 2014), and crippled the public health program (Khan and van den Heuvel 2007). Large amounts of American military aid have likely exacerbated the problem. Islam (2004) attributes Pakistani corruption to its collectivist administrative culture and masculinist adherence to hierarchy. These observations underscore the point that corruption is not simply an economic phenomenon, but deeply embedded in moral and cultural values as well. Not surprisingly, the country's high level of corruption has impeded its economic progress (Farooq et al. 2013).

Similarly, impoverished Bangladesh has witnessed its share of corruption. Thus, corrupt practices have contaminated public procurements (Mahmood 2010), health and education services (Knox 2009), and cyclone preparedness (Mahmud and Prowse 2012). As a result, citizen distrust of the state is high and the country's economic growth has been impaired. Typically, it is the poor who suffer the most. E-government – the use of the internet to deliver some government services and dispense funds – has been widely touted as a solution to Bangladesh's corruption (Hossan and Bartram 2010; Bhuiyan 2011), a dubious strategy given the country's internet penetration rate in 2018 was only 48 percent.

In Indonesia, graft and bribery became deeply institutionalized under Suharto's New Order regime (1965–98), in which elaborate patronage networks were used to reward supporters. Suharto's successor, Abdurrahman Wahid, was himself implicated in three corruption scandals. This centralized system of cronyism gradually became more decentralized over time as administrative autonomy was shifted to local districts. More recently, President Joko Widodo launched a weak, ineffectual anti-corruption campaign. The official Corruption Eradication Commission targeted relatively low-level officials who were less able to conceal their crimes. As *The Economist* (2015a) notes, "Indonesians of all stripes gripe about sticky-fingered officials." For example, 66 percent of Indonesians report having bribed a member of the judiciary. Indeed, popular protests in Indonesia often include the chant "KKN," which stands for corruption (*korupsi*), collusion (*kolusi*), and nepotism (*nepotisme*). Corruption has undermined environmental protection laws and fostered illegal logging in Kalimantan (Smith et al. 2003), increased the uncertainty of doing business (Kuncoro 2006), suppressed dissent (Robertson-Snape 1999), and hampered economic development (Collins 2007). Micro-economic evidence suggests that corruption leads to missing government expenditures, a practice that can be reduced through enhanced auditing (Olken 2007).

Despite multiple government campaigns against it, corruption in Malaysia remains stubbornly entrenched (Siddiquee 2010). Although the government appointed the Malaysian Anti-Corruption Commission (MACC), it has had relatively little effect. Prime minister Najib Razak has been embroiled in a mushrooming scandal in which more than $700 million was deposited into his personal bank account, funds, the MACC concluded, which were donated by unidentified private donors and did not come from the state development fund, 1Malaysia Development Berhad. However, foreign sources, including the *Wall Street Journal*, insisted otherwise, and linked the funds to the debt-laden agency. Upon investigating the scandal, attorney general Abdul Gani Patail was removed from office. These events illustrate how intractable corruption can be even in relatively modern, wealthy, and globalized Asian societies.

Other examples of corruption abound. In Thailand, vote-buying casts doubt on the validity of the electoral process (Callahan 2005). Petty corruption is endemic in Myanmar, such as bribes of tax collectors and customs officials; monies intended for NGOs are often siphoned off (Saw 2015). In Vietnam, corruption has hampered the growth of the private sector but not that of the state (Nguyen and van Dijk 2012). Filipino elections are marred by fraud and vote-buying (Quimpo 2007); corruption has seriously impaired the country's public health system, including delays in vaccinations and increased waiting times at clinics, all of which affect poor and rural areas the most (Azfar and Gurgur 2008).

Figure 5.4 shows the distribution of the CPI score across Asia.

AFRICA: CORRUPTION WITHOUT END

Corruption is a highly visible aspect of African politics. For example, Mobutu Sese Seko, long-time tyrant of Zaire (now Democratic Republic of the Congo), amassed a fortune of $5 billion, equal to the country's entire external debt, before he was ousted in 1997 (Thomas 2001; Svensson 2005). The widespread corruption overseen by Kenya's Daniel arap Moi saw millions of dollars lost in "massive cash subsidies for fictitious exports of gold and diamonds" in the Goldenberg scandal (Vasagar 2006). Nigeria's Sani Abacha (Pallister and Capella 2000) and South Africa's Jackie Selebi (Schwella 2013) are also among public officials implicated in major corruption scandals. Occupy Nigeria arose later to protest removal of an oil subsidy that undergirded an uneasy peace between parts of Nigerian society and the corrupt state (Agbedo 2012). Kofele-Kale (2006, p. 697) summarizes the dismal state of African corruption succinctly:

> Corruption is a punishable offense under the laws of nearly every African state, and it is expressly prohibited in several of their constitutions and in various regional and pan-African anti-corruption instruments. In fact, Africa's leadership is so concerned about the problem of corruption that hardly a day goes by without some government entity criticizing corruption and its cancerous effects on African society. Yet, for all the bombast about eradicating corruption, Africa has made little progress on this front.

Compared to many parts of the world, African states are particularly prone to severe corruption, with low average incomes, low literacy levels,

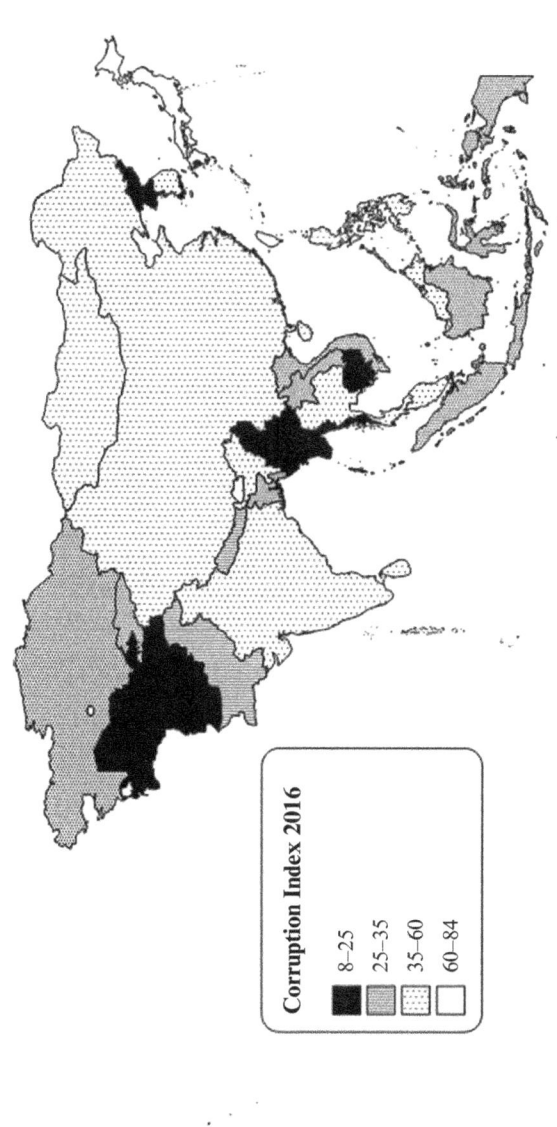

Source: Author, using data from https://www.transparency.org/news/feature/corruption_perceptions_index_2016 (accessed May 14, 2018).

Figure 5.4   Transparency International corruption scores across Asia, 2016

and numerous repressive governments. Nigeria, for example, has been widely cited as a state enveloped by a withering scourge of corruption (Fagbadebo 2007; Smith 2007; Ologbenla 2008; Agbiboa 2012), which has deeply destabilized the government, led to inadequate provision of services such as electricity, hampered foreign investment, and helped to fuel insurgencies such as Boko Haram. Similar conditions apply to the Democratic Republic of the Congo (Matti 2010), Uganda (Tangri and Mwenda 2006), Ethiopia (Plummer 2012), and Kenya (Mwangi 2008). Gettleman (2015, p. 10) notes that

> In Kenya, police corruption starts even before officers are hired. Analysts say it is so stubbornly ingrained that it begins with young men raising money from their villages, usually around $2,000, to bribe recruitment officials just to get the job. The young officers, who make only $200 a month, then have a stiff debt to repay. Hapless motorists and passers-by then become their quarry. Superiors in the police department are widely believed to demand a cut, sustaining a system in which countless men and women in uniform are on the take.

In Somalia, which has effectively lacked a functional government for two decades, the government consists essentially of spoils obtained in a Darwinian struggle for power and survival (Menkhaus 2007). Like Sudan, South Sudan is an institutionalized kleptocracy (de Waal 2014). Similarly, prolonged conflict in Angola over its oil and diamond wealth, including tribal wars waged by contending factions of the elite class, have rendered the state unable to contain systemic and widespread corruption (Malaquias 2001). Kristof (2015) notes of Angola, which has the world's highest infant mortality rate, "Under the corrupt and autocratic president, José Eduardo dos Santos, who has ruled for 35 years, billions of dollars flow to a small elite – as kids starve." Eritrea, obsessed with security since its independence from Ethiopia in 1993, is so corrupt that there is severely inadequate funding for public bureaucracies and minimal provision of services by civil servants hampered by low salaries and morale (Habtom 2014; Tronvoll and Mekonnen 2014).

Not surprisingly, Africans are concerned about corruption. As a survey of selected countries in 2014 by the Pew Charitable Trust reveals, large majorities of people in diverse countries rate it as a "very serious problem." In some countries only crime was rated as a worse concern. Roughly 22 percent of Africans acknowledge having paid a bribe within the last year (The Economist 2015b). For example, police officers in Kenya often stop motorists, and in lieu of citations, which require a court appearance, ask for "fees."

Several factors conspire to create high levels of corruption in Africa. Corruption in Africa in no small part reflects its long and tragic history of colonialism (Acemoglu et al. 2001), and by its occurrence within "a very specific historical, material, and moral global framework involving a variety of actors – as opposed to the standard image of a single despot siphoning off vast quantities of funds" (Scher 2005, p. 18). Pre-independence territorial boundaries have been largely preserved intact, an entrenched political geography that reinforces tribal conflicts and fratricidal civil wars, undermining the growth of healthy civil society. Post-independence political conditions did not help in reducing corruption. For example, although foreign aid undoubtedly plays a role in facilitating growth (Hansen and Tarp 2000), aid to Africa is often misappropriated, resulting in outright theft (Thomas 2001), retarded rates of development (Gyimah-Brempong 2002), the consolidation of power among corrupt elites (Tangri and Mwenda 2006), or the ossification of emerging patterns of social organization and related identities (e.g., among refugees; see Duffield 2002).

Moreover, neoliberal structural adjustment policies foisted on Africa by the International Monetary Fund (IMF) and World Bank impoverished millions, reduced government resources (e.g., to pay competitive public salaries), and amplified anti-democratic political factions that took advantage of unrest and uncertainty to enrich themselves. Neoliberal reforms that sought to downsize the size of the state and its monopolistic powers often only generated new opportunities for corruption instead. Pressure by international financial institutions and other states to adopt particular institutional forms, among other impacts, led to an institutional framework that facilitated Egypt's Hosni Mubarak's movement of vast wealth around the world (Fadel 2011).

De Sardan (1999) locates the phenomenon within a normalized "corruption complex" that includes the absence of effective sanctions, widespread gift-giving, predatory forms of public authority, and redistributive regimes of accumulation. The majority of Africa's states are characterized by weak public institutions, anemic private sectors, high degrees of dependence on foreign aid, and a history of restructuring guided by international institutions such as the IMF and World Bank (Gyimah-Brempong 2002). Development in the African context is particularly vulnerable to corruption, while clientelism, patronage, and patrimonialism are seemingly indelible. African politics tend to exhibit high degrees of neopatrimonialism, a blend between personal rule, patrimonial politics, and bureaucratic forms of domination (Médard 2001; Erdmann and Engel 2007; Bach and Gazibo 2013). Public office is often treated as a route to personal gain by elites and the public, and research on corruption often defines away the overlap between the private and public sectors that can inhibit efforts to combat it (Adebanwi and Obadare 2011).

Despite challenges to the concept's utility (see Di John 2010), it remains influential, as does Bratton and van de Walle's (1994) claim that neopatrimonialism is a core feature of African politics, although it is certainly not confined to Africa. Because it centralizes power among secretive and well-connected political elites, corruption tends to undermine fragile democracies and encourage the growth of despots, almost always at the expense of the poor, as the case of the Democratic Republic of the Congo amply demonstrates (Matti 2010). As Chabal and Daloz (1999) note in their volume *Africa Works*, rigorously structured political systems may work to the disadvantage of entrenched elites, which often prefer to conduct the business of running the state informally, contributing to an environment in which corruption thrives. In this reading, patrimonial structures and tribal conflicts all help to mitigate against a democratic and efficient state. However, as Bayart (2009) argues, in this respect Africa is hardly an unusual exception, and despite its unique historical context, overemphasis on the particularities of the region serves to demonize the African state unfairly. While neopatrimonialism is a useful heuristic for analysis of the region, Bratton and van de Walle (1994) presented a typology of regimes into which African states could be sorted, and it has long been recognized that neopatrimonialism arises in different ways in different geographic contexts (Allen 1995).

Africa's abundance of natural resources also contributes to its corruption. One poignant example is the Niger Delta in Nigeria, where revenues from petroleum exports have left the vast majority of residents impoverished (Watts 2004). Arezki and Gylfason (2013) found a strong relationship between resource exports and corruption in Africa; rents from resources may also be used to quell public dissent. Indeed, part of the "resource curse" of many African states seems to be corrupt, indifferent, tyrannical, and ineffective governments (McFerson 2009).

The presence or absence of natural resources is beyond a state's control, and even a deterministic understanding of the impact of natural resources on a country would therefore see reason for substantial differences within Africa, specifically between nations with and without such resources. A concentration of foreign direct investment in resource extraction can lead to a reduction in the benefits it is expected to produce (Asiedu 2006), which may depend on a more functional institutional context and more open trade policies (Asiedu 2004). While the link between natural resources, civil war, and authoritarian governance is not unquestioned (e.g., Di John 2007), a concentration of investment by multinational corporations in resource extraction can destabilize local institutions and spur conflict (Montague 2002; McFerson 2009). Moreover, resource-rich states are not always affected in the same way. Natural resources comprise the vast majority of Botswana's exports (McFerson 2009), for example, yet Botswana is considered the least corrupt nation in the continent.

Although it varies in type and severity, there can be no doubt that the lives of hundreds of millions of Africans are negatively impeded by the corruption in their states. One example is health care. Transparency International (2015) identified a report by Audit Service Sierra Leone (ASSL) as the only audit to date of local efforts to combat the 2014 Ebola outbreak within hard-hit West African nations. The ASSL report found that allocation of these funds was inadequately documented and that as much as one-third may have been used improperly. The misuse of funds slated for public health and other social services can be a particularly damaging form of corruption (Thomas 2001; Svensson 2005), and this problem is rampant in Africa, including pervasive bribery, sales of counterfeit drugs, misuse of funds for other purposes, and diversion of resources, among other abuses. Although African nations are not alone in their struggles with corruption in health care, the lack of oversight in a closely watched emergency situation reflects the severity of corruption in this case.

Corruption also enters into African wildlife management efforts. For example, it hinders attempts to preserve East Africa's dwindling stocks of herbivores (Jones 2006), often tolerating if not exacerbating poaching, such as the recent massacres of elephant populations in Uganda, Kenya, and Somalia. In the same vein, widespread bribery has reduced the utility of fishing management in South Africa (Sundström 2013). Similarly, corruption often accelerates rates of illegal deforestation, as the case of Benin sadly demonstrates (Siebert and Elwert 2004). Corruption may also enhance environmental damage when it fails to curtail illegal logging or similar forms of resource abuse.

Botswana, Africa's least corrupt country, is a regional leader in infrastructure, education, and public services, and has a history of comparatively strong democratic institutions (Manga Fombad 1999; McFerson 2009). While anti-corruption efforts and attempts to repatriate stolen funds have enjoyed varying levels of success (and failure; see Scher 2005; Lawson 2009; Adebanwi and Obadare 2011), Botswana's anti-corruption efforts exist in a context that permits no comparison. Although the nation's Directorate on Corruption and Economic Crime (DCEC) was founded in response to a series of corruption scandals in the early 1990s, Botswana already had relatively low levels of corruption; further, the nation benefited from a number of inputs that make its situation unique: a relatively healthy economy, available funds for public services, a small (and relatively ethnically homogeneous) population, a lack of bloody conflicts, and no history of significant political power wielded by the military

(Theobald and Williams 1999; McFerson 2009). Further, the DCEC was based on, and initially staffed by veterans of, Hong Kong's Independent Commission Against Corruption (Theobald and Williams 1999). Manga Fombad (1999) cites the DCEC as an example of what a multi-pronged approach to corruption emphasizing prevention and reinforcing the rule of law can be, even in Africa, although in other African nations such an agency would be tasked with combating greater corruption with fewer resources (Theobald and Williams 1999). Botswana's background and successes highlight the variation between African nations and reinforce the need to consider their historical, political, and economic situations.

Figure 5.5 displays African corruption cartographically. In broad terms, the highest levels of corruption may be observed in states such as Sudan, South Sudan, Eritrea, and Somalia, which are marginally functional and often unable to provide minimal government services. Libya, too, falls in this category given the disintegration of the state there following the fall

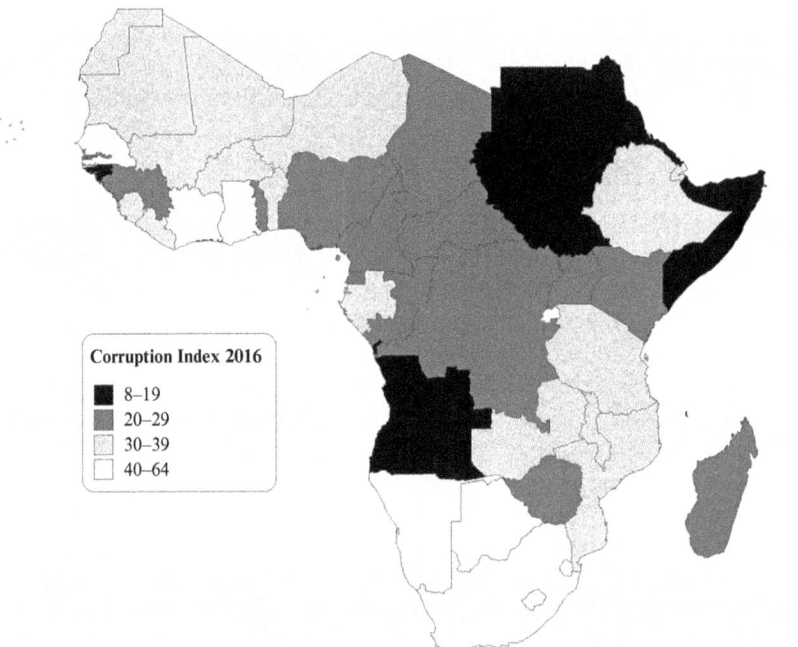

Source: Author, using data from https://www.transparency.org/news/feature/corruption_perceptions_index_2016 (accessed May 14, 2018).

Figure 5.5   Transparency International corruption scores across Africa, 2016

of Qaddafi. Angola emerges as an extremely corrupt petro-state in this context. The next-highest levels of corruption are found in a broad swath of states, including Chad, Nigeria, Congo, the Democratic Republic of Congo, and Kenya, as well as Zimbabwe to the south. Moderately corrupt states are located largely in the northern and western portions of the continent, such as Niger, Mali, Algeria, and Maghreb countries such as Morocco, as well as Ethiopia and several East African nations such as Tanzania, Zambia, and Mozambique. Africa's least corrupt states are found in the south, notably Botswana and to a lesser extent South Africa and Namibia, although Tunisia and Ghana also fall into this group.

CORRUPTION ACROSS LATIN AMERICAN LANDSCAPES

Corruption exhibits a wide variety of degrees of severity and substantive forms throughout Latin America (Morris and Blake 2010). Popular attitudes toward corruption also vary within the region (Licht et al. 2007; Lavena 2013). Latin America's corruption scores are displayed spatially in Figure 5.6. Chile and Uruguay emerge as the least corrupt. Brazil and Cuba form a secondary tier. Guatemala and Honduras have been swept by recent corruption scandals (Malkin 2015). Venezuela and Haiti are seen as the region's two most corrupt states. In many states corruption has hindered attempts to control deforestation (Bulte et al. 2007)

Chile stands out as the success of the past 40 years. Like Brazil, Chile experienced relative stability soon after independence (North et al. 2000) and had an outspoken Church leadership during the build-up to democratization, unlike Argentina (Philpott 2004). Like Uruguay – another country with relatively little corruption – Chile emerged from military rule to return to a relatively well-functioning democracy (Mainwaring and Scully 2008), with the attendant institutions that limit corruption. Schamis (1991) further suggests that the authoritarian regimes in Chile and Uruguay, which emerged in the 1970s, stood in contrast to the authoritarian rule that had emerged elsewhere in earlier decades. And unlike much of the region, Chile is relatively ethnically homogeneous and was somewhat developed already in the 1980s (Alesina et al. 2003). Thus, Chile's unique trajectory and characteristics suggest limits to policy prescriptions for battling corruption.

Venezuela is the most egregious case of Latin American corruption. The phenomenon long pre-dated the Chavez regime, and undermined democratic institutions there, notably mainstream political parties (Gates 2014). However, under Chavez corruption reached new heights as petroleum

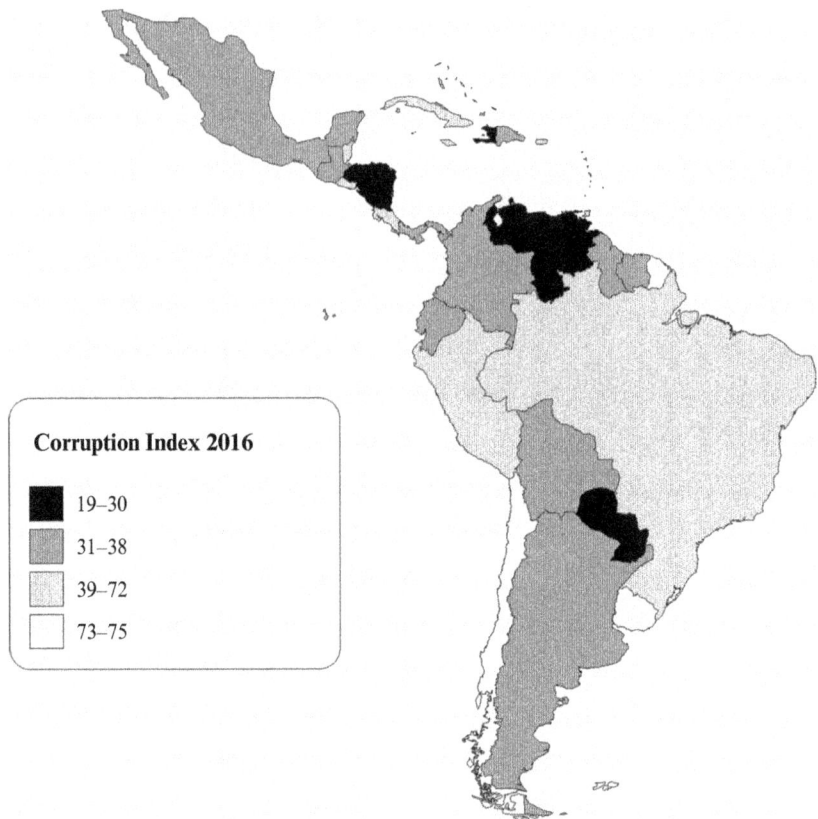

Source: Author, using data from https://www.transparency.org/news/feature/corruption_
perceptions_index_2016 (accessed May 14, 2018).

*Figure 5.6   Transparency International corruption scores for Latin
America, 2016*

exports created massive incentives for rent-seeking, severely impairing the
government's ability to function effectively (Corrales and Penfold 2007).
Little of the petro-revenues trickled down to the middle class (Dietz and
Myers 2007), in part because of the lack of effective political opposition.
Venezuela thus suffers from the classic "resource curse" that plagues
many African countries (Hammond 2011; Kott 2012). A clearer case of
the linkages between corruption and authoritarian governance is difficult
to find. Some Latin American countries with histories of authoritarianism
have become democratic, while Venezuela – which had been relatively
democratic for decades despite the presence of oil – has declined in this

respect (Di John 2007, 2010). While the Chavez regime offered the appearance of democracy, upon closer inspection it more closely resembled a Bolivarian state in which multiple interest groups struggled for power (Gates 2014), setting the stage for widespread abuse of public authority. Not surprisingly, given this climate, collusion between public officials and private managers has long been a major concern of the business community (Perdomo 1990).

In marked contrast to Venezuela, corruption in Colombia is relatively contained, a reflection of its more decentralized political system, stable democracy, thriving markets, and robust political opposition (Langbein and Sanabria 2013). The incidence of bribery is manageable, but varies among cities. Although drug lords, such as those in the Cali cartel, engaged in widespread bribery and violence (Chepesiuk 2003), even this threat was ultimately contained, a situation quite different from that in Mexico. Much of the variation in corruption across the Colombian landscape reflects varying levels of economic development among its departments, notably reliance on minerals and drug trafficking (Poseda 2013).

Brazil also exemplifies severe degrees of corruption, which reflect a long, entrenched institutional history. Neither the transition to democracy nor the ambitions of successive presidential administrations have curtailed it (Adorno 2013). In 2002, a corruption scandal led to the resignation of several ministers of President Luiz Inácio Lula da Silva and damaged the reputation of the ruling Workers' Party (PT) (Flynn 2005). More recently, *The Economist* reported that the PT – if not then-president Dilma Rousseff herself – was implicated in a colossal scandal involving state-run oil giant Petrobras, which apparently paid roughly \$5 billion in bribes. The investigation of sitting politicians, including arrests, has severely shaken the country's elite (Romero 2015), including Rousseff's resignation due to her role in the company in the 2000s, when massive bribery and money laundering are suspected to have occurred. Petrobas is responsible for a considerable share of Brazil's fixed investments, and the scandal has highlighted long-standing demands for heightened accountability (see Taylor and Buranelli 2007). In Brazil, as in many countries undergoing corruption scandals, accountability is divided into oversight, investigations, and sanctions, all of which are necessary for the establishment and maintenance of public trust. Latin American voters are often faced with a dilemma, namely, whether to support powerful presidents whose regimes are fragile when faced with political crises or to vote for legislators and multiparty regimes that may be unstable and dysfunctional.

Argentina is an interesting laboratory to observe Latin American corruption, exhibiting forces that both enable and constrain it. Argentine corruption takes a variety of forms, ranging from bribery among high-ranking officials to widespread allegations of police on the take (Hinton 2005). Interestingly, corruption varies spatially within the country; as Kaufmann et al. (2000, p. 10) note,

> In Argentina, corruption in procurement and budget allocation was found to be common in the province of Corrientes. In contrast, in the city of Buenos Aires, a participatory program to enhance transparency in procurement is bringing about major improvements.

The frequency of scandals among the Argentine elite has led to "scandal fatigue," resignation that inspires little public moral outrage (Waisbord 2004), a situation amplified by the decline in the country's watchdog press (Pinto 2008). Argentina's Anti-Corruption Office, which has had mixed success, has played a role in dampening the abuse of public offices there (De Michele 2001).

Corruption has long been deeply engrained in Mexican political culture (Ionescu 2011). Morris (2008) conducted an extensive study of Mexican corruption, concluding that it served the interests of elites, who pay bribes in order to conduct their affairs, but among the public led to low levels of trust in the government. Widespread suspicions of corruption among the Mexican public have generated substantial distrust of the state (Morris and Klesner 2010). While the country's gradual transition to more democratic political regimes may be expected to mitigate corruption (Bailey and Paras 2006), the enormously destructive and violent drug wars, the rise of powerful cartels and *narcotrafficantes* who have infiltrated many local state apparatuses, and the government's authoritarian response have undermined any progress in this regard (Freeman 2006).

Guatemalan corruption has effectively left large sections of the country under the control of drug lords and crime syndicates (Brands 2010). The country's elites routinely enrich themselves at public expense, a process that has given rise to organized crime and frequently threatens its fragile peace accord (Gavigan 2009). In many respects the country is so riddled with corruption it is on the brink of becoming a failed state (Isaacs 2010). Such was the level of mass dissatisfaction that in 2015 popular demonstrations led to the resignation and imprisonment of President Otto Pérez Molina following an anti-corruption investigation of customs fraud by the International Commission against Impunity in Guatemala (Dada 2015).

Peru offers a rich well of anecdotes about corruption, with a lengthy history stretching to the colonial era, including contraband silver, endemic patronage networks, misused guano revenues, military graft, and war profiteering (Quiroz 2008). Under the notoriously corrupt Fujimori administration (1990–2000), rigged elections and bribes of congressmen became commonplace (Conaghan 2005; Schulte-Bockholt 2013). Torture and drug trafficking were not unknown. Despite his faux populism, substantive government decisions were made out of sight of the public, a political climate ripe for abuse. Although Peru established a special Anti-Corruption System to enforce accountability, it was undermined by an obsession with expeditiousness at the expense of legal evidence, or *legalidad de la prueba* (Navarro 2006). Because it undermines the rule of law, corruption in Peru has negatively affected illegal logging (Smith et al. 2006) and health care (Hunt 2007).

Finally, corruption is also readily evident in the Caribbean, hampering development (Collier 2005). Haiti stands out as the worst offender. Long run as a private cash machine by the Duvalier dynasty, recent attempts at democratization have done little to reign in graft. Billions of dollars in international aid over several decades have had essentially zero impact. International aid that flowed in following the disastrous January 12, 2010 earthquake, for example, was often diverted by corrupt officials and military officers (Oliver-Smith 2010). Widespread shoddy construction amplified the country's vulnerability and resulting casualties (Ambraseys and Bilham 2011). In sharp contrast, Barbados and the Bahamas exhibit relatively low levels of corruption, which calls to mind the argument put forth by Treisman (2007), viz., that corruption is inversely related to a legacy of British rule and Protestantism.

CONCLUDING THOUGHTS

The severity of corruption varies enormously among countries, and often within them (as studies from China, Russia, and the United States attest). The most severe levels of corruption tend to be in poorer countries that have highly centralized political systems, low levels of literacy, violent civil conflicts, and lack an effective independent media, such as Eritrea and North Korea. Conversely, the lowest levels are invariably found in rich, democratic societies with high rates of literacy and a free press, notably Canada and Scandinavia. The nature and impacts of corruption also vary as well: it means quite different things to an Italian business executive and a Javanese peasant. Such observations should refute simplistic "one-size-fits-all" models of corruption that ignore geographic contexts.

Most countries undertake periodic anti-corruption campaigns. In a rare geographical contribution, Brown and Cloke (2004) situate this phenomenon within the broader context of neoliberal governance, including economic liberalization and the institutional reforms that accompany it. Thus, anti-corruption movements are often funded by external donors, including the World Bank and NGOs; after all, corrupt actors can hardly be expected to monitor themselves. Such undertakings usually have few long-term impacts, and often amount to little more than pogroms or vendettas against perceived domestic political opponents. Typically, such efforts amount to little more than hollow rhetoric, the punishment of a few sacrificial lambs, and little substantive change (Bukanovsky 2006). In China, episodic crusades to stop it (including the latest by Xi Jinping), which may lead to executions of Communist Party officials, have failed to stem the tide. Similarly, anti-corruption campaigns have failed visibly in Italy (Wilson 2007), Uganda (Tangri and Mwenda 2006), Thailand (Mutebi 2008), and Indonesia (Butt 2011).

More meaningful measures taken to minimize corruption include: well-defined career paths for public officials, with frequent rotations of staff among offices; a fiercely independent media; encouragement of whistle-blowers; and improved transparency in government appointments, contracts, and expenditures. In the long run, the geography of corruption mirrors that of uneven spatial development. The most effective antidote, therefore, is rising standards of living, improved literacy, greater equality, and democratic political reforms.

REFERENCES

Websites accessed 14 May 2018.
Abdulraheem, A. 2009. Corruption in India: an overview. *Social Action* **59**, 351–63.
Acemoglu, D., S. Johnson, and J. Robinson. 2001. The colonial origins of comparative development: an empirical investigation. *American Economic Review* **91**, 1369–401.
Adebanwi, W. and E. Obadare. 2011. When corruption fights back: democracy and elite interest in Nigeria's anti-corruption war. *Journal of Modern African Studies* **49**, 185–213.
Adorno, S. 2013. Democracy in progress in contemporary Brazil: corruption, organized crime, violence and new paths to the rule of law. *International Journal of Criminology and Sociology* **2**, 409–25.
Agbedo, C. 2012. Placards as a language of civil protest in Nigeria: a systemic-functional analysis of the fuel subsidy crisis. *IOSR Journal of Humanities and Social Sciences* **6**, 17–26.
Agbiboa, D. 2012. Between corruption and development: the political economy of state robbery in Nigeria. *Journal of Business Ethics* **108**(3), 325–45.
al-Ali, Z. 2014. Iraq's rot starts at the top. *New York Times*, August 10. http://www.nytimes.com/2014/08/11/opinion/iraq-s-rot-starts-at-the-top.html
Alamdari, K. 2005. The power structure of the Islamic Republic of Iran: transition from populism to clientelism, and militarization of the government. *Third World Quarterly* **26**(8), 1285–301.
Alesina, A., A. Devleeschauwer, W. Easterly, S. Kurlat, and R. Wacziarg. 2003. Fractionalization. *Journal of Economic Growth* **8**(2), 155–94.
Allen, C. 1995. Understanding African politics. *Review of African Political Economy* **22**, 301–20.
Alley, A. 2010. The rules of the game: unpacking patronage politics in Yemen. *Middle East Journal* **64**(3), 385–409.
Ambraseys, N. and R. Bilham, 2011.Corruption kills. *Nature* **469**, 153–55.
Anjaria, J. 2011. Ordinary states: everyday corruption and the politics of space in Mumbai. *American Ethnologist* **38**(1), 58–72.

Arafa, M. 2011. Towards a culture for accountability: a new dawn for Egypt. *Phoenix Law Review* **1**, 2–20.

Arango, T. 2013. Corruption scandal is edging near Turkish premier. *New York Times*, December 25. http://www.nytimes.com/2013/12/26/world/europe/turkish-cabinet-members-resign.html?_r=0

Arezki, R. and T. Gylfason. 2013. Resource rents, democracy, corruption and conflict: evidence from sub-Saharan Africa. *Journal of African Economies* **22**(4), 552–69.

Asiedu, E. 2004. The determinants of employment of affiliates of US multinational enterprises in Africa. *Development Policy Review* **22**, 371–9.

Asiedu, E. 2006. Foreign direct investment in Africa: the role of natural resources, market size, government policy, institutions and political instability. *World Economy* **29**, 63–77.

Aybar, S. and C. Lapavistsas. 2001. The recent Turkish crisis: another step toward free market authoritarianism. *Historical Materialism* **8**, 279–308.

Azfar, O. and T. Gurgur. 2008. Does corruption affect health outcomes in the Philippines? *Economics of Governance* **9**(3), 197–244.

Bach, D. and M. Gazibo (eds). 2013. *Neopatrimonialism in Africa and Beyond*. London: Routledge.

Bågenholm, A. 2013. Throwing the rascals out? The electoral effects of corruption allegations and corruption scandals in Europe 1981–2011. *Crime, Law and Social Change* **60**(5), 595–609.

Bailey, J. and P. Paras. 2006. Perceptions and attitudes about corruption and democracy in Mexico. *Mexican Studies* **22**(1), 57–82.

Bardhan, P. 1997. Corruption and development: a review of issues. *Journal of Economic Literature* **35**, 1320–46.

Bayart, J.-F. 2009. *The State in Africa*. Cambridge: Polity.

Bedirhanoğlu, P. 2007. The neoliberal discourse on corruption as a means of consent building: reflections from post-crisis Turkey. *Third World Quarterly* **28**(7), 1239–54.

Bertrand, M., S. Djankov, R. Hanna, and S. Mul-Lainathan. 2007. Obtaining a driver's license in India: an experimental approach to studying corruption. *Quarterly Journal of Economics* **122**(4), 1639–76.

Bhuiyan, S. 2011. Modernizing Bangladesh public administration through e-governance: benefits and challenges. *Government Information Quarterly* **28**, 54–65.

Borshchevskaya, A. 2010. Sponsored corruption and neglected reform in Syria. *Middle East Quarterly*, Summer, 41–50.

Braendle, U., T. Gasser, and J. Noll. 2005. Corporate governance in China – is economic growth potential hindered by g*uanxi? Business and Society Review* **110**(4), 389–405.

Brands, H. 2010. *Crime, Violence, and the Crisis in Guatemala: A Case Study in the Erosion of the State*. Washington, DC: Strategic Studies Institute.

Bratton, M. and N. van de Walle. 1994. Neopatrimonial regimes and political transitions in Africa. *World Politics* **46**, 453–89.

Brown, E. and J. Cloke. 2004. Neoliberal reform, governance and corruption in the South: assessing the international anti-corruption crusade. *Antipode* **36**(2), 272–94.

Brunetti, A. and B. Weder. 2003. A free press is bad news for corruption. *Journal of Public Economics* **87**(7–8), 1801–24.

Bukanovsky, M. 2006. The hollowness of anti-corruption discourse. *Review of International Political Economy* **13**(2), 181–209.

Bulte, E., R. Damania, and R. López. 2007. On the gains of committing to inefficiency: corruption, deforestation and low land productivity in Latin America. *Journal of Environmental Economics and Management* **54**(3), 277–95.

Butt, S. 2011. Anti-corruption reform in Indonesia: an obituary? *Bulletin of Indonesian Economic Studies* **47**(3), 381–94.

Byman, D. and J. Lind. 2010. Pyongyang's survival strategy: tools of authoritarian control in North Korea. *International Security* **35**(1), 44–74.

Callahan, W. 2005. Social capital and corruption: vote buying and the politics of reform in Thailand. *Perspectives on Politics* 3, 495–508.

Carmody, P. 2009. Cruciform sovereignty, matrix governance and the scramble for Africa's oil: insights from Chad and Sudan. *Political Geography* **28**(6), 353–61.

Chabal, P. and J.-P. Daloz. 1999. *Africa Works: Disorder as Political Instrument.* Bloomington, IN: Indiana University Press.

Chae, W., H. Sohn, and O. Kim. 2006. A study of the realities and causes of corruption of North Korean bureaucrats: a survey of perceptions of North Korean refugees. *Journal of the Korean Association for Governance* **13**(1), 297–321.

Chakrabarti, R., W. Megginso, and P. Yadav. 2008. Corporate governance in India. *Journal of Applied Corporate Finance* **20**(1), 59–72.

Chepesiuk, R. 2003. *The Bullet or the Bribe: Taking Down Colombia's Cali Drug Cartel.* New York: Praeger.

Cole, M., R. Elliott, and J. Zhang. 2009. Corruption, governance and FDI location in China: a province-level analysis. *Journal of Development Studies* **45**(9), 1494–512.

Collier, M. 2005. *Political Corruption in the Caribbean Basin: Constructing a Theory to Combat Corruption.* London and New York: Routledge.

Collins, K. 2006. *Clan Politics and Regime Transition in Central Asia.* Cambridge: Cambridge University Press.

Collins, S. 2007. *Indonesia Betrayed: How Development Fails.* Manoa, Hawai'i: University of Hawai'i Press.

Conaghan, C. 2005. *Fujimori's Peru: Deception in the Public Sphere.* Pittsburgh, PA: University of Pittsburgh Press.

Corbridge, S. and S. Kumar. 2002. Community, corruption, landscape: tales from the tree trade. *Political Geography* **21**(6), 765–88.

Corrales, J. and M. Penfold. 2007. Venezuela: crowding out the opposition. *Journal of Democracy* **18**(2), 99–113.

Dada, C. 2015. Corruption charges turn Guatemala upside down. *The New Yorker*, September 4. http://www.newyorker.com/news/news-desk/corruption-charges-turn-guatemala-upside-down

De Michele, R. 2001.The role of the Anti-Corruption Office in Argentina. *Journal of Public Inquiry*, Fall/Winter, 17–20. http://www.acauthorities.org/sites/aca/files/countrydoc/The%20Role%20of%20the%20Anti-Corruption%20Office%20in%20Argentina.pdf

De Sardan, J. 1999. A moral economy of corruption in Africa? *Journal of Modern African Studies* **37**(1), 25–52.

De Waal, A. 2014. When kleptocracy becomes insolvent: brute causes of the civil war in South Sudan. *African Affairs* **113**(452), 347–69.

Della Porta, D. and Y. Mény (eds). 1997. *Democracy and Corruption in Europe.* London: Pinter.

Di John, J. 2007. Oil abundance and violent political conflict: a critical assessment. *Journal of Development Studies* **43**, 961–86.

Di John, J. 2010. *From Windfall to Curse? Oil and Industrialization in Venezuela, 1920 to the Present.* College Station, PA: Pennsylvania State University Press.

Dietz, H. and D. Myers. 2007. From thaw to deluge: party system collapse in Venezuela and Peru. *Latin American Politics and Society* **49**(2), 59–86.

Doig, A. 2010. Asking the right questions? Addressing corruption and EU accession: the case study of Turkey. *Journal of Financial Crime* **17**(1), 9–21.

Domoro, O. and S. Agil. 2012. Factors influencing police corruption in Libya – a preliminary analysis. *International Journal of Economics and Management Sciences* **2**(2), 25–35.

Duffield, M. 2002. Aid and complicity: the case of war-displaced Southerners in the Northern Sudan. *Journal of Modern African Studies* **40**, 83–104.

Engler, S. 2015. Corruption and electoral support for new political parties in Central and Eastern Europe. *West European Politics* **39**(2), 278–304.

Erdmann, G. and U. Engel. 2007. Neopatrimonialism reconsidered: critical review and elaboration of an elusive concept. *Commonwealth and Comparative Politics* **45**, 95–119.

Fadel, M. 2011. Public corruption and the Egyptian Revolution of January 25: can emerging international anti-corruption norms assist Egypt recover misappropriated public funds? *Harvard International Law Journal* **52**. https://tspace.library.utoronto.ca/ bitstream/1807/78202/1/Fadel%20-%20Public%20Corruption%20and%20Egyptian%20 Revolution%20HILJ.pdf

Fagbadebo, O. 2007. Corruption, governance and political instability in Nigeria. *African Journal of Political Science and International Relations* **1**(2), 28–37.

Fan, Y. 2002. Guanxi's consequences: personal gains at social cost. *Journal of Business Ethics* **38**, 371–80.

Farooq, A., M. Shahbaz, M. Arouri, and F. Teulon. 2013. Does corruption impede economic growth in Pakistan? *Economic Modelling* **35**, 622–33.

Finkel, A. 2000. Who guards the Turkish press? A perspective on press corruption in Turkey. *Journal of International Affairs* **54**(1), 147–66.

Flynn, P. 2005. Brazil and Lula, 2005: crisis, corruption and change in political perspective. *Third World Quarterly* **26**(8), 1221–67.

Freeman, L. 2006. *State of Siege: Drug-related Violence and Corruption in Mexico: Unintended Consequences of the War on Drugs*. WOLA Special Report. http://www.wola. org/sites/default/files/downloadable/Mexico/past/state_of_siege_06.06.pdf

Gates, L. 2014. Interest groups in Venezuela: lessons from the failure of a "model democracy" and the rise of a Bolivarian democracy. *Journal of Public Affairs* **14**(3–4), 240–53.

Gavigan, P. 2009. Organized crime, illicit power structures and Guatemala's threatened peace process. *International Peacekeeping* **16**(1), 62–76.

Gettleman, J. 2015. Kenya struggles over best response to attack. *New York Times*, April 9, p. 10.

Ghalwash, T. 2014. Corruption and economic growth: evidence from Egypt. *Modern Economy* **5**(10), 1001–9.

Gillespie, K. 2006. The Middle East's corruption conundrum. *Current History* **1**, 40–6.

Gleason, G. 1995. Corruption, decolonization and development in Central Asia. *European Journal on Criminal Policy and Research* **3**(2), 38–47.

Gold, T., D. Guthrie, and D. Wank. 2002. *Social Connections in China: Institutions, Culture and the Changing Nature of Guanxi.* Cambridge: Cambridge University Press.

Gong, T. 2002. Dangerous collusion: corruption as a collective venture in contemporary China. *Communist and Post-Communist Studies* **35**(1), 85–103.

Gounev, P. and F. Ruggiero (eds). 2012. *Corruption and Organised Crime in Europe: Illegal Partnerships.* London: Routledge.

Green, P. 2005. Disaster by design: corruption, construction and catastrophe. *British Journal of Criminology* **45**(4), 528–46.

Grimaldi, J. and R. Harrow. 2011. In Egypt, corruption cases had an American root. *Washington Post*, October 19. http://www.washingtonpost.com/investigations/in-egypt-corruption-had-an-americanroot/2011/10/07/gIQAApWoyL_story.html

Guetat, I. 2006. The effects of corruption on growth performance of the MENA countries. *Journal of Economics and Finance* **30**(2), 208–21.

Guhan S. and S. Paul (eds). 1997. *Corruption in India: Agenda for Action.* New Delhi: Vision Books.

Guo, Y. 2008. Corruption in transitional China: an empirical analysis. *China Quarterly* **194**, 349–68.

Gyimah-Brempong, K. 2002. Corruption, economic growth, and income inequality in Africa. *Economics of Governance* **3**, 183–209.

Haber, S. and V. Menaldo. 2011. Do natural resources fuel authoritarianism? A reappraisal of the resource curse. *American Political Science Review* **105**(1), 1–26.

Habtom, G. 2014. Public administration reform in Eritrea: past trends and emerging challenges. *Journal of Public Administration and Policy Research* **6**(3), 44–58.

Hadjadj, D. 2007. Algeria: a future hijacked by corruption. *Mediterranean Politics* **12**(2), 263–77.

Haggard, S. and M. Noland. 2010. Reform from below: behavioral and institutional change in North Korea. *Journal of Economic Behavior & Organization* **73**, 133–52.

Hammond, J. 2011. The resource curse and oil revenues in Angola and Venezuela. *Science & Society* **75**(3), 348–78.

Hansen, H. and F. Tarp. 2000. Aid effectiveness disputed. *Journal of International Development* **12**, 375–98.

He, Z. 2000. Corruption and anti-corruption in reform China. *Communist and Post-Communist Studies* **33**, 243–70.

Heineman, B. 2011. In China, corruption and unrest threaten autocratic rule. *The Atlantic*, June 29. http://www.law.harvard.edu/programs/corp_gov/articles/Heineman_Atlantic_0 6-29-11.pdf

Hertog, S. 2011. *Princes, Brokers, and Bureaucrats: Oil and the State in Saudi Arabia*. Ithaca, NY: Cornell University Press.

Heston, A. and V. Kumar. 2008. Institutional flaws and corruption incentives in India. *Journal of Development Studies* **44**(9), 1243–61.

Hill, G., P. Salisbury, L. Northedge, and J. Kinninmont. 2013. *Yemen: Corruption, Capital Flight and Global Drivers of Conflict*. A Chatham House Report. http://www.hiwar-watani.org/uploads/1/5/2/3/15238886/0913r_yemen.pdf

Hinton, M. 2005. A distant reality: democratic policing in Argentina and Brazil. *Criminal Justice* **5**(1), 75–100.

Hirsch, A. 2010. WikiLeaks cables: Sudanese president "stashed $9bn in UK banks". *Guardian*, December 17. http://www.theguardian.com/world/2010/dec/17/wikileaks-sudanese-preside nt-cash-london

Holmes, L. 2009. Crime, organised crime and corruption in post-communist Europe and the CIS. *Communist and Post-Communist Studies* **42**(2), 265–87.

Hossan, C. and T. Bartram. 2010. The battle against corruption and inefficiency with the help of egovernment in Bangladesh. *Electronic Government, An International Journal* **7**(1), 89–100.

Hunt, J. 2007. Bribery in health care in Peru and Uganda. National Bureau of Economic Research Working Paper No. 13034. Washington, DC.

Ionescu, L. 2011. Mexico's pervasive culture of corruption. *Economics, Management, and Financial Markets* **2**, 182–7.

Isaacs, A. 2010. Guatemala on the brink. *Journal of Democracy* **21**(2), 108–22.

Islam, N. 2004. Sifarish, sycophants, power and collectivism: administrative culture in Pakistan. *International Review of Administrative Sciences* **70**(2), 311–30.

Ismail, O. 2011. The failure of education in combating corruption in Sudan: the impact on sustainable development. *OIDA International Journal of Sustainable Development* **2**(11), 43–50.

Jackson, J., M. Asif, B. Bradford, and M. Zakar. 2014. Corruption and police legitimacy in Lahore, Pakistan. *British Journal of Criminology* **54**(6), 1067–88.

Jain, A. 2001. Corruption: a review. *Journal of Economic Surveys* **15**(1), 71–121.

Jeffrey, C. 2002. Caste, class, and clientelism: a political economy of everyday corruption in rural North India. *Economic Geography* **78**(1), 21–41.

Johnston, M. and Y. Hao. 1995. China's surge of corruption: delayed political development, markets, and political reform. *Journal of Democracy* **6**(4), 80–94.

Jones, S. 2006. A political ecology of wildlife conservation in Africa. *Review of African Political Economy* **33**(109), 483–95.

Kamba I. and M. Sakdam. 2012. Reality assessment of the corruption in Libya and search for causes and cures. *Journal of Business and Economics* **3**(5), 357–68.

Kaufmann, D., A. Kraay, and P. Zoido-Labatón. 2000. Governance matters: from measurement to action. *Finance & Development*, June, 10–13.

Khan, F. 2007. Corruption and the decline of the state in Pakistan. *Asian Journal of Political Science* **15**(2), 219–47.

Khan, M. and W. van den Heuvel. 2007. The impact of political context upon the health policy process in Pakistan. *Public Health* **121**(4), 278–86.

Kirkpatrick, D. 2014. Graft hobbles Iraq's military in fighting ISIS. *New York Times*, November 23. http://www.nytimes.com/2014/11/24/world/middleeast/graft-hobbles-iraqs-military-in-fighting-isis.html

Knox, C. 2009. Dealing with sectoral corruption in Bangladesh: developing citizen involvement. *Public Administration and Development* **29**, 117–32.

Ko, K. and C. Weng. 2012. Structural changes in Chinese corruption. *China Quarterly* **211**, 718–40.

Kofele-Kale, N. 2006. Change or the illusion of change: the war against official corruption in Africa. *George Washington International Law Review* **38**, 697–748.

Kostadinova, T. 2012. *Political Corruption in Eastern Europe: Politics after Communism.* Boulder, CO: Lynne Rienner Publishers.

Kott, A. 2012. Assessing whether oil dependency in Venezuela contributes to national instability. *Journal of Strategic Security* **5**(3), 69–86.

Kristof, N. 2015. Deadliest country for kids. *New York Times*, March 18. http://www.nytimes.com/2015/03/19/opinion/nicholas-kristof-deadliest-country-for-kids.html?_r=0

Kugelman, M. and R. Hathaway (eds). 2009. *Running on Empty: Pakistan's Water Crisis.* Washington, DC: Woodrow Wilson International Center for Scholars.

Kuncoro, A. 2006. Corruption and business uncertainty in Indonesia. *ASEAN Economic Bulletin* **23**(1),11–30.

Kwong, J. 2015. *The Political Economy of Corruption in China.* Armonk, NY: M.E. Sharpe.

Langbein, L. and P. Sanabria. 2013. The shape of corruption: Colombia as a case study. *Journal of Development Studies* **49**(11), 1500–13.

Lavena, C.F. 2013. What determines permissiveness toward corruption? A study of attitudes in Latin America. *Public Integrity* **15**(4), 345–66.

Lawson, L. 2009. The politics of anti-corruption reform in Africa. *Journal of Modern African Studies* **47**, 73–100.

Le Billon, P. 2005. Corruption, reconstruction and oil governance in Iraq. *Third World Quarterly* **26**(4–5), 685–703.

Leenders, R. 2012. *Spoils of Truce: Corruption and State-building in Postwar Lebanon.* Ithaca, NY: Cornell University Press.

Leenders, R. and J. Sfakianakis. 2002. *Middle East and North Africa.* Transparency International annual report. http://unpan1.un.org/intradoc/groups/public/documents/AP CITY/UNPAN008450.pdf

Li, L. 2011. Performing bribery in China: guanxi-practice, corruption with a human face. *Journal of Contemporary China* **20**(68), 1–20.

Li, S. and J. Wu, 2007. Why China thrives despite corruption. *Far Eastern Economic Review* **170**(3), 24–8.

Licht, A., C. Goldschmidt, and S. Schwartz. 2007. Culture rules: the foundations of the rule of law and other norms of governance. *Journal of Comparative Economics* **35**(4), 659–88.

Looney, R. 2005. Profiles of corruption in the Middle East. *Journal of South Asian and Middle Eastern Studies* **28**(4), 1–20.

Looney, R. 2008. Reconstruction and peacebuilding under extreme adversity: the problem of pervasive corruption in Iraq. *International Peacekeeping* **15**(3), 424–40.

Luo, Y. 2002. Corruption and organization in Asian management systems. *Asia Pacific Journal of Management* **19**, 405–22.

Luo, Y. 2008. The changing Chinese culture and business behavior: the perspective of intertwinement between guanxi and corruption. International Business Review **17**(2), 188–93.

Ma, S. 2008. The dual nature of anti-corruption agencies in China. *Crime, Law and Social Change* **49**(2), 153–65.

Mahmood, S. 2010. Public procurement and corruption in Bangladesh confronting the challenges and opportunities. *Journal of Public Administration and Policy Research* **2**(6), 103–11.

Mahmud, T. and M. Prowse. 2012. Corruption in cyclone preparedness and relief efforts in coastal Bangladesh. *Global Environmental Change* **22**(4), 933–43.

Mainwaring, S. and T. Scully. 2008. Latin America: eight lessons for governance. *Journal of Democracy* **19**(3), 113–27.

Malaquias, A. 2001. Making war & lots of money: the political economy of protracted conflict in Angola. *Review of African Political Economy* **28**(90), 521–36.

Malkin, E. 2015. Wave of protests spreads to scandal-weary Honduras and Guatemala. *New York Times*, June 14, p. 9.

Manga Fombad, C. 1999. Curbing corruption in Africa: some lessons from Botswana's experience. *International Social Science Journal* **51**, 241–54.

Manion, M. 2004. *Corruption by Design: Building Clean Government in Mainland China and Hong Kong*. Cambridge, MA: Harvard University Press.

Martini, M. 2011. Corruption and anti-corruption in Sudan. http://www.transparency.org/files/content/corruptionqas/342_Corruption_and_anti-corruption_in_Sudan.pdf

Mashali, B. 2012. Analyzing the relationship between perceived grand corruption and petty corruption in developing countries: case study of Iran. *International Review of Administrative Sciences* **78**(4), 775–87.

Matti, S. 2010. The Democratic Republic of the Congo? Corruption, patronage, and competitive authoritarianism in the DRC. *Africa Today* **56**(4), 42–61.

McFerson, H. 2009. Governance and hyper-corruption in resource-rich African countries. *Third World Quarterly* **30**(8), 1529–47.

Médard, J. 2001. Corruption in the neopatrimonial states of sub-Saharan Africa. In A. Heidenheimer and M. Johnston (eds), *Political Corruption: Concepts and Contexts*, pp. 379–402. Rutgers, NJ: Transactions Publishing.

Menkhaus, K. 2007. Governance without government in Somalia: spoilers, state building, and the politics of coping. *International Security* **31**(3), 74–106.

Miller, W., Grødeland, Å, and T. Koshechkina. 2001. *A Culture of Corruption? Coping with Government in Post-communist Europe*. Budapest: Central European University Press.

Montague, D. 2002. Stolen goods: Coltan and conflict in the Democratic Republic of the Congo. *SAIS Review* **22**, 103–18.

Morris, S. 2008. Disaggregating corruption: a comparison of participation and perceptions in Latin America with a focus on Mexico. *Bulletin of Latin American Research* **27**(3), 388–409.

Morris, S. and C. Blake (eds). 2010. *Corruption and Politics in Latin America: National and Regional Dynamics*. Boulder, CO and London: Lynne Rienner Publishers.

Morris, S. and J. Klesner. 2010. Corruption and trust: theoretical considerations and evidence from Mexico. *Comparative Political Studies* **43**(10), 1258–85.

Mutebi, A. 2008. Explaining the failure of Thailand's anti-corruption regime. *Development and Change* **39**(1), 147–71.

Mwangi, O. 2008. Political corruption, party financing and democracy in Kenya. *Journal of Modern African Studies* **46**(2), 267–85.

Navarro, N. 2006. Fighting corruption: the Peruvian experience. *Journal of International Criminal Justice* **4**(3), 488–509.

Nguyen, T. and M. van Dijk. 2012. Corruption, growth, and governance: private vs. state-owned firms in Vietnam. *Journal of Banking & Finance* **36**, 2935–48.

Niehaus, P. and S. Sukhtankar. 2013. The marginal rate of corruption in public programs: evidence from India. *Journal of Public Economics* **104**, 52–64.

North, D., W. Summerhill, and B. Weingast. 2000. Order, disorder, and economic change: Latin America versus North America. In B. de Mesquita and H. Root (eds), *Governing for Prosperity*, pp. 17–58. New Haven, CT: Yale University Press.

Oliver-Smith, A. 2010. Haiti and the historical construction of disasters. *NACLA Report on the Americas* **43**(4), 32–6.

Olken, B. 2007. Corruption and the costs of redistribution: micro evidence from Indonesia. *Journal of Public Economics* **90**, 853–70.

Ologbenla, D. 2008. Leadership, governance and corruption in Nigeria. *Economic and Policy Review* **14**(1), 36–44.

Pallister, D. and P. Capella. 2000. British banks set to freeze dictator's millions. *Guardian*, July 7. http://www.theguardian.com/

Patey, L. 2010. Crude days ahead? Oil and the resource curse in Sudan. *African Affairs* **109**(437), 617–36.

Peisakhin, L. 2012. Transparency and corruption: evidence from India. *Journal of Law and Economics* **55**(1), 129–49.

Peisakhin, L. and P. Pinto 2010. Is transparency an effective anti-corruption strategy? Evidence from a field experiment in India. *Regulation & Governance* **4**, 261–80.

Perdomo, R. 1990. Corruption and business in present day Venezuela. *Journal of Business Ethics* **9**(7), 555–66.

Peyrouse, S. 2012. *Turkmenistan: Strategies of Power, Dilemmas of Development*. Armonk, NY: M.E. Sharpe.

Philpott, D. 2004. The Catholic wave. *Journal of Democracy* **15**(2), 32–46.

Pinto, J. 2008. Muzzling the watchdog: the case of disappearing watchdog journalism from Argentine mainstream news. *Journalism* **9**(6), 750–74.

Plummer, J. 2012. *Diagnosing Corruption in Ethiopia: Perceptions, Realities, and the Way Forward for Key Sectors.* Washington, DC: World Bank.

Polk, J., J. Rovny, R. Bakker et al. 2017. Explaining the salience of anti-elitism and reducing political corruption for political parties in Europe with the 2014 Chapel Hill Expert Survey data. *Research and Politics*, January–March, 1–9.

Pollack, K. 2006. Iran: three alternative futures. *Middle East Review of International Affairs* **10**(2), 1–12. http://www.naba.org.uk/Content/news/Daily/MERIA/MERIA_60612.pdf

Poseda, A. 2013. Corruption, economic development, and insecurity in Colombia. In I. Osman (ed.), *Handbook of Research on Strategic Performance Management and Measurement Using Data Envelopment Analysis*, pp. 373–87. New York: IGI Global Press.

Quah, J. 2008. Curbing corruption in India: an impossible dream? *Asian Journal of Political Science* **16**(3), 240–59.

Quimpo, N. 2007. The Philippines: political parties and corruption. *Southeast Asian Affairs*, 277–94.

Quiroz, A. 2008. *Corrupt Circles: A History of Unbound Graft in Peru*. Washington, DC: Woodrow Wilson Center Press.

Robbins, P. 2000. The rotten institution: corruption in natural resource management. *Political Geography* **19**(4), 423–43.

Robertson-Snape, F. 1999. Corruption, collusion and nepotism in Indonesia. *Third World Quarterly* **20**(3), 589–602.

Romero, S. 2015. An anticorruption drive throws Brazil's leadership into disarray. *New York Times*, August 13, pp. 1, 7.

Rosenburg, M. and G. Bowley. 2012. Intractable Afghan graft hampering U.S. strategy. *New York Times*, March 7. http://www.nytimes.com/2012/03/08/world/asia/corruption-remains-intractable-in-afghanistan-under-karzai-government.html?pagewanted=all

Ross, M. 2011. Will oil drown the Arab Spring? *Foreign Affairs* **90**(5), 2–7.

Sadowski, Y. 1987. Patronage and the Ba'th: corruption and control in contemporary Syria. *Arab Studies Quarterly* **9**(4), 442–61.

Sajo, A. 1998. Corruption, clientelism, and the future of the constitutional state in Eastern Europe. *East European Constitutional Review* **7**(2), 54–63.

Saw, K.S. 2015. Tackling Myanmar's corruption challenge. Munich Personal RePEc Archive Paper 63764. https://mpra.ub.uni-muenchen.de/63764/1/MPRA_paper_63764.pdf

Sayan, S. (ed.). 2009. *Economic Performance in the Middle East and North Africa*. London: Routledge.

Schamis, H. 1991. Reconceptualizing Latin American authoritarianism in the 1970s: from bureaucratic-authoritarianism to neoconservatism. *Comparative Politics* **23**(2), 201–20.

Scher, D. 2005. Asset recovery: repatriating Africa's looted billions. *African Security Review* **14**, 17–26.

Schulte-Bockholt, A. 2013. *Corruption as Power: Criminal Governance in Peru during the Fujimori Era (1990–2000)*. New York: Peter Lang.

Schwella, E. 2013. Bad public leadership in South Africa: the Jackie Selebi case. *Scientia Militia, South African Journal of Military Studies* **41**, 65–90.

Shah, A. 2009. Morality, corruption and the state: insights from Jharkhand, eastern India. *Journal of Development Studies* **45**(3), 295–313.

Siddiquee, N. 2010. Combating corruption and managing integrity in Malaysia: a critical overview of recent strategies and initiatives. *Public Organization Review* **10**, 153–71.

Siebert, U. and G. Elwert. 2004. Combating corruption and illegal logging in Bénin, West Africa: recommendations for forest sector reform. *Journal of Sustainable Forestry* **19**(1–3), 239–61.

Smith, D. 2007. A Culture of Corruption: Everyday Deception and Popular Discontent in Nigeria. Princeton, NJ: Princeton University Press.

Smith, J., K. Obidzinski, Subarudi, and J. Suramengala. 2003. Illegal logging, collusive corruption and fragmented governments in Kalimantan, Indonesia. *International Forestry Review* **5**(3), 293–302.

Smith, J., V. Colan, C. Sabogal, and L. Snook. 2006. Why policy reforms fail to improve logging practices: the role of governance and norms in Peru. *Forest Policy and Economics* **8**(4), 458–69.

Soliman, H. and S. Cable. 2011. Sinking under the weight of corruption: neoliberal reform, political accountability and justice. *Current Sociology* **59**(6), 735–53.

Sun, Y. and M. Johnston. 2009. Does democracy check corruption? Insights from China and India. *Comparative Politics* **42**(1), 1–19.

Sundström, A. 2013. Corruption in the commons: why bribery hampers enforcement of environmental regulations in South African fisheries. *International Journal of the Commons* **7**(2), 454–72.

Svensson, J. 2005. Eight questions about corruption. *Journal of Economic Perspectives* **19**(3), 19–42.

Tangri, R. and M. Mwenda. 2006. Politics, donors, and the ineffectiveness of anti-corruption institutions in Uganda. *Journal of Modern African Studies* **44**, 101–24.

Tänzler, D., K Maras, and A. Giannakoupoulos (eds). 2012. *The Social Construction of Corruption in Europe*. London: Routledge.

Taylor, M. and V. Buranelli. 2007. Ending up in pizza: accountability as a problem of institutional arrangement in Brazil. *Latin American Politics and Society* **49**, 59–87.

Tekin-Coru, A. 2006. Corruption and the ownership composition of the multinational firm at the time of entry: evidence from Turkey. *Journal of Economics and Finance* **30**(2), 251–69.

The Economist. 2015a. Corruption in Indonesia: a damnable scourge. *The Economist*, June 6. http://www.economist.com/news/asia/21653671-jokowis-arduous-task-cleaning-up-government-damnable-scourge

The Economist. 2015b. The scale of corruption in Africa. *The Economist*, December 3. http://www.economist.com/news/middle-east-and-africa/21679473-gloomy-news-transparency-international-scale-corruption-africa

Theobald, R. and R. Williams. 1999. Combating corruption in Botswana: regional role model or deviant case? *Commonwealth and Comparative Politics* **37**, 117–34.

Thomas, M. 2001. Getting debt relief right. *Foreign Affairs* **80**, 36–45.

Transparency International. 2015. Ebola: corruption and aid. February 27. http://www.transparency.org

Transparency International. 2016. People and corruption: Europe and Central Asia 2016. https://www.transparency.org/whatwedo/publication/people_and_corruption_europe_and_central_asia_2016

Treisman, D. 2007. What have we learned about the causes of corruption from ten years of cross-national empirical research? *Annual Review of Political Science* **10**, 211–44.

Tronvoll, K. and D. Mekonnen. 2014. *The African Garrison State: Human Rights and Political Development in Eritrea*. New York: Boydell and Brewer.

Tummala, K. 2009. Combating corruption: lessons out of India. *International Public Management Review* **10**(1), 34–58.

Urinboyev, R. and M. Svensson. 2013. Corruption in a culture of money: understanding social norms in post-Soviet Uzbekistan. In M. Baier (ed.), *Social and Legal Norms*, pp. 267–84. Aldershot: Ashgate.

Vasagar, J. 2006. Charges in Kenya corruption scandal. *Guardian*, March 16. http://www.theguardian.com/

Volker, P. 2004. *Syria under Bashar al-Asad: Modernisation and the Limits of Change*. Oxford: Oxford University Press.

Waisbord, S. 2004. Scandals, media, and citizenship in contemporary Argentina. *American Behavioral Scientist* **47**(8), 1072–98.

Wallace, C. and R. Latcheva. 2006. Economic transformation outside the law: corruption, trust in public institutions and the informal economy in transition countries of Central and Eastern Europe. *Europe-Asia Studies* **58**(1), 81–102.

Walsh, D. 2014. At Afghan border, graft is part of the bargain. *New York Times*, November 11. http://www.nytimes.com/2014/11/12/world/asia/in-afghanistan-customs-system-corrup tion-is-part-of-the-bargain.html?_r=0

Warner, C. 2007. *The Best System Money Can Buy: Corruption in the European Union.* Ithaca, NY: Cornell University Press.

Watts, M. 2004. Resource curse? Governmentality, oil and power in the Niger Delta, Nigeria. *Geopolitics* **9**(1), 50–80.

Wedeman, A. 2004. The intensification of corruption in China. *China Quarterly* **180**, 895–921.

Wedeman, A. 2012. *Double Paradox: Rapid Growth and Rising Corruption in China.* Ithaca, NY: Cornell University Press.

Wedeman, A. 2015. Anticorruption campaigns and the intensification of corruption in China. *Journal of Contemporary China* **14**(42), 93–116.

Werner, C. 2000. Gifts, bribes, and development in post-Soviet Kazakstan. *Human Organization* **59**(1), 11–22.

Whyte, D. 2007. The crimes of neo-liberal rule in occupied Iraq. *British Journal of Criminology* **47**(2), 177–95.

Williams, P. 2009. Organized crime and corruption in Iraq. *International Peacekeeping* **16**(1), 115–35.

Wilson, A. 2007. Corruption and anti-corruption: the political defeat of "Clean Hands" in Italy. *West European Politics* **30**(4), 830–53.

Wu, Y. and J. Zhu. 2011. Corruption, anti-corruption, and inter-county income disparity in China. *Social Science Journal* **48**(3), 435–48.

Zhu, J. 2012. The shadow of the skyscrapers: real estate corruption in China. *Journal of Contemporary China* **21**(74), 243–60.

6. The consequences of corruption
Dominik H. Enste and Christina Heldman

Over the last decades, there has been a heated discussion on the impact of corruption on a country's economy. Corrupt structures allow for circumventing rules and laws and privilege a small number of people of particular interest groups. Thus, from a moral perspective corruption is unambiguously condemned. When it comes to the economic consequences, however, there are two views: some scholars argue that corruption is "grease in the wheels" of a country by allocating scarce resources toward those who have the highest willingness to pay. Furthermore, corruption helps to avoid inefficient and time-consuming bureaucratic processes. The opposing side views corruption as "sand in the wheels." Corruption might help some business in the bureaucratic process, but at the same time hinders others. In addition, corruption causes an economic loss which excels the few gains by far, as the supporters argue (Enste and Heldman 2017).

The empirical evidence favors the second argument, but the net effect of corruption on a country's economy has still not yet been identified ultimately. This chapter will contribute to the discussion by examining the central consequences of corruption identified by the literature published over the past decades. Although endogeneity and causality cannot always be ruled out, looking at the consequences more closely can help to grasp the overall effect and to design accurate policy measures to deal with corruption.

PRIVATE INVESTMENT

The first variable we will take a closer look at are the total investments made by private entities. The way corruption impacts private investment undertaken in an economy is straightforward in theory: when planning a project, investors have to account for the bribes that need to be paid in the process. The anticipated costs increase and projects, which would have been profitable, become unattractive and are not realized. This is supported by several studies which prove a negative correlation between corruption and investment, for example, Mauro (1995), who shows that the ratio of investment on gross domestic product (GDP) is negatively affected by corruption. Further support is provided by Brunetti et al. (1998).

Several economists advise caution though, claiming that the relationship between investments and corruption is more complex. They argue that it is not the absolute level itself but rather the predictability of corruption and thus the institutionalization that affects investment. If investors know in advance that they have to pay a certain extra and if they can be relatively sure to get a service in return, the negative effect of corruption is reduced (World Bank 1997; Campos et al. 1999). Lambsdorff (2007) adds that the form also influences the decision to invest. There are two types of corruption: petty and grand. Petty corruption happens at the low level and typically includes small payments, while grand corruption describes the abuse of high-level power. Grand corruption can be accompanied by top-down corruption, because the bribed official often has to make payments himself to fulfill the demands on him (Rose-Ackerman 1999). Petty corruption is said to cause less harm to society than grand, because it is small scaled, but in the context of private investments, grand corruption is preferred, as Lambsdorff (2007) argues. The intuition is that investors only have to deal with one powerful official instead of many, which increases the predictability of the payment that has to be made and the service he will get. Even though it is possible that the corrupt official tries to extort more money, his opportunism is restricted by the threat of getting a reputation for being unreliable which can put off future investments and his fear of reciprocal action taken by the bribe-payer, for example, withdrawing from the deal or even uncovering the official's illegitimate activities.

FOREIGN DIRECT INVESTMENT AND CAPITAL INFLOWS

The same mechanism that deters total investment by domestic entities prevails in capital inflows and foreign direct investments (FDI). This is supported by several studies, for example, an analysis of 21 Organisation for Economic Co-operation and Development (OECD) and 59 non-OECD countries by Egger and Winner (2006). They find evidence that corruption deters FDI, with the effect being stronger in non-OECD countries indicating that the growth of FDI in those regions depends on other factors than corruption, such as economic growth and factor endowments. Habib and Zurawicki (2002) argue that corruption raises the costs of doing business, increases uncertainty and leads to distortions. This makes entering a corrupt country less attractive for international firms as they demonstrate empirically. In another analysis using a dataset of 111 countries, the authors furthermore provide support for the theory that corruption has a stronger impact on foreign than on local investors. In addition, their datasets indicate that investors from countries that are corrupt themselves are less deterred by corruption in a host country. This is intuitive, because if a firm is already used to corrupt practices they have better knowledge on how to deal in such a business environment and furthermore the psychological distance is lower (Habib and Zurawicki 2001).

In a study of 20 OECD source and 52 host countries from 1996 to 2003, Barassi and Zhou (2012) also show that the probability of FDI taking place is reduced in corrupt countries. When controlling for the location effects, however, the authors find evidence that corruption increases the level of FDI stock, indicating that once a multinational enterprise (MNE) has set a location, corruption increases the amounts invested. To explain this result, the authors refer to Wei (2000a, 2000b) who argues that MNEs that are willing to invest abroad are less averse to corruption than those who do not.

Not only the amount of investment by foreign firms, but also the form is influenced by corruption: foreign investors prefer joint ventures over owning a whole firm abroad, because they need the expertise and insider information of a local partner who is accustomed to the practices of doing business. This only applies to simple production processes though; in the case of high-tech products, this relationship vanishes as MNEs have to protect their knowledge from intellectual property theft (Smarzynska and Wei 2000; Uhlenbruck et al. 2006).

Another interesting result of the empirical literature concerns the distortions of foreign capital inflows caused by corruption. Wei and Wu (2001) observe a movement away from FDI toward international bank loans in countries with high corruption rates. One possible explanation is that while direct investors have to face a high level of risk of losing their investment in countries with unpredictable ways of doing business, banks are less exposed to that risk, because they are protected by a possible bail-out of the financial system (Wei and Wu 2001). This relationship between corruption and the composition of capital inflows negatively affects the economic performance of a host country, because bank loans are more flexible than direct investment and can be withdrawn more quickly (Lambsdorff 2007). This makes corrupt countries more vulnerable to currency crises and helps explain why the economic growth of the Asian Tiger States did not suffer from high corruption levels at first, but eventually had to face a severe destabilization starting at the end of the 1990s accelerated by the rapid withdrawal of foreign capital.

FOREIGN TRADE

Corruption does not only affect international economic relationships via FDI and capital inflows, but also has an influence on foreign trade. Foreign trade is prone to corruption because the government has a strong interest in controlling it and because it is of high value to international companies to enter a new market, so they are willing to accept extra costs if it gives them a competitive advantage. Regulations on foreign trade are controlled by customs agents, who ensure that importers comply with prohibitions, import restrictions or tariffs for importing goods. The officials have ample scope for collecting bribes since their actions are hard to monitor: once a product has entered the host country, tracing it can be difficult and furthermore, depending on the nature of the deal, there is no paperwork.

Corrupt practices at the border can cause competitive disadvantages for international firms, because either their costs of doing business increase due to the necessary bribe-payments or, if they refuse to pay because it is too expensive or because they condemn bribery, they are kept out of the market. The latter is indicated by a regression of 18 exporting countries and 87 importing countries performed by Lambsdorff (1998) and reproduced by the same author in 2000. He shows that companies from Australia, Malaysia and Sweden only have little involvement in corrupt countries, while the exporters Belgium, France, Italy, the Netherlands and South Korea have higher market shares there. These differences can be explained by cultural and moral reasons that lead corruption to be more condemned in the societies where the first group comes from. There might also be legal reasons insofar as punishment for corrupt behavior is lower in the second group. If a company is more inclined to pay bribes in order to obtain a contract, high corruption levels give them a competitive advantage compared to those companies that shy away from such behavior. This also sheds some light on the responsibilities regarding corruption. Since countries have different ways to deal with corrupt business partners, the notion that exporting countries just have to accept the code of conduct in the host countries and adjust to it does not hold. The decision on whether to accept corrupt practices and pay bribes in order to obtain contracts is rather an individual choice made by the exporting firm. Blaming the corrupt host countries for their practices is therefore shortsighted, as the exporters also carry some responsibility (Lambsdorff 1998).

GOVERNMENT EXPENDITURES AND SERVICES

While the theoretical relationship between corruption and private investment is relatively unambiguously supported by the empirical literature, the results on size, direction and quality of government spending are contradictory and difficult to interpret.

The intuition behind the link between corruption and government spending is the following: corrupt officials want to maximize their total income, which consists of their regular income and the bribes collected. This gives them incentive to direct government expenditure toward areas where the expected income is highest, that is, where there is much room to extort money. These areas are characterized by high levels of discretion in allocating public resources and by strong particular interest which are associated with a high willingness to pay for favors. Opportunities for officials are also higher in capital-intense public projects, as Rose-Ackerman (1999) concludes. In line with this research is Hessami (2013), who hypothesizes that corruption distorts public spending in the direction of areas that involve public commissions and away from those that do not, such as social protection and culture. In sectors involving procurement, businesses compete for these rents by paying bribes to politicians in order to influence this process. This theory is supported by the evaluation of 29 OECD countries and their health sector and the extent of environmental protection such as waste management. These areas are characterized by public procurement, oligopolistic structures and high-tech equipment and thus give room to extract bribes. The share of expenditure on recreation, culture and religion on the other hand decreases with corruption (Hessami 2013).

Another attractive area for corrupt officials are projects that are hard to monitor by the public (Mauro 1998). One of these is military equipment, which is very heterogeneous since it differs depending on its operational area. The decision on products and prices is thus complex and offers a lot of leeway for a corrupt official to choose a project or product where his rent is highest, even though it might not be the most lucrative one from an objective standpoint. While older studies cannot support this theory (e.g., Mauro 1998), more recent research proves a significant and robust correlation between corruption and military spending showing that in corrupt countries military spending is high either in total amounts or as a share of GDP (Gupta et al. 2001a). Areas that are more visible in the public eye and where decisions are more comprehensible, for example, because the technology is less complex, give less leeway for officials which makes them unattractive. Exemplary in this context is education. Empirical research shows that in countries with high levels of corruption, the government directs less resources toward it. This results in significantly higher dropout rates (Mauro 1998; Gupta et al. 2001b, 2002). Health care is also notable in this context: while high class medical equipment profits from corruption because the room to collect bribes is large, it is less so in the quality of health care which is strongly regulated. The empirical results are not strong though, but they do hint at expenditures being lower in corrupt countries (Mauro 1998). Gupta et al. (2001b) find that child mortality rates are significantly higher when corruption levels are high, which supports this notion.

Corruption not only affects the composition of government expenditure, but also the quality of the service. Kaufman and Wei (1999), for example, find that the time managers deal with bureaucracy is significantly higher in corrupt countries. Another serious consequence of corruption is that it reduces trust in the civil service and the acceptance of the political system as documented by Anderson and Tverdova (2003) based on survey data. The lack of trust in the civil service can lead to a destabilization of the country, because the legitimacy of the political authority is challenged.

GROSS DOMESTIC PRODUCT

The comparison of corruption levels and the size of GDP provides a clear image: in countries with high levels of corruption the GDP per head is lower (Figure 6.1). Even though this seems intuitive considering that corruption lowers investments as shown before, the relationship is not unambiguously causal. While corruption lowers GDP to some extent, a low GDP also fosters corruption because it reduces the resources to fight it. Therefore, we are most likely dealing with simultaneity (Paldam 2002).

In order to isolate the effect of corruption on GDP, several solutions have been suggested. One of them is an Instrumental Variables (IV) approach, but since adequate instruments are hard to find, especially since corruption and GDP are so closely connected, the majority of the research has focused on other variables altogether (Lambsdorff 2007).

Many scholars have focused on the effect of corruption on growth of GDP, but the results are very ambivalent, depending on the dataset used for the corruption level and the explanatory variables included in the model (see Campos et al. 2010 for a meta-analysis of 41 empirical studies). These studies show that the relationship between corruption and growth is highly complex and difficult to interpret.

Some studies provide evidence for a negative impact (Tanzi and Davoodi 2000). Other studies cannot support the relationship (Abed and Davoodi 2002). Mo (2001) produces significant results showing at

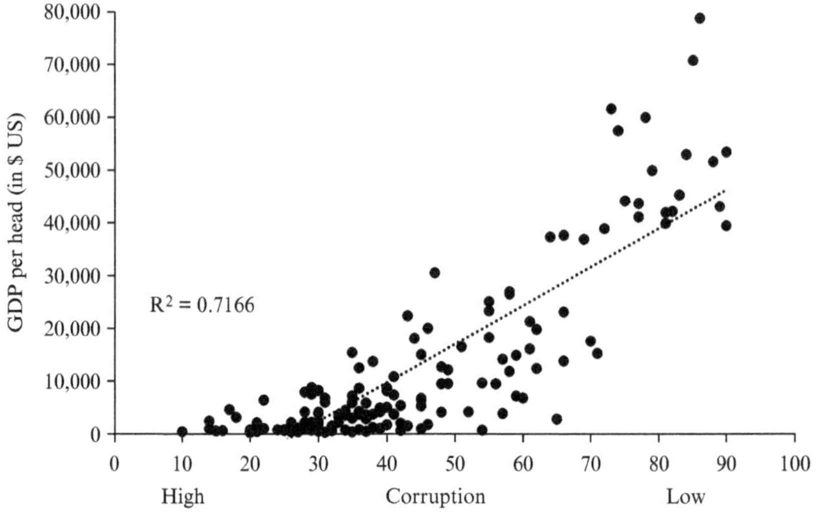

Source: Authors' calculations using the World Bank (2016) and Transparency International (2014).

Figure 6.1   Correlation between corruption and GDP per capita

first that a 1 percent increase in corruption reduces growth by about 0.72 percent. After adding several control variables, the significance cannot be sustained though. The most important channel through which corruption affects growth is political stability, making up for 53 percent of the effect. Other channels are the ratio of private investment on GDP and human capital formation (Mo 2001). Méon and Sekkat (2005) on the other hand produce significant results on a negative effect of corruption and growth even after controlling for the ratio of investment. The effect depends on the quality of governance however.

The role of institutional quality has also been addressed by Aidt et al. (2008) who take account of the fact that corruption and growth are determined jointly within a particular regime which makes the relationship dependent on that regime. In a dataset of 67 to 71 countries, the authors find that corruption indeed lowers growth of GDP. The effect is only observed in countries with good political institutions though, where a one-point decrease in corruption increases growth rates by 0.5–0.6 percentage points in the short run and about 0.4 percent in the long run. Countries that suffer from low institutional quality are not affected by the negative relationship, meaning that even if countries improve their corruption levels, growth would not increase accordingly unless institutions are improved. This result can be interpreted as support for the "grease in the wheels" hypothesis, which states that corruption helps individuals to circumvent bad institutions and thus increases efficiency. The authors caution against that inference though because it is unclear what causes weak institutions and how corruption plays into it.

The use of GDP as an indicator for a country's prosperity has been a topic of public debate for years. Consequently, research on corruption and its effects on the wealth of countries has questioned if GDP is the best indicator in this context. Aidt (2011), for example, claims that GDP does not capture important determinants of well-being and using it as an indicator for prosperity might underestimate the harmful effects of corruption. In addition, as shown before, previous research has not produced unambiguous results, which is why Aidt (2011) proposes another measure, sustainable development, measured by growth of genuine wealth per capita. While growth of GDP does not imply long-run stability, sustainable development describes a country's capability to keep up the living standards of its citizens through time and is therefore a better concept for defining policy measures, as the author claims. In a study of 110 countries from 1996 to 2007 he finds robust proof of a negative relationship between corruption and sustainable development. Aidt connects his findings to the resource curse which states that resource-rich countries grow slower than resource-scarce countries. The reasons are not fully explored yet, but it is widely assumed that a high endowment of resources leads to overinvestment and rent-seeking in particular industries while others, such as human and manufactured capital, are neglected. The author concludes that policy measures regarding corruption but also other political fields should focus on reducing rent-seeking and corruption in areas like natural assets in order to prevent the underinvestment in the capital base crucial for sustainable development.

INEQUALITY

Corruption is a phenomenon that does not occur equally in all social classes, but rather benefits the well endowed. Bribes are usually directed at the privileged members of society because they have the power to give something in return. If legal fees are substituted by bribes, the government treasury and thus the wealth of the community do not grow whereas the incomes of the civil servants who control and distribute the public good are enriched (Tanzi 1998). This relationship has led to the hypothesis that corruption increases inequality in a country. The empirical proof has been provided by Gupta et al. (2002), who find that corrupt countries show higher income inequality and poverty measured by the Gini coefficient and higher inequality in land distribution and education. The authors explain the results by arguing that corruption reduces economic growth, the effectiveness of social spending and human capital formation, biases the tax system, and furthermore corroborates inequality in asset ownership and access to education. This negatively affects the core functions of a government, namely, the allocation of resources, the stabilization of the economy and the redistribution of income, all of which decide income distribution and poverty. Thus, corruption increases income inequality and poverty (Gupta et al. 2002).

Other studies provide evidence on the reverse relationship, claiming that inequality also increases corruption, such as You and Khagram (2005). According to the authors there are a few well-endowed people in unequal societies, who use corruption to obtain their power and position. The poor on the other hand are badly organized and have neither the means nor the political power to monitor the upper class. Furthermore, the poor are more likely to be extorted when trying to access public services which are usually scarce in unequal societies. You and Khagram note though that the relationship between inequality and corruption works both ways.

The empirical and theoretical research does not provide a clear answer on the relationship between inequality and corruption and it is most likely that they affect each other and stem from the same cultural determinants. Husted (1999), for example, finds that corrupt countries are characterized by high uncertainty avoidance, high masculinity and high power distance, which have also been associated with inequality acceptance in a society (Karstedt 2002; Malinoski 2012).

SHADOW ECONOMY

The last consequence of corruption analysed in this chapter is the shadow economy. Plotting the level of corruption as measured by the Corruption Perception Index by Transparency International against the size of the shadow economy in percent of GDP hints at a relationship between the two indices: high corruption levels correspond to a larger shadow economy and vice versa (Figure 6.2). This correlation is also found in a regression of Johnson et al. (1998) who analyse the informal economy of 43 countries and the effect of corruption using three indices for corruption.

It is not clear, however, in what direction the causality runs. Corruption and the shadow economy might be substitutes, since a large unofficial sector negatively impacts the power of public officials and therefore their opportunities to ask for bribes (e.g., Choi and Thum 2005). It is also possible that the two variables are complements: high levels of corruption

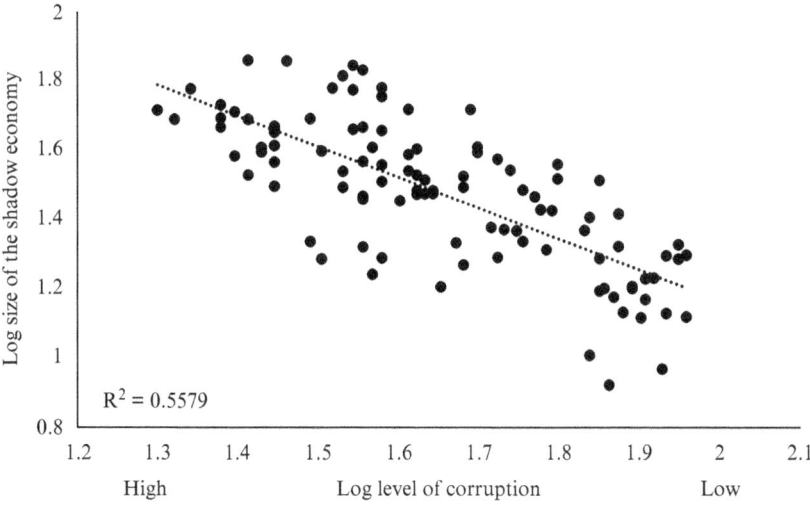

Note: Data from 2013.

Source: Authors' calculations using Schneider and Medina (2017) and Transparency International (2013).

Figure 6.2 Correlation between corruption and the shadow economy

can cause firms to shift their business underground. One reason for this effect is that corruption decreases the efficiency of the government services as explained before. If the services are insufficient, the unofficial sector becomes more attractive. Corruption also increases the costs of doing business and thus works like an extra tax. This again encourages firms to go underground, as Friedman et al. (2000) show.

The relationship of corruption and the shadow economy is examined on the basis of empirical data from 98 countries collected by Dreher and Schneider (2010). The authors make a distinction between high and low income countries, arguing that the relationship between corruption and the unofficial sector differs. In high income countries, firms usually do not fully operate underground since the official sector functions and provides them with contract enforcement or rule of law. There, bribes are not paid to escape the public sector, but rather to achieve benefits or speed up a process within. A large proportion of the revenues gained in the shadow economy are in turn spent in the official sector, which increases the government revenue again. An increase in government revenue leads to a higher institutional quality and lower corruption (Schneider and Enste

2000; Dreher and Schneider 2010). In low income countries, institutions are often insufficient in protecting and regulating business, and companies are expected to shift underground completely. Bribes are then used in order to hide the activities in the shadow economy from public officials and avoid punishment. Corruption and the size of the shadow economy are therefore expected to be complements here, since more illegal activities require corruption to stay concealed and also because bribes are common in the shadow economy. The data shows that corruption and the shadow economy are indeed complements in low income countries, while in high income countries no significant effect can be found (Dreher and Schneider 2010). Interestingly, when using a perceptions-based index to measure corruption, no correlation can be found at all. Dreher and Schneider refer to recent critique claiming that perceived corruption cannot adequately measure corruption because respondents are unable to separate the variable from the perception of the overall institutional environment. Thus, perceptions of corruption might in fact be perceptions of the intuitional quality. Using an alternative index based on a measurement of the most common consequences of corruption (developed by Dreher et al. 2007) yields strongly significant results. Obtaining clear results in this field is hard to impossible though, as the authors conclude, because both the shadow economy and corruption are concealed and thus there is an obvious lack of data (Dreher and Schneider 2010).

In summary, there is empirical support for a connection between corruption and the shadow economy, but the direction of the causality is not clear. Policy measures against corruption can still have a positive effect on the shadow economy because they stabilize institutions and increase economic and sustainable growth which lowers the attractiveness of illegal activities (Schneider and Enste 2015).

CONCLUSION

Identifying the impact corruption has on a country's economy is a difficult task. Corruption cannot be measured precisely since it takes place out of the public eye. Scholars therefore have to rely mostly on the perception a society has of corruption, which has been criticized as biased by other factors such as institutional quality. Indices that are based on the most likely consequences of corruption face a similar problem and might be biased by simultaneity. When talking about the consequences of corruption, it is therefore of value to examine various studies of different approaches. Even though there are some contradictions depending on the datasets and variables, they mostly have a similar tenor: while corruption might help to avoid inefficient regulations in some cases, the negative consequences prevail. Corruption increases the costs of doing business which makes a country less attractive for both local and foreign investors, it reduces the GDP and distorts government spending, increases inequality and is associated with a larger unofficial sector. In total, corruption reduces the competitiveness of a country, its wealth and the quality of institutions with tensions and instability as possible results. This environment is in turn a breeding ground for corruption and so a vicious circle develops. This cycle needs to be broken in order to stabilize affected countries and ensure the well-being of its citizens. Policy measures should be directed at corruption itself and at the consequences at the same time. They furthermore need the cooperation of the international community, since country leaders often do not have the necessary means and knowledge and are in some cases not willing to change a corrupt system.

REFERENCES

Abed, G. and Davoodi, H. 2002. Corruption, structural reforms, and economic performance. IMF Working Paper No. 00/132, Washington, DC.

Aidt, T. 2011. Corruption and sustainable development. In S. Rose-Ackerman and T. Soreide (eds), *International Handbook on the Economics of Corruption*, Vol. 2, pp. 3–51 Cheltenham, UK and Northampton, MA, USA: Edward Elgar.

Aidt, T., Dutta, J., and Sena, V. 2008. Governance regimes, corruption and growth: theory and evidence. *Journal of Comparative Economics* **36**(2), 195–220.

Anderson, C. and Tverdova, Y.V. 2003. Corruption, political allegiances, and attitudes toward government in contemporary democracies. *American Journal of Political Science* **47**(1), 91–109.

Barassi, M. and Zhou, Y. 2012. The effect of corruption on FDI: a parametric and non-parametric analysis. *European Journal of Political Economy* **28**(3), 302–12.

Brunetti, A., Kisunko, G., and Weder, B. 1998. Credibility of rules and economic growth: evidence from a worldwide survey of the private sector. *World Bank Economic Review* **12**(3), 353–84.

Campos, E., Lien, D., and Pradhan, S. 1999. The impact of corruption on investment: predictability matters. *World Development* **27**(6), 1059–67.

Campos, N., Dimova, R., and Saleh, A. 2010. Whither corruption? A quantitative survey of the literature on corruption and growth. CEPR Discussion Paper No. DP8140.

Choi, J.-P. and Thum, M. 2005. Corruption and the shadow economy. *International Economic Review* **46**(3), 817–36.

Dreher, A. and Schneider, F. 2010. Corruption and the shadow economy: an empirical analysis. *Public Choice* **144**(1), 215–38.

Dreher, A., Kotsogiannis, C., and McCorriston, S. 2007. Corruption around the world: evidence from a structural model. *Journal of Comparative Economics* **35**(3), 443–66.

Egger, P. and Winner, H. 2006. How corruption influences foreign direct investment: a panel data study. *Economic Development and Cultural Change* **54**(2), 459–86.

Enste, D. and Heldman, C. 2017. *Causes and Consequences of Corruption – an Overview of Empirical Results*. IW Report 02/17. Cologne: Cologne Institute for Economic Research. https://www.iwkoeln.de/fileadmin/publikationen/2017/323508/IW-Report_2_2017_Corruption.pdf.

Friedman, E., Johnson, S., Kaufmann, D., and Zoido-Lobaton, P. 2000. Dodging the grabbing hand: the determinants of unofficial activity in 69 countries. *Journal of Public Economics* **76**(3), 459–93.

Gupta, S., de Mello, L., and Sharan, R. 2001a. Corruption and military spending. *European Journal of Political Economy* **17**(4), 749–77.

Gupta, S., Davoodi, H., and Tiongson, E. 2001b. Corruption and the provision of health care and education services. IMF Working Paper 00/116, Washington, DC.

Gupta, S., Davoodi, H., and Alonso-Terme, R. 2002. Does corruption affect income inequality and poverty? *Economics of Governance* **3**(1), 23–45.

Habib, M. and Zurawicki, L. 2001. Country-level investments and the effect of corruption: some empirical evidence. *International Business Review* **10**(6), 687–700.

Habib, M. and Zurawicki, L. 2002. Corruption and foreign direct investment. *Journal of International Business Studies* **33**(2), 291–307.

Hessami, Z. 2013. Corruption, public procurement, and the budget composition: theory and evidence from OECD countries. Working Paper Series 2013-27, Universität Konstanz.

Husted, B. 1999. Wealth, culture, and corruption. *Journal of International Business Studies* **30**(2), 339–60.

Johnson, S., Kaufmann, D., and Zoido-Lobaton, P. 1998. Regulatory discretion and the unofficial economy. *American Economic Review* **88**(2), 387–92.

Karstedt, S. 2002. The culture of inequality and corruption: a cross-national study of corruption. Working Paper, Department of Criminology, Keele University.

Kaufmann, D. and Wei, S.-J. 1999. Does "grease money" speed up the wheels of commerce? National Bureau of Economic Research Working Paper 7093, Cambridge, MA.

Lambsdorff, J. 1998. An empirical investigation of bribery in international trade. *European Journal for Development Research* **10**(1), 40–59.

Lambsdorff, J. 2007. Consequences and causes of corruption: what do we know from a cross-section of countries? In S. Rose-Ackerman (ed.), *International Handbook on the Economics of Corruption*, pp. 3–51. Cheltenham, UK and Northampton, MA, USA: Edward Elgar.

Malinoski, M. 2012. On culture and income inequality: regression analysis of Hofstede's international cultural dimensions and the Gini coefficient. *Xavier Journal of Politics* **3**(1), 32–48.

Mauro, P. 1995. Corruption and growth. *Quarterly Journal of Economics* **110**(3), 681–712.

Mauro, P. 1998. Corruption and the composition of government expenditure. *Journal of Public Economics* **69**(2), 263–79.

Méon, P.-G. and Sekkat, K. 2005. Does corruption grease or sand the wheels of growth? *Public Choice* **122**(1), 69–97.

Mo, P.-H. 2001. Corruption and economic growth. *Journal of Comparative Economics* **29**(1), 66–79.

Paldam, M. 2002. The big pattern of corruption, economics, culture and the seesaw dynamics. *European Journal of Political Economy* **18**(2), 215–40.

Rose-Ackerman, S. 1999. *Corruption and Government: Causes, Consequences and Reform.* Cambridge: Cambridge University Press.

Schneider, F. and Enste D. 2000. Shadow economies: size, causes, and consequences. *Journal of Economic Literature*, **38**(1), 77–114.

Schneider, F. and Enste, D. 2015. *The Shadow Economy: An International Survey.* Cambridge: Cambridge University Press.

Schneider, F. and Medina, L. 2017. Shadow economies around the world: new results for 158 countries over 1991–2015 Working Paper. http://www.economics.jku.at/mem bers/Schneider/files/publications/2017/JointPaper_LeandroMedina_158countries.pdf (accessed May 16, 2018).

Smarzynska, B. and Wei, S.-J. 2000. Corruption and the composition of foreign direct investment: firm-level evidence. World Bank Discussion Paper Series No. 2360, Washington, DC.

Tanzi, V. 1998. Corruption around the world: causes, consequences, scope and cures. *IMF Staff Papers* **45**(4), 559–94.

Tanzi, V. and Davoodi, H. 2000. Corruption, growth, and public finances. IMF Working Paper 00/182, Washington, DC.

Transparency International. 2013. Corruption Perceptions Index 2013, Berlin.

Transparency International. 2014. Corruption Perceptions Index, 2014, Berlin.

Uhlenbruck, K., Rodriguez, P., Doh, J., and Eden, L. 2006. The impact of corruption on entry strategy: evidence from telecommunication projects in emerging economies. *Organizational Science* **17**(3), 402–14.

Wei, S.-J. 2000a. How taxing is corruption on international investors. *Review of Economics and Statistics* **82**(1), 1–11.

Wei, S.-J. 2000b. Local corruption and global capital flows. *Brooking Papers on Economic Activity* **2**, 303–54.

Wei, S.-J. and Wu, Y. 2001. Negative alchemy? Corruption, composition of capital flows, and currency crises. National Bureau of Economic Research Working Paper 8187, Cambridge, MA.

World Bank. 1997. *World Development Report 1997: The State in a Changing World.* Washington, DC: World Bank.

World Bank. 2016. GDP per capita (current US$). http://data.worldbank.org/indicator/NY.GDP.PCAP.CD (accessed June 30, 2016).

You, J.-S. and Khagram, S. 2005. A comparative study of inequality and corruption. *American Sociological Review* **70**(1), 136–57.

7. E-government and corruption: a review
Nasr G. Elbahnasawy

A growing body of evidence considers corruption as one of the greatest obstacles to economic and social development.[1] It distorts the allocation of resources, reduces the efficiency of public expenditures, and intensifies economic inefficiency in general. It weakens formal institutions and the protection of property rights, and erodes the incentives system that boosts economic growth. It also lowers economic growth via its negative impact on investment. In addition, it hinders the success of the United Nations sustainable development goals.[2] It also fuels social exclusion and inequalities, and violates human rights, especially in cases of systemic grand corruption. In their recent joint report, the United Nations Department of Economic and Social Affairs and the International Organization of Supreme Audit Institutions (2013) emphasized the ability of corruption to corrode states from within and bring about a collapse of institutions, as has occurred in some cases.

Because corruption undermines economic development efforts in developing countries, it is not surprising it has received considerable attention from policymakers, researchers, and numerous international organizations. Several developing countries have recently drafted various anti-corruption strategies to coordinate national anti-corruption efforts, but many of them were less effective (United Nations Development Programme 2014). Although the fight against corruption remains difficult and a major challenge to various nations,[3] e-government has been recently introduced as a promising tool in the battle against corruption, due to the disincentives it generates to engage in corrupt behavior.[4] It reduces

---

[1]   World Bank (2017), Transparency International (2016), OECD (2013), Gupta et al. (2002), Mauro (1995), and Murphy et al. (1991). It is estimated that businesses and individuals pay about $1.5 trillion in bribes each year, which represents 2 percent of global gross domestic product and about ten times the value of overseas development assistance (World Bank 2017).

[2]   Target 16.5 of the sustainable development goals calls for nations to "substantially reduce corruption and bribery in all their forms" (United Nations 2015).

[3]   For the various causes of corruption, see Dimant and Tosato (2018), Elbahnasawy and Revier (2012), Jain (2001), and Treisman (2000, 2007).

[4]   E-government can also promote Target 16.6 of the sustainable development goals that calls for countries to "develop effective, accountable, and transparent institutions at all levels" (United Nations 2015).

the discretionary power of public officers and induces greater levels of accountability and transparency in the economy, eliminating numerous opportunities for corrupt rent-seeking activities. Yet, it is important to apply e-government systems properly in order to generate the positive impact on corruption reduction, as e-government implementation may have weaknesses that allow corruption to persist.

The main purpose of this chapter is to review the current literature on the linkages between e-government and corruption. The chapter proceeds as follows. The first section presents the definitions and measures of both e-government and corruption. The second section explores the potential connection between e-government and corruption. The third section surveys the current empirical literature. The last section concludes.

DEFINITIONS AND MEASURES

E-government refers to the government use of information and com-munications technology (ICT), such as the internet and other digital media, to enhance the quantity and quality of services delivered to the public, share more information, promote equality and social inclusion, and to improve government efficiency in general (United Nations 2016; United Nations Development Programme 2006; World Bank 2015). A widely used measure of e-government is the United Nations e-government development index (EGDI), which measures the willingness and capacity of governments to utilize online and mobile technology while performing their functions (United Nations 2010, 2016). The EGDI is a weighted average of three various dimensions of e-government, including the scope and quality of online services, the development of telecommunications infrastructure, and the extent of human capital.[5] It ranks countries on a scale from zero to one, with higher values denoting better e-government. Based on the EGDI levels, countries can be grouped into four main categories: very high EDGI (more than 0.75 EGDI value), high EDGI (between 0.50 and 0.75), middle EDGI (between 0.25 and 0.50), and low EDGI (less than 0.25 EGDI value).

The Department of Economic and Social Affairs of the United Nations publishes the results of the e-government development survey every two years. In the 2016 e-government development survey, 29 countries were in the very high EDGI group, compared to only ten countries in 2003, while 51 percent of countries in 2016 had a low to medium score, compared to

---

[5]　For the definition and composition of each index, see United Nations (2016).

73 percent in 2003 (United Nations 2016). Despite such improvement, African countries on average continue to lag globally, with a significantly low average of 0.2882, which is below the global average of 0.4922 and far below the average score for European countries, 0.7241.[6] Moreover, e-government remains a major challenge to least developed countries in 2016, with an average EGDI of 0.2350, about half of the global average.[7] The least developed countries, mainly from Africa, remain at the bottom of the index, where 29 out of 32 total countries in the low EGDI group are least developed countries.[8]

Furthermore, the gaps in e-government development among various regions of the world have persisted in the 2016 survey. In the very high EDGI group, 19 out of 29 countries in this group (representing 66 percent) are from Europe. In the low EDGI group, 26 countries are from Africa (constituting 81.2 percent of this group), three from Asia (9.4 percent), two from Oceania (6.3 percent), and one from the Americas (3.1 percent).[9] The average EGDI varies from one region to another, with 0.2882 in Africa, 0.4154 in Oceania, 0.5132 in Asia, 0.5245 in the Americas, and 0.7241 in Europe.

In the 2016 survey, the United Kingdom comes in the top rank with a 0.9193 EDGI value, followed by Australia and South Korea.[10] At the bottom of the EDGI are Eritrea, the Central African Republic, Niger, and Somalia, with EDGI values of 0.0902, 0.0789, 0.0593, and 0.0270, respectively.

Greater access to the internet improves the efficacy of e-government. Yet, globally, only about 43 percent of people had access to the internet,

---

[6]   It is worth noting that five African countries were in the high EDGI group in the 2016 survey: Mauritius (the top rank in Africa), Tunisia, South Africa, Morocco, and Seychelles. The remaining African countries fall in the lower two tiers of e-government development; the low EDGI and the middle EDGI (United Nations 2016).

[7]   The term "least developed countries" (LDCs) is a United Nations classification for countries that exhibit the lowest indicators of socioeconomic development. Countries included in this group meet three main criteria: low gross national income per capita, low human assets, and high economic vulnerability. For more details and a complete list of those countries, see United Nations (2017).

[8]   On the other hand, in the 2016 survey, three countries graduated from low to middle EGDI levels: Nepal, Togo, and Zambia, while three other countries dropped from middle to low EGDI: Congo, Madagascar, and Yemen. In addition, four countries fell from high EDGI to medium EDGI levels: Antigua and Barbuda, Egypt, and Fiji, while ten countries graduated from middle to high EGDI, such as the Philippines, South Africa, and Vietnam. See United Nations (2016) for more details.

[9]   See figure 5.2 in United Nations (2016). E-government gaps within regions also exist. See United Nations (2016) for more details.

[10]   It is worth mentioning that South Korea was ranked first in the 2014 survey. The United States ranks 12th in the 2016 survey and Canada ranks 14th, where both are the only two countries in the Americas in the very high EDGI group.

as of 2015 (United Nations 2016). The gap between developing and developed countries in internet usage remains, with only 35 percent of people in developing countries that use the internet, compared with 82 percent of people in developed countries (United Nations Department of Economic and Social Affairs 2015).

Corruption, on the other hand, denotes the misuse of public authority or office for private gains (Transparency International 2016; United Nations Development Programme 2006, 2008).[11] Various measures of perceived corruption exist. The corruption perception index (CPI) by Transparency International measures perceived corruption in the public sector worldwide, based on expert and business surveys from around the world (Transparency International 2016). It ranks countries on a scale from zero to 100, where zero denotes the highest level of perceived corruption and 100 indicates the lowest level. In the 2016 CPI, the lowest-ranked countries with the highest-perceived corruption are Iraq, Guinea-Bissau, Afghanistan, Libya, Yemen, Sudan, Syria, North Korea, South Sudan, and Somalia.[12] Conversely, the highest-ranked countries with the lowest-perceived corruption are New Zealand, Denmark, Finland, Sweden, Switzerland, Norway, Singapore, Netherlands, Canada, and Germany.[13]

The control of corruption (CC) index, from the World Bank's Worldwide Governance Indicators (WGI),[14] measures "perceptions of the extent to which public power is exercised for private gain, including both petty and grand forms of corruption, as well as capture of the state by elites and private interests" (Kaufmann et al. 2010, p. 4). The CC measure is a weighted average of a large number of individual data sources, incorporated into a single aggregate measure by using the unobserved components model, with greater weights given to sources with higher correlation with each other.[15] It ranges from −2.5 to 2.5, with higher values indicating lower perceived corruption.[16] The correlation between the CC measure and the CPI is typically very high, above 0.9.

---

[11]   The focus in this chapter is on public corruption.
[12]   The CPI values for these countries are 17, 16, 15, 14, 14, 14, 13, 12, 11, and 10, respectively. See Transparency International (2016) for more details.
[13]   The CPI values for these countries are 90, 90, 89, 88, 86, 85, 84, 83, 82, and 81, respectively (Transparency International 2016). The Unites States was ranked 18th in this index.
[14]   Available online at: http://info.worldbank.org/governance/wgi/#home. Accessed November 1, 2017.
[15]   For more details on the CC methodology, see Kaufmann et al. (2010).
[16]   In the 2016 CC measure, the most corrupt countries are Equatorial Guinea, Somalia, Yemen, Sudan, South Sudan, Syria, Libya, Afghanistan, Guinea-Bissau, and Turkmenistan, respectively. The least corrupt countries are New Zealand, Finland, Denmark, Sweden, Norway, Luxembourg, Singapore, Liechtenstein, Switzerland, Iceland, and Canada, respectively.

The International Country Risk Guide (ICRG) also constructs an index of perceived corruption within the political system. This index ranges between zero and six risk points, with higher values indicating a likely lower risk of corruption.[17]

THE E-GOVERNMENT AND CORRUPTION NEXUS

Klitgaard (1988) used a principal-agent-client framework to describe corruption, where elected government officials that represent the state and its citizens (the principals) have to hire public servants (the agents) to deliver public services to citizens (the clients) on their behalf, since they are unable to supply those services on their own. In this scenario, the agents know more about the administration than the principals and the clients, and that asymmetric information is present. This dilemma may induce agents to act more opportunistically in their own interest, taking advantage of the entrusted power and engaging in corrupt acts. The probability of illicit rent-seeking behavior is amplified by weak accountability of agents to principals, greater discretionary power delegated to agents, and larger monopoly power conceded to agents over clients.

In this context, e-government can be regarded as an effective tool in restructuring the principal-agent-client association by diminishing the extent of discretionary power entrusted to agents and improving their accountability to principals, in addition to enhancing transparency in general (Elbahnasawy 2014). E-government lessens the direct interaction between government agents and the public, and standardizes the delivery of services to the public, which reduces the discretionary power of government agents. It publicizes greater quantity and quality of information in the economy, which induces the public to question unreasonable procedures and decisions of government agents, leading to greater accountability and transparency. It also allows maintaining extensive data on various transactions, easing the process of tracking actions and decisions made by agents, and hence augmenting the rate of detection of corrupt acts. Additionally, it expands access to information and simplifies rules and procedures, which also promotes transparency and accountability (Bhatnagar 2003; Elbahnasawy 2014). In general, e-government is likely to eliminate various opportunities to supply or demand bribes and create disincentives for government agents to engage in corrupt acts. Therefore,

---

[17]   For the ICRG methodology, see The PRS Group (2017).

it can significantly curb corruption levels and thus be seen as an effective instrument in anti-corruption efforts.

On the other hand, e-government may fail to curb corruption and can even be counterproductive. Corrupt government agents may learn ways to beat the new e-government systems and take advantage of any potential weaknesses of these systems, which enables corrupt behavior to persist and even to grow faster (Bhatnagar 2003). In addition, e-government may cause corruption to migrate somewhere else in the economy and may create new rent-seeking opportunities (Andersen 2009; Pacific Council on International Policy 2002).[18] Therefore, it is important to apply e-government systems appropriately to produce favorable results on corruption reduction.

E-government project failure is quite common in many developing countries (Aladwani 2016; United Nations Development Programme 2014). High corruption levels may well induce this failure, as unqualified technical employees may lead e-government projects due to various forms of corruption, such as nepotism (Aladwani 2016). Accordingly, creating a clear vision that targets lower corruption levels is necessary when implementing e-government projects, along with a proper legislative and institutional framework and strong political commitment. Furthermore, constant monitoring and revision of e-government project implementations is useful. Each country may find solutions to e-government implementation problems that fit its own national context, while the private sector can provide considerable assistance in this respect.

EMPIRICAL EVIDENCE CONCERNING E-GOVERNMENT AND CORRUPTION

A number of case studies using micro-level data, such as Kim et al. (2009), Chawla and Bhatnagar (2004), and Bhuiyan (2011), report a favorable impact of e-government implementation on curbing corruption. Nevertheless, a few other studies found that e-government, and information technology in general, may create new opportunities for corruption (Heeks 1998).[19]

---

[18]   It may generate greater rent-seeking opportunities for technology leaders, initiating intergenerational shift in corruption towards younger government officials who are more tech-literate (Pacific Council on International Policy 2002).

[19]   For a list of case studies and their results, see Weerakkody et al. (2015) and Neupane et al. (2012). Also, for further discussion on the application of e-government principles to lower corruption, see Neupane et al. (2017).

Even though some anecdotal evidence finds that e-government is a successful tool in anti-corruption efforts, it is not necessarily the case at the macro level, since corruption may shift elsewhere in the economy with e-government implementations. Using a large panel dataset, Elbahnasawy (2014) provides macro-level evidence that e-government is a useful tool in anti-corruption programs, which is reinforced by greater internet adoption. Likewise, Andersen (2009) finds support to the claim that e-government is a valuable tool in the global efforts to curb corruption.

In a similar vein, some studies have examined the influence of internet adoption, measured by the number of people with access to the world-wide network, on corruption. Lio et al. (2011) find that internet usage lowers corruption, but with a small magnitude. Andersen (2009) argues that while e-government is a useful instrument to reduce corruption only in non-OECD[20] countries, the internet use is more important than e-government in this regard. Using data on the number of internet hits on corruption and bribery, generated by Google and Yahoo search engines in a given country at a certain time, Goel et al. (2012) argue that internet usage augments access to information and the speed of information dissemination, which escalates the level of corruption awareness and raises the risk of corruption detection. Relatedly, Jha and Sarangi (2017) found a significant negative relationship between Facebook penetration and corruption.[21] Yet, Elbahnsawy (2014) finds that the influence of internet adoption on depressing corruption is ambiguous and sensitive to model specifications. Because corruption is contagious, especially when corrupt acts go unpunished or the punishment is not a deterrent, social media and internet usage that intensively disseminate information about corrupt behavior from individuals' perspectives may give rise to the feeling that "everyone is corrupt" and hence may stimulate further corruption. Nevertheless, the extent of internet adoption complements the favorable impact of e-government on corruption reduction, and that a larger share of the population with access to the internet strengthens the influence of e-government on reducing corruption (Elbahnasawy 2014).

Some recent research investigates the influence of culture on the diffusion of e-government and on the effectiveness of e-government on corruption reduction.[22] Zhao et al. (2014) and Khalil (2011) find that greater uncertainty avoidance and higher power distance are negatively associated

---

[20]　Organisation for Economic Co-operation and Development.

[21]　They indicate that this relationship is stronger in countries where press freedom is limited. See Jha and Sarangi (2017) for more details.

[22]　Culture denotes the set of beliefs and values that distinguishes people of a specific society from others (Hofstede 1984; Zhao et al. 2014).

with e-government diffusion. Uncertainty avoidance refers to the scope to which societies rely on social norms and bureaucratic measures to ease the unpredictability of future events, while power distance denotes the extent to which society members agree and expect power to be shared unequally (Hofsted 1984; House et al. 2004).[23] Zhao et al. (2014) and Khalil (2011) also argue that greater in-group collectivism lowers e-government diffusion, whereas higher future orientation increases it. In-group collectivism refers to the extent to which individuals express their pride, loyalty, and cohesiveness in their families,[24] while future orientation indicates the extent to which societies engage in future-oriented behavior, such as planning (Hofstede 1984; Zhao et al. 2014). In a related vein, Zhao et al. (2017) argue that e-government has greater influence on corruption reduction in cultures with lower uncertainty avoidance and power distance levels.

CONCLUSION

This study provides an overview of the current literature on the potential relationship between e-government and corruption. While the battle against corruption remains very difficult and a key challenge to many countries, especially developing countries, e-government is a promising tool in anti-corruption efforts, as it provides disincentives to engage in corrupt activities. It also lowers the discretionary power of government agents and generates greater accountability and transparency in the economy. Yet, e-government may fail to reduce corruption and can also be counterproductive. It may create new opportunities for corrupt rent-seeking activities, and corrupt government agents may take advantage of potential e-government system weaknesses. Therefore, it is vital to implement e-government systems properly in order to produce the favorable effect on curbing corruption. Developing a clear vision and a long-term target of corruption reduction, while implementing e-government projects, increases the probability of e-government success in lowering corruption. Persistent monitoring and revision of e-government project implementations seem necessary to address their potential weaknesses. Each country needs to find solutions to e-government implementation problems, and

---

[23] Societies exhibiting greater uncertainty avoidance tend to maintain strict codes of behavior and beliefs and implement more rules, laws, and regulations to minimize the occurrence of unknown future events. They may also be less tolerant of change. Cultures exhibiting greater power distance are more likely to conform to hierarchy. For more details, see Hofstede (1984) and House et al. (2004).

[24] Societies with more in-group collectivism may prefer face-to-face communication to maintain relationships.

to corruption in general, that fit its own national context, and there is no one-size-fits-all solution that applies to each society. Furthermore, it may be useful to reach out to the private sector for innovative strategies to enhance the positive impact of e-government on accountability, transparency, and corruption reduction.

As fighting corruption requires a strong political commitment with collaborative leadership and an appropriate legislative and institutional framework, this will also help the success of e-government as a useful instrument in anti-corruption programs. Future research should explore various conditions under which e-government will be more effective in restraining corruption. It should also investigate the capacity of e-government to restrict the various forms of corruption.

REFERENCES

Aladwani, A. 2016. Corruption as a source of e-government projects failure in developing countries: a theoretical exposition. *International Journal of Information Management* **36**(1), 105–12.

Andersen, T. 2009. E-government as an anti-corruption strategy. *Information Economics and Policy* **21**(3), 201–10.

Bhatnagar, S. 2003. *E-government and Access to Information*. United Nations, Global Corruption Report 2003, pp. 24–32. http://unpan1.un.org/intradoc/groups/public/documents/apcity/unpan008435.pdf. Accessed July 17, 2017.

Bhuiyan, S. 2011. Modernizing Bangladesh public administration through e-governance: benefits and challenges. *Government Information Quarterly* **28**(1), 54–65.

Chawla, R. and Bhatnagar, S. 2004. Online delivery of land titles to rural farmers in Karnataka, India. The International Bank for Reconstruction and Development, the World Bank. http://web.worldbank.org/archive/website00819C/WEB/PDF/INDIA_BH.PDF. Accessed August 5, 2017.

Dimant, E. and Tosato, G. 2018. Causes and effects of corruption: what has the past decade's empirical research taught us? A survey. *Journal of Economic Surveys* **32**(2), 335–56.

Elbahnasawy, N. 2014. E-government, internet adoption, and corruption: an empirical investigation. *World Development* **57**, 114–26.

Elbahnasawy, N. and Revier, C. 2012. The determinants of corruption: cross-country-panel-data analysis. *Developing Economies* **50**(4), 311–33.

Goel, R., Nelson, M., and Naretta, M. 2012. The internet as an indicator of corruption awareness. *European Journal of Political Economy* **28**(1), 64–75.

Gupta, S., Davoodi, H., and Alonso-Terme, R. 2002. Does corruption affect income inequality and poverty? *Economics of Governance* **3**(1), 23–45.

Heeks, R. 1998. Information technology and public sector corruption. Information Systems for Public Sector Management Working Paper Series No. 4. Institute for Development Policy and Management, Manchester. http://unpan1.un.org/intradoc/groups/public/documents/NISPAcee/UNPAN015477.pdf. Accessed June 15, 2017.

Hofstede, G. 1984. *Culture's Consequences: International Differences in Work-related Values*, Vol. 5. Cross Cultural Research and Methodology Series. London: Sage.

House, R., Hanges, P., Javidan, M., Dorfman, P., and Gupta, V. 2004. *Culture, Leadership, and Organizations: The GLOBE Study of 62 Societies*. London: Sage.

Jain, A. 2001. Corruption: a review. *Journal of Economic Surveys* **15**(1), 71–121.

Jha, C. and Sarangi, S. 2017. Does social media reduce corruption? *Information Economics and Policy* **39**, 60–71.

Kaufmann, D., Kraay, A., and Mastruzzi, M. 2010. The worldwide governance indicators: methodology and analytical issues. Policy Research Working Paper 5430. World Bank, Washington, DC.

Khalil, O. 2011. E-government readiness: does national culture matter? *Government Information Quarterly* **28**(3), 388–99.

Kim, S., Kim, H., and Lee, H. 2009. An institutional analysis of an e-government system for anti-corruption: the case of OPEN. *Government Information Quarterly* **26**(1), 42–50.

Klitgaard, R. 1988. *Controlling Corruption*. Berkeley, CA: University of California Press.

Lio, M.C., Liu, M.C., and Ou, Y.P. 2011. Can the internet reduce corruption? A cross-country study based on dynamic panel data models. *Government Information Quarterly* **28**(1), 47–53.

Mauro, P. 1995. Corruption and growth. *Quarterly Journal of Economics* **110**(3), 681–712.

Murphy, K., Shleifer, A., and Vishny, R. 1991. The allocation of talent: implications for growth. *Quarterly Journal of Economics* **106**(2), 503–30.

Neupane, A., Soar, J., Vaidya, K., and Yong, J. 2012. Role of public e-procurement technology to reduce corruption in government procurement. In *Proceedings of the 5th International Public Procurement Conference (IPPC5)*, pp. 304–34. Lagos: Public Procurement Research Center.

Neupane, A., Soar, J., Vaidya, K., and Aryal, S. 2017. Application of e-government principles in anti-corruption framework. In R. Shakya (ed.), *Digital Governance and E-government Principles Applied to Public Procurement*, pp. 56–74. Hershey, PA: IGI Global.

OECD. 2013. Issues paper on corruption and economic growth. https://www.oecd.org/g20/topics/anti-corruption/Issue-Paper-Corruption-and-Economic-Growth.pdf. Accessed September 10, 2017.

Pacific Council on International Policy. 2002. Roadmap for e-government in the developing world: 10 questions e-government leaders should ask themselves. The Working Group on E-government in the Developing World. http://www.itu.int/wsis/docs/background/themes/egov/pacific_council.pdf. Accessed July 10, 2017.

The PRS Group. 2017. International country risk guide, ICRG methodology. https://www.prsgroup.com/about-us/our-two-methodologies/icrg. Accessed October 24, 2017.

Transparency International. 2016. Corruption perceptions index 2016. https://www.transparency.org/news/feature/corruption_perceptions_index_2016. Accessed June 10, 2017.

Treisman, D. 2000. The causes of corruption: a cross-national study. *Journal of Public Economics* **76**(3), 399–457.

Treisman, D. 2007. What have we learned about the causes of corruption from ten years of cross-national empirical research? *Annual Review of Political Science* **10**, 211–44.

United Nations. 2010. United Nations e-government survey 2010: leveraging e-government at a time of financial and economic crisis. Department of Economic and Social Affairs, New York. http://www.unpan.org/egovkb/global_reports/08report.htm. Accessed July 17, 2017.

United Nations. 2015. Transforming our world: the 2030 agenda for sustainable development. A/RES/70/1. Accessed May 15, 2018. https://sustainabledevelopment.un.org/content/documents/21252030%20Agenda%20for%20Sustainable%20Development%20web.pdf. Accessed June 25, 2017.

United Nations. 2016. United Nations e-government survey 2016: e-government in support of sustainable development. Department of Economic and Social Affairs, New York.

United Nations. 2017. Least developed countries (LDCs). Development Policy & Analysis Division. https://www.un.org/development/desa/dpad/least-developed-country-category.html. Accessed October 25, 2017.

United Nations Department of Economic and Social Affairs. 2015. National adoption, governance and institutions of MDGs: lessons and implications for post-2015. https://wess.un.org/wp-content/uploads/2015/02/WESS-2015-BP-Adoption-Governance-Institutions-24-02-15-unedited.pdf. Accessed October 1, 2017.

United Nations Department of Economic and Social Affairs and the International Organization of Supreme Audit Institutions. 2013. Audit and advisory activities by SAIs: opportunities and risks, as well as possibilities for engaging citizens. Report of the expert group meeting. ST/ESA/PAD/SER.E/192. http://www.intosai.org/fileadmin/downloads/downloads/5_events/symposia/2013/EN_22_UN_INT_SympReport.pdf. Accessed July 5, 2017.

United Nations Development Programme. 2006. Fighting corruption with e-government applications. APDIP e-note 8. http://www.unapcict.org/ecohub/resources/apdip-e-note-8-fighting-corruption-with-e. Accessed June 17, 2017.

United Nations Development Programme. 2008. *Tackling Corruption, Transforming Lives: Accelerating Human Development in Asia and the Pacific.* New Delhi: Macmillan Publishers India.

United Nations Development Programme. 2014. *Anti-corruption Strategies: Understanding What Works, What Doesn't and Why. Lessons Learned from the Asia-Pacific Region.* New York: United Nations.

Weerakkody, V., Irani, Z., Lee, H., Osman, I., and Hindi, N. 2015. E-government implementation: a bird's eye view of issues relating to costs, opportunities, benefits and risks. *Information Systems Frontiers* **17**(4), 889–915.

World Bank. 2015. E-government. http://www.worldbank.org/en/topic/ict/brief/e-government. Accessed September 5, 2017.

World Bank. 2017. Combating corruption. http://www.worldbank.org/en/topic/governance/brief/anti-corruption. Accessed October 30, 2017.

Zhao, F., Shen, K.N., and Collier, A. 2014. Effects of national culture on e-government diffusion – a global study of 55 countries. *Information & Management* **51**(8), 1005–16.

Zhao, H., Ahn, M.J., and Manoharan, A. 2017. E-government, corruption reduction and culture: a study based on panel data of 57 countries. In *Proceedings of the 18th Annual International Conference on Digital Government Research*, pp. 310–18. New York: ACM.

PART II

NATIONAL CASE STUDIES

8. Corruption in Mexico: continuity amid change
Stephen Morris

Long depicted as a fundamental component of the Mexican political system, corruption spans virtually all levels of government and affects every policy arena, business and household in the country. It costs Mexico between 2 to 10 percent of gross domestic product (Casar 2015; Petersen 2015), weakens and distorts investment affecting economic growth, undermines public confidence in state institutions and faith in democracy, worsens the already high rates of inequality, and prevents the country from addressing poverty and unemployment, violence and drug trafficking. From a comparative perspective, Mexico offers an intriguing case, raising theoretical questions and empirical challenges. It exhibits higher levels of corruption than one would expect based on its level of development, and the levels of corruption have stubbornly persisted and arguably even increased in recent years despite broad political changes over the past few decades. These include a shift from one-party dominant authoritarianism to competitive elections (what many refer to as democratization), pluralistic opening, growing social opposition to corruption and the construction of normative and institutional mechanisms designed to address it, and a dreadful and dramatic rise in drug trafficking-related violence. Not only does the literature suggest that greater democracy should curb corruption especially after a period of time, but many of the reforms put in place since the onset of democracy in 2000 reflected much of the anticorruption thinking of the time. This does not mean, of course, that the types, forms or patterns of corruption have remained unchanged in response to the broader political changes, though gauging this presents methodological challenges. Even so, how did democratization, now almost two decades old, the growing pluralism, and even anticorruption reforms fail to reign in corruption? How did these changes alter the patterns or forms of corruption? And will the current integral reforms known as the National Anticorruption System (SNA for the initials in Spanish) taking effect from 2017 finally begin to address the issue?

This chapter seeks to explore these questions while offering a broad overview of corruption in Mexico. It highlights the presence of and the predominant aspects of corruption in the country, the changing political and social landscape, trends in corruption, and reform efforts. As such, it tries to account for the failure of democracy broadly and past reforms to address corruption, and how the broader political changes contributed to changes in the patterns of corruption. The chapter also, in the final section, seeks to underscore the impact of corruption on the nation's political culture and how the lack of confidence and expectations stemming from corruption, impunity, and the weak rule of law craft the vicious cycle facilitating corruption and undermining reform efforts.

CORRUPTION IN MEXICO OVER THE YEARS

Despite the methodological challenges related to measuring corruption (see, e.g., Anderson and Heywood 2009; Razafindrakoto and Roubaud 2010; Soreide 2006), data show corruption levels in Mexico have changed very little over the past two decades. Of course, this says nothing about the underlying patterns or forms of corruption – one of the challenges in gauging corruption – a point I will return to later. Still, data from Transparency International's oft-cited Corruption Perception Index (CPI) and the World Bank Governance Indicator Control of Corruption presented in Figures 8.1 and 8.2 show this minimal degree of change. Clearly, neither indicates a notable decrease in the levels of corruption over the years. In

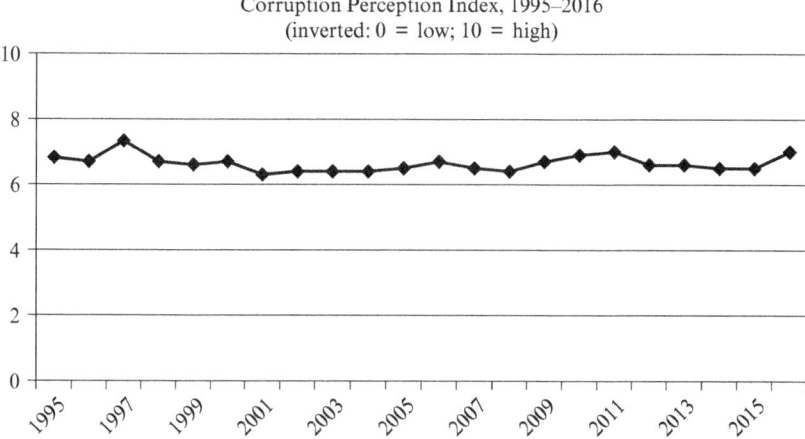

Source: Transparency International.

Figure 8.1 Levels of perceived corruption in Mexico, 1995–2016

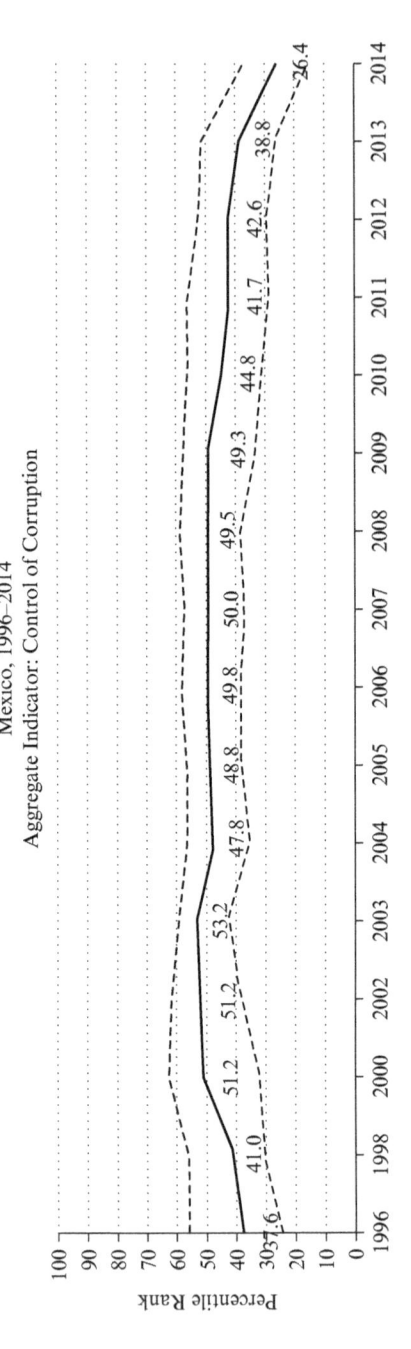

Source: World Bank Group, Worldwide Governance Indicators.

Figure 8.2   Control of Corruption, 1996–2014

Table 8.1   Progress in the reduction of corruption in state institutions over
            the prior two years

	2004	2007	2010	2013	2015
A lot	3	8	5	8	5
Some	21	30	27	23	17
A little	46	37	37	31	38
None	30	24	29	36	38

Source: Latinobarómetro.

the 2016 CPI, the most recent measure, Mexico ranked 123 among the 176 countries, alongside such nations as Azerbaijan, Djibouti, and Honduras.

Whereas these two measures tap expert and business opinion, polls gauging public opinion tell a similar story, highlighting not only the exceedingly high perceived levels of corruption in the country, but also recognition of corruption as a national problem and the stunning lack of progress in fighting it despite rhetorical promises and reforms to the contrary. While 83 percent in 2004 considered corruption "very or somewhat generalized," for example, 84 percent in 2010 and 75 percent in 2014 expressed that view (Global Corruption Barometer, Transparency International). In 1995 corruption ranked fifth as the most important problem facing the country (behind unemployment, inflation, low salaries, and poverty), the same spot it would occupy 20 years later (behind crime and insecurity, unemployment, political problems, and economic problems). Similarly, as shown in Table 8.1, 76 percent of respondents believed there had been "little" or "no" reduction in the levels of corruption over the prior two years in 2004 and in 2015.

Beyond perceptions regarding the extent of corruption, another means of measuring corruption asks respondents about their actual participation in corruption by making extralegal payments for assorted government services. Calculated in multiple years by Transparencia Mexicana, the index of national corruption and good government presented in Figure 8.3 shows 2010 measures near those recorded back in 2001. Indeed, INEGI's (2015) National Survey of Quality and Governmental Impact uncovered a total of 9.9 million such extralegal payments in 2015, while the Global Fraud Survey 2016 found that 82 percent of businesses surveyed considered corruption habitual practice in doing business in Mexico (Ethos 2017).

By itself, the high levels of corruption in Mexico beg for explanation; but this noted continuity over recent decades presents perhaps an even

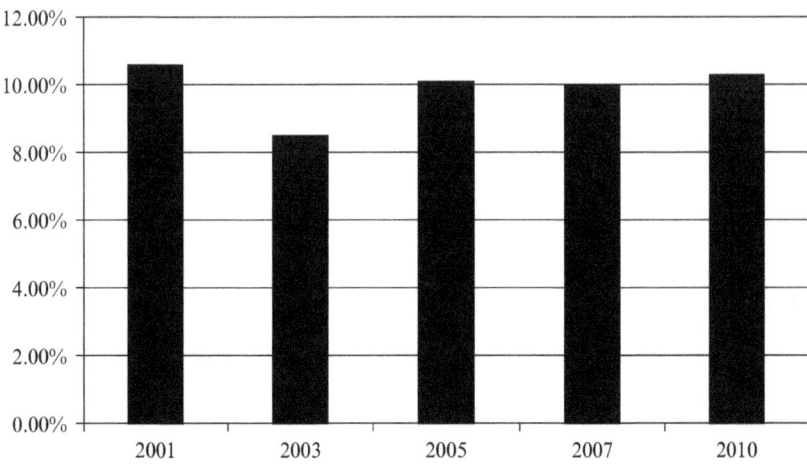

Source: Transparencia Mexicana, 2001, 2003, 2005, 2007, 2010.

Figure 8.3 *Participation in corruption: INCBG (payment of bribes in use of government services), 2001–10*

bigger analytical challenge: how can we explain the lack of changes in the levels of corruption amid what has been significant political changes in other areas? As the following discussion will indicate, Mexico of today is quite different politically and socially than the Mexico of two decades or more ago, yet this pillar of the system seems to persist.

CAUSES

Earlier analyses of corruption in Mexico pointed squarely to the nature and structure of the one-party authoritarian system (Morris 1991). Ruled by the Institutional Revolutionary Party (PRI) since the party's formation from the seat of power by President Elias Calles in 1929, the party won virtually all elections and monopolized state power until the waning years of the century. In this rather unique system – once referred to by Peruvian novelist Mario Vargas Llosa as the perfect dictatorship – the president exercised sweeping, extra-constitutional powers, largely because of the party's dominance within the legislature, the judiciary, and at state and local levels. Within a centralized structure, the president played the dominant role in determining appointments within the vast state bureaucracy and the PRI's nominees for the many "elected" positions throughout the system. Coupled with a strict and sweeping ban on reelection, this ensured a mechanism to discipline those within the system and created a set of informal rules and boundaries. The no reelection principle also ensured that presidents would rule only for their six-year term, thus guaranteeing a fluidity of leadership. Through import substitution industrialization economic policies, a largely neo-patrimonial state also controlled major levers of the economy which the state and more specifically the president used effectively to maintain discipline, co-opt or punish opponents, and to engage in rent-seeking endeavors.

Corruption and the abuse of power by state officials not only grew out of this authoritarian system largely owing to the absence of vertical or horizontal mechanisms of accountability, a weak and co-opted society, but in many ways the corruption actually contributed to the nation's rather remarkable record of political stability, especially compared to that of other Latin American countries. Politicians and bureaucratic officials, labor union and peasant organizations incorporated into the corporatist structure of the PRI, and others not only enjoyed the spoils of the system and the rent-seeking opportunities of a neo-patrimonial state, but by allowing officials to partake in the spoils proved critical in cementing loyalty to the president, the party, and the political system. The informal boundaries, largely determined and enforced by the president, meant the president could, when needed, use the law and even anticorruption measures to go after opponents or political enemies, or remove governors or other high-level officials who either overstepped the bounds of acceptable conduct or became caught up in scandal. At the same time, presidents periodically crafted rhetorical anticorruption campaigns. Nestled within a democratic narrative, such efforts, while clearly unsuccessful at having anything more than a temporary impact on corruption and never threatening the pillars of the system or political stability, nevertheless helped the government, the PRI, and particularly the incoming president garner popular legitimacy, distance himself from the prior president who of course hailed from his own party, and maintain a certain democratic facade.

Yet, Mexican politics have changed dramatically since the authoritarian period of presidentialism, corporatism, and one-party rule: the heyday of what was known as the *PRI-gobierno*. Though opposition and even crisis to the authoritarian regime amid uninterrupted elections had long existed, it is really only with the economic downturn of the 1980s "lost decade" that the PRI-led system began to unravel. As discontent over the economy and the corruption spread, opposition parties, particularly the center-right PAN (National Action Party), began to make inroads in local elections. While President Miguel de la Madrid (1982–88) allowed PAN victories in local races to stand in the early 1980s in response to social pressures, this trend was reversed a few years later, relying on fraud to deny further opposition advances. Such a response, however, had the effect of further mobilizing the opposition's demands for free and fair elections. This not only further delegitimized the PRI-led system, but opposition victories even at the local level provided them opportunities to govern and further pressure the PRI-led government for change. By the end of the decade, the PAN had secured one gubernatorial seat outright: a trend it would continue into the 1990s, with some state executive seats gained following post-electoral negotiations with the government. Further erosion of the political system occurred prior to and following the historic 1988 presidential election. Prominent leaders of the party's nationalistic left-wing, led by former PRI governor and the son of the country's most popular president, Cautéhmoc Cárdenas, and the former head of the PRI, Porfirio Muñoz Ledo, resigned from the party following disputes over the traditional process of the president hand-picking the party's presidential nominee. Soon thereafter, Cárdenas challenged the PRI's nominee in the 1988 contest. Amid widespread accusations and credible evidence of massive voter fraud, the PRI held on to the presidency, electing Carlos Salinas de Gortari (1988–94). But the events further undermined the regime's legitimacy. They also left the PRI with a small minority in congress (not enough to alter the Constitution) and, for the first time, facing major opposition from the right (PAN) and the Cárdenas-led left, which by 1989 had coalesced around the newly created PRD (Party of the Democratic Revolution).

Accompanying and nurturing these political changes was an ever growing degree of pluralism and maturing of civil society that increasingly escaped PRI's control. Facilitated by a combination of Mexico's growing middle class as a result of decades of solid economic growth (the Mexican Miracle) as well as government austerity measures, the government's poor response to the 1985 earthquake, the dismantling of import substitution industrialization, and the embrace of neoliberal economic reforms beginning in the mid to late 1980s, the state had fewer resources in which to co-opt opponents or even reward loyalists. Economic troubles also tend to highlight the levels of corruption. Whereas the traditional PRI-led corporatist system structurally incorporated societal organizations into the system and rewarded activists and key leaders with positions in government (along with the opportunities for corrupt gain), by the 1990s society increasingly began to escape these controls and assert a degree of autonomy vis-à-vis the government. Democratic labor movements struggled to break free from state-controlled labor unions, a more assertive business community began to actually lend support to the PAN, and the emergence of independent civil society organizations all began to pressure government for needed changes. This included demands for free and fair elections, greater respect for human rights, accountability and transparency.

With these growing pressures, the PRI-led government had few options but to negotiate with its opponents, while struggling to maintain its power and privilege. This included a series of electoral reforms that further opened the electoral system, transferring control by 1996 away from the PRI government to a truly citizen-controlled electoral council. These moves, in turn, set the stage for even further erosion of PRI's political control, resulting in the loss of its majority in the lower chamber of congress in the 1997 mid-term election, followed by the historic loss of the presidency in 2000 to the PAN's Vicente Fox. Yet despite the dramatic loss, the PRI remained during the ensuing two terms (12 years) of PAN presidential rule, the largest national political party, controlling the largest bloc of seats in the national congress, in state houses and at the local level. Nonetheless, civil society organizations continued to grow and mature as opposition parties captured more and more states and localities. Largely in response to President Felipe Calderon's (2006–12) policy of deploying the military to fight the growing power of drug trafficking organizations (DTOs) beginning in 2006, the level of violence and crime in Mexico also skyrocketed. Amid the political changes, Mexico emerged as one of the most violent countries in the world with military forces exercising police functions. With the 2012 election of Enrique Peña Nieto, the PRI recaptured the presidency, but was now ruling over a different political system than it had masterfully crafted during the twentieth century.

CHANGING PATTERNS OF CORRUPTION

While the level of corruption remained relatively unchanged throughout this period, as indicated earlier, the changing political landscape nonetheless altered the forms and patterns of corruption. Five areas stand out (Morris 2009).

First, by losing the presidency, the PRI-led political system lost much of its centralized control over state-level officials. Governors who once bowed to presidential authority were now virtually free to develop and exploit their own authoritarian political fiefdoms. This trend, which also saw a dramatic increase in funding from the federal government to the states, led to an explosion of corruption among state governors who have largely controlled state congress and local governments, thereby rendering accountability mechanisms, already weak at the state and local levels, ineffective. In recent years, a number of state governors have been charged with the diversion of state funds to election campaigns, to ghost companies receiving lucrative state contracts, to conflict of interest, the abuse of power to acquire land and properties, money laundering, ties to drug traffickers, and so on. In fact, the Infolatina platform notes 303 political scandals involving 62 governors between 2000 and 2013 (cited in Montes Mendoza 2017). With the former governor of Quitana Roo already in prison for laundering money for drug traffickers, a handful of others, including Javier Duarte of Veracruz, Roberto Borge of Quintata Roo, and Tomas Yarrington of Tamaulipas, are currently on trial or facing extradition charges.[1]

A second trend also related to decentralization has been the rapid growth in drug trafficking-related corruption. While drugs have been moving from and through Mexico to the insatiable US market for a century, it was only in the late 1990s to early 2000s when this became closely tied to the emergence of both powerful DTOs (drug trafficking organizations), violence, and to a growing level of drug-related corruption among law enforcement and government officials at all levels. For many, the decentralization of the system fed this process as DTOs now faced a breakdown in centralized control and thus had to seek alliances with local and state law enforcement and government officials of different political stripes, while battling Calderon's war and disputes with rival organizations (Grayson 2010; Morris 2013). By way of campaign contributions, bribes, extortion, threats, intimidation, and violence, DTOs have effectively captured many local governments, municipal, state, and even federal police, resulting in what Edgardo Buscaglia calls an "impunity pact" (*Proceso* 2047). In fact, in early 2011, Buscaglia estimated that 73 percent of Mexico's cities are "captured or under the control" of organized crime (quoted in *Justice in Mexico News Report*, January 2011, p. 9). This goes beyond the police to include the capturing of governors, attorney generals, and other top officials. This is clearly a highly destabilizing form of state capture since rather than uniting individuals to pursue a state objective, it undermines the state itself and pits state forces against one another.

A third and growing pattern of corruption in Mexico centers on elections. Shaped by the political and institutional changes since the late twentieth century, particularly the growing competitiveness of elections

---

[1]   The former governor of Quintana Roo, Roberto Borge, allegedly diverted over 50 million pesos to a nonexistent company by way of a closed bidding process (animalpolitico.com/2017/02/borge-contrato-empresa-fantasma-atlas-riesgo/, accessed May 16, 2018). The former governor of Chihuahua, Cesar Duarte, among other charges, misused more than 100 million pesos in the state purchase of medicines and medical supplies by overpaying more than 20 times the value authorized. Other governors facing charges include the former governor of Chihuahua and former head of the PRI, Humberto Moreira, actually arrested in Spain for illicit enrichment before the Mexican government intervened to achieve his release; Guillermo Padrés of Sonora facing charges for fraud and money laundering; Tomas Yarrington, the former governor of Tamaulipas, recently arrested in Italy, is charged with money laundering, and Fausto Vallejo of Michoacan for diversion of government funds.

and the spoils the system has to offer, this category of corruption includes the illegal use of state resources to fund campaigns, the conditioning of the provision of state services for votes (or voter credentials), the clandestine and illegal channeling of funds from outside sources like organized crime, businesses, and foreign entities, private contributions and campaign spending beyond legal limits, and various forms of electoral fraud. Examples abound. Noteworthy cases range from campaign spending far beyond legal limits in the gubernatorial campaign of Roberto Madrazo in the state of Tabasco in 1994 and the presidential campaign of Peña Nieto in 2012,[2] the channeling of funds from the state-owned oil firm Pemex to PRI's presidential candidate in the 2000 presidential campaign (known as Pemexgate),[3] illegally channeled private, foreign, and corporate donations to the PAN's 2000 presidential campaign (Amigos de Fox scandal), the triangulation of private funds through Monex Bank during Peña Nieto's 2012 campaign, or the secret agreement between his campaign and Mexico's near television monopoly *Televisa* to ensure favorable coverage years ahead of the contest. Obscured perhaps by the more high-profile cases, some of which actually resulted in fines being levied against individuals and political parties but without altering electoral results, lie numerous unconfirmed allegations of the channeling of funds from organized crime into electoral campaigns, including the campaign of the President Peña Nieto.[4] According to Buscaglia, military intelligence sources allege that between 55 and 65 percent of the electoral campaigns in Mexico have been infiltrated by criminal groups pumping money into the campaigns (cited in *Justice in Mexico News Report*, January 2011, p. 9).

A fourth pattern of corruption that reflects recent political changes centers on corruption within legislatures. Under the old PRI regime, legislatures (federal, state, and local) had very little effective power. Since most members owed their current and future positions to leading PRI officials in the executive branch, the legislatures operated largely as a rubber stamp to executive initiatives. And just as they played virtually

---

[2] Though initially alleged to have spent 238 million pesos (about $79 million) on his campaign – or about 50 times the legal limit – it was subsequently confirmed by the attorney general that the campaign spent 128 million pesos. Though still well above the spending cap, Roberto Madrazo was never formally charged (*Los Angeles Times*, October 3, 1999).

[3] In 2015, the Mexican government levied fines of $255 million – twice the amount believed to have been transferred from Pexmex to the PRI presidential campaign – against the PRI, the former director of Pemex and five other officials involved in finance, labor relations, and budget administration.

[4] The special committee in the Chamber of Deputies exploring the Monex case estimated that the Peña Nieto campaign surpassed the spending cap by a factor of 14. Reports also suggest that the financing for the Monex cards came via the firm Grupo Comercializador Cónclave with ties to the Juárez drug cartel (see Ackerman 2017; Cártel de Juárez 2017).

no role in shaping policy, they did little in terms of holding the executive in check or exercising oversight powers. Yet, as executives began to lose majorities in congresses at all levels, it became increasingly necessary to negotiate with members of congress, providing members with certain benefits and a degree of autonomy in exchange for their support. Legislators increased their control over state resources, generally increasing funding for the parties and even creating large discretionary funds that individual members could use for their own political purposes. This gave rise to a form of corruption known as *moches*, where legislators provide funding to their states or local areas in exchange for kickbacks or a percentage of the budgeted transfer. It was also part of an initiative in which the executive (president, governor, and mayor) could effectively capture congress, thus undermining or weakening its oversight capacity.

A final pattern of corruption in recent years centers on a seemingly new type of state-business relationship. While historically, certain areas of the Mexican state were captured by state-affiliated labor unions and large corporate interests, much of the corruption taking place in recent years seems to involve government concessions and contracts to particular firms connected to public officials receiving anywhere from 10 percent to as high as 25 percent of the value of the contract in the form of kickbacks and related conflict of interest issues (Ugalde 2015). This dates back to privatization schemes in the late 1980s tied to secret agreements and corruption, which made a few entities extremely wealthy (Salas-Porras 2014; Sandoval Ballesteros 2009; World Bank 2007). More recent examples of collusion have exploded. The White House scandal early in the Peña Nieto administration, for example, involved the construction and special financing of a home for the First Lady by a government contractor with a history of contracts with Peña Nieto while serving as governor in the state of Mexico. The former governor of Veracruz, Javier Duarte, currently on trial, reportedly channeled millions of dollars in state funds into a number of ghost companies and phantom projects seemingly merely to enhance his own wealth. One such project included a highway extension valued at over US$7 million on which construction was never actually begun (see SourceMex 2017). Indeed, at the municipal level the preferred method to make money in the three short years of the term is in the granting of permits and concessions for use of land to business establishments. One recent example involves Walmart paying US$52,000 in bribes for use of land rights (Ugalde 2015).

While such patterns are relatively new or more pronounced, other more traditional forms of corruption have largely remained. They include the highly ubiquitous pattern involving the routine payment of bribes at lower levels of government. As highlighted in the surveys of actual participation in bribery noted earlier, Mexicans frequently make extra-legal payments (bribes) to acquire routine government services. This phenomenon is particularly common in quotidian dealings with the police (*la mordida*, literally, the bite). In fact, 61 percent of Mexican respondents in Transparency International's Global Corruption Barometer reported having paid a bribe to the police just within the prior 12-month period. Such bribes, according to polls, are also commonly found in interactions with the judiciary and state agencies and in the provision of business licenses and permits. As with the other forms of corruption noted above, this pattern also reflects an abuse of state power by state officials rooted in the discretion that enables them to structure the provision of services in such a way as to demand or solicit bribes (creating red tape in order to "do you a favor" by cutting through it) coupled with a low risk of detection and sanction. Officials also tend to have the support and cooperation of colleagues and are often part of a more organized system or network of corruption with quotas and payments often moving up the chain of command. With loyalty largely trained upwards toward patrons rather than to the written rules and regulations of the institution, bureaucrats abuse state power to enjoy the rents provided by Mexico's neo-patrimonial state (Collins 2011). This pattern also reflects the failure to professionalize the bureaucracy or implement the 2003 civil service reforms (on civil service reforms see Arrellano Gault and Guerrero Amparán 2003; Dussauge Laguna 2011; World Bank 2007).

The depths of corruption, the emerging patterns in recent years, and an assortment of scandals have wrought a growing outcry among the public. Certainly, corruption contributed to the PRI's delegitimization during the waning years of the twentieth century and to Fox's appeal during the historic 2000 election. Fox not only campaigned on fighting corruption, but made it a top priority of his administration. As one of his first acts as president, he created an inter-secretarial commission to develop and implement a broad anticorruption strategy. It engaged in detailed diagnostics, sought to streamline and open bureaucratic procedures, strengthened internal controls over the bureaucracy, created a world-renowned, open competitive, public procurement process known as Compranet, passed a transparency law and system in 2002 that pried open the government for the first time, passed electoral reforms that placed stricter controls on campaign financing and political parties, and even incorporated civil society into oversight functions. An active civil society pushed for many of these changes, particularly the historic transparency law in 2003 (Michener 2005). Since then, civil society organizations have continued to take the lead in the fight to promote transparency and accountability.

FAILURE OF DEMOCRACY TO CURB CORRUPTION

Despite cross-national research showing democracy tied to lower levels of corruption, at least after a period of time (Rock 2009; Triesman 2007), Mexican democratization has not led to a reduction of corruption and may, instead, have brought about higher levels of corruption (Morris 2009). As the above discussion suggests, rather than closing opportunities for corruption, democratization in Mexico seemingly opened new ones. Growing electoral competition amid strict formal controls on campaign finance and spending, for example, led many to seek alternative means to channel state and other illegal funds into the campaigns to ensure victory. While the weakening of the executive vis-à-vis the legislatures (at all levels) within a democratic context should lead to greater oversight and stronger checks and balances, in Mexico it seemed to facilitate corruption among state governors, within the legislatures, and even within the judiciary. Rather than exercising oversight, the now more powerful legislatures vie for control over state budgets to bolster their political support and/or line their own pockets. Executives, in turn, provide resources in return for legislative support on policy initiatives while undermining and crippling oversight, thus ensuring impunity. In this sense, as Ugalde (2015) notes, the executive has effectively co-opted congress into the corruption.

Arguably the key component behind the persistent corruption, even amid the higher levels of transparency and democratization, involves Mexico's exceedingly high levels of impunity (Bohórquez et al. 2016) and its ineffective rule of law institutions. As Ugalde (2015) contends, what is required for the checks and balances related to democracy to curb corruption is a culture of the rule of law, and in Mexico, this is lacking. Indeed, few high-level officials have ever been prosecuted for corruption in Mexico. Rather than offer a cure for the corruption, the judiciary suffers from it. Studies not only show high perceived levels of corruption within the judiciary, but also tremendously high impunity rates estimated at 94 to 97 percent, with many crimes not reported (Bargent 2016; Zepeda Lecuona 2004). Impunity further opens the opportunities for corruption as more and more officials recognize the low likelihood of being caught and punished. In sum, the many changes taking place over the years have largely failed to fundamentally alter the authoritarian way the system operates.

ANTICORRUPTION REFORMS: THEN AND NOW

In addition to the apparent failure of democracy to curb corruption, the persistent and growing levels of corruption also point to the failure of the numerous anticorruption reforms of the past. Though anticorruption reforms date back to the PRI period – President de la Madrid (1982–88), for instance, championed a "moral renovation" campaign – President Fox (2000–06), the first opposition president, developed and put in place a broad, systematic anticorruption program. Yet, the reforms did little to alter the overall levels of corruption. Part of the reason lies in the political situation that forced Fox to work with and deal with the PRI, the nation's largest political party which at the time controlled the largest bloc in congress, in order to achieve his economic and political reforms. As a result, serious efforts to overhaul the system and/or prosecute corrupt government officials never took place, despite the historic opportunity. The implementation of the professionalization of the civil service system put in place under Fox also eventually slowed and has made little progress since (Dussauge Laguna 2011). In the end, even the Fox team itself became caught up in high-level allegations of corruption with government contracts benefitting companies tied to the president's wife and children. The PAN, as well, now in many positions of power, seemed to succumb to corruption and scandal. Adding to this issue, the new transparency regime, which successfully opened up the government, seemed only to reveal more corruption rather than really help curb it. Set within the prevailing pattern of widespread impunity, such revelations further undermined government legitimacy and eroded people's confidence in state institutions. Many of the political reforms during the period were also undermined by the ability of the political parties to capture oversight institutions like the federal electoral institution, undermining their autonomy and potential role in ensuring accountability.

Many hold that the reforms of the past represented mere simulations and failed because of the lack of political will, coordination, and integration, and for failing to expand the role of civil society. In many cases, including the early reforms under PRI presidents and even Fox's reforms, the measures indeed sought to empower the government to exercise oversight over itself while doing little to promote the rule of law.

Recent years, however, have brought a major overhaul in the area of transparency and anticorruption. Though the original Access to Information Law dates back to 2002 (taking effect in 2003), constitutional changes in 2014 coupled with the passage of secondary laws a year later greatly strengthened access to information and the autonomy of the National Institute of Access to Information (INAI). It also harmonized rules and procedures within the states, and expanded the reach of the law to include trusts, political parties, labor unions, state and local governments, and all entities dealing with the government.

More significantly, over the course of the past few years, the government and civil society have worked – sometimes together, sometimes not – to create the SNA (National Anticorruption System). This broad, consolidated anticorruption system encompassed constitutional changes swiftly passed in May 2015 and seven secondary laws passed in June 2016 with the system taking effect in July 2017. Touted as a broad and integrated reform, the SNA seeks to ensure greater coordination among strengthened oversight institutions and creates new institutions dedicated solely to fighting corruption. There are four main pillars in the institutional architecture: (a) a revised and strengthened Secretariat of Public Function in charge of internal audits and controls over the federal administration with the power to impose administrative sanctions for minor offenses; (b) a strengthened external auditing system, the Auditoria Superior de la Federación (ASF) with greater autonomy and now with the power to audit federal spending in real time (as opposed to two years after the spending occurred) and to audit the spending of trusts and state/local governments, and the power to investigate and sanction officials for irregularities in public spending; (c) a newly created Special Prosecutor for Crimes related to Corruption housed within the Attorney General's office and empowered to investigate and prosecute more significant charges of corruption before (d) a newly created Federal Court of Administrative Justice that will hear the cases and have the autonomy to sanction public officials and private citizens for serious administrative violations and corruption. Coordination and direction of the SNA centers on a technical secretariat which, among other tasks, will create a national digital information system on corruption; a Coordinating Committee, an inter-secretarial body composed of the heads of the major departments including the head of the Access to Information Agency, the Judicatura, the head of the National Electoral Institute and the president of the Citizen Committee; and, finally, a Citizen Committee charged with overall coordination, policy direction and evaluation of the performance of the different components of the system. In addition, the reforms revised the Penal Code and the Law of Responsibilities to broaden definitions of corruption, create harsher penalties, require greater disclosure, and even protect whistleblowers. Even more significantly in the long run, the reforms are also national in scope and thus require all state governments to develop and implement a similar anticorruption system.

Behind the SNA reforms lie a growing attention, frustration, and anger on the part of the population and civil society organizations to corruption and insecurity. Though such concerns have long existed as documented earlier, they have seemingly gone beyond prior levels reaching what Mauricio Merino (2014) describes as a tipping point. Much of this emerged from the "White House" conflict of interest scandal involving the construction and preferential financing of a house provided to the First Lady by a government contractor with ties to the president (he was later exonerated by his appointed Secretary of Public Function) and by the disappearance of 43 students in Ayotzinapa, Guerrero at the hands of local and most likely federal forces working in conjunction with organized crime, both in 2014. This outcry not only pushed President Peña Nieto to agree to sponsor the constitutional reforms for the SNA – reforms that he had tabled – but the public mobilized behind the efforts of civil society organizations like Causa Común, IMCO (Mexican Institute for Competitiveness), Transparencia Mexicana, and others to obtain 634,000 signatures for a proposed #3x3 law. The effort far surpassed the threshold needed to require legislative action. Incorporated into the Federal Law on Administrative Responsibilities, the initiative requires public officials to release tax returns and disclose property and conflicts of interest. Of particular importance was the role of civil society organizations in pushing and even influencing the reforms. It is significant that facing stiff opposition from parties and civil society organizations, the president largely failed to dictate the terms of the SNA reform. Through Open Parliament procedures organizations such as Transparencia Mexicana were even allowed to directly consult with Senate committees during the drafting of the laws (see Rios 2017).

Overall, the new system seeks to provide greater autonomy, more teeth, is broader in scope, and includes a prominent role for civil society. Yet, it remains as of this writing a work in progress. While at the federal level much has been done, the Senate has yet to name the Special Prosecutor or appoint magistrates to the new tribunal. States are even further behind the timeline in their efforts despite the passage of the one-year deadline to comply with the law. Again as of this writing, only two of Mexico's 32 states have fully complied with some registering virtually no progress to date. More problematic, some states have been accused of trying to undermine the reforms at birth by placing allies of the governor in key positions, thus ensuring a continuation of impunity. In addition, many criticize the new system for being weak and unlikely to change things. The Mexican think tank CIDAC, for example, highlights SNA's failure to provide mechanisms for the prevention of corruption and for a diagnostic structure; the think tank FUNDAR cites the law's inability to deal with corruption within the judicial system, while Buscaglia, an expert on organized crime, points to SNA's inability to control the role of organized crime money in elections (see Jesús Cantú 2016 on the institutional failures of the SNA). Similarly, Ugalde (2015) contends that the new system fails to address how budgets are actually made, strengthen legislative oversight, ensure broader citizen participation, reform campaign finance, or do much to promote a culture of rule of law. In the end, Eduardo Bohórquez (2016), the executive director of Transparencia Mexicana, calls SNA a guide or a roadmap, and that if it functions will require at least a decade of systematic and continual work.

POLITICAL CULTURE AND ITS IMPACT

Even while discounting the notion that culture causes corruption, it is nonetheless important to recognize how corruption shapes the contours of Mexico's political culture and that by forging collective dilemmas and vicious circles, people's attitudes, perceptions, and expectations tend to facilitate and perpetuate corruption while rendering reform difficult. This, as Uslaner (2008, p. 26) points out, is what makes corruption "sticky."

First, corruption instills within the political culture the view that politicians serve their own interests and fail to abide by the rule of law, thereby undermining the legitimacy of and confidence in state institutions. When asked in one national survey of the political culture (ENCUP 2012) whose interests are served when a member of congress approves a law, 67 percent said it served the interests of their political party or their own interests; only 14 percent said the interests of the people. In the same poll 73 percent expressed the view that government officials do not comply with the law, and when asked who violates the law the most, 51 percent pointed to politicians and public officials and another 15 percent to the police. It is not surprising then that the 2012 World Values Survey shows 70 percent of respondents in Mexico with "little" or "no confidence" in political parties, congress, and the police and more than 60 percent expressing "little" of "no confidence" in government, the president, and other public institutions. This lack of confidence extends to elections with 73 percent characterizing vote buying as "very common" or "common," 64 percent considering it "very common" or "common" that TV news favors the party in power, and 47 percent finding it "very common" or "common" that votes are not properly counted. Among other things, this translates into a generalized dissatisfaction with democracy. According to Latinobarometro (2015), 78 percent were not satisfied with democracy in Mexico, while 70 percent believed that the people have "little" to "no" influence over the decisions of government: the very definition of democracy.

Arguably, such corruption-produced perceptions set the example for society, undermining any sort of a culture of legality. Witnessing and understanding that the law is not applied to those to whom the law grants authority (state) gives rise to and tends to justify illegal conduct by other members of society, including participating in corruption where feasible for one's own benefit. The political culture – again, the product of a history of corruption – teaches people that it is not necessary to comply with the law, not only because others rarely do so or because few are ever sanctioned for failing to do so, but also because alternative means – informal institutions – are available and efficacious in arranging matters with government officials. In short, the political culture teaches one to utilize and abide by the informal rules of the system instead of the formal rules. For most, this means abide by the law when it works to your favor or when you have no other choice – pay the bribe when there is no other rational option or even seek a legal injunction (*amparo*) from a legal system one does not fully trust – and ignore it when it is not in one's favor. Part of the political culture is thus the idea of personally taking advantage of the system when the opportunity exists which, according to the prevailing views, everyone else does, particularly those in the government.

Second, and in a similar manner, corruption weakens confidence, not just in the public institutions, but also among members of society. As the 2012 ENCUP survey shows, 69 percent of respondents lack confidence in the majority of people, 72 percent believe that the majority are only concerned about themselves and that a majority fails to abide by the law. Such lack of confidence in others further nurtures the expectation of corruption – the view that everyone does it – which prompts people to participate in the practice. Indeed, empirical studies show the negative relationship linking trust in both public institutions and interpersonal trust and corruption (Morris and Klesner 2010).

The Mexican political culture also tends to cast anticorruption reforms on the part of the government as mere simulation. If there is no confidence in the system and the people do not believe the politicians and their public officials, they are hardly likely to believe promises to fight against corruption. As a result, as Aguilar Zinser (2000) notes, "the official versions are rejected beforehand and the promises to carry out the investigations to their ultimate consequences are received with general skepticism." It is within this context – where impunity is seen as the rule rather than the exception – that even the occasional prosecution of a public official is interpreted not as an example of the proper application of the rule of law, but as part of the same corrupt system, motivated by the pursuit of political gain. Recent moves against governors and the capture of fugitive governors outside of Mexico have been interpreted as a component of the president's political strategy, particularly the arrest of Javier Duarte in Panama coming just days prior to the critical election in the state of Mexico (see also Jesús Cantú 2016). Such a cynical view, moreover, also undermines the people's participation in fighting corruption: a necessary ingredient in the effective implementation of any anticorruption efforts.

Despite some, including President Peña Nieto, claiming that culture, specifically the people's tolerance for corruption, is to blame for the nation's problem, data show that Mexicans are actually staunchly opposed to corruption. According to polls, an average of 75 percent of respondents disapproved of 21 different acts of corruption (Transparencia Mexicana 2005); 73 percent felt that accepting a bribe was never justified (World Values Survey 2012); 83 percent believed that politicians should be held accountable (CIDAC 2010); and 77 percent believed that "the law should be applied equally to all" (Encuesta Nacional de Cultural Constitucional 2011). Of course, if the people were truly tolerant of corruption, they would hardly consider it a national problem as alluded to earlier, or support anticorruption organizations.

In the end, corruption in Mexico represents a form of conduct that is expected, but not accepted; a course of action that is morally rejected and considered wrong, but frequently utilized as a survival tactic within a corrupt system; a device that is publicly condemned, but not formally denounced to the authorities (whom they do not trust). As in a classic collective action dilemma, the people recognize that everyone would be better off without corruption, but it is and remains rational to participate in corruption, and, too often, irrational to fight against it (see Persson et al. 2013).

CONCLUSION

Mexico has a long history of corruption that has seemingly evolved and adapted to changing political and social times. Once a byproduct of presidentialism, one-party dominance, authoritarianism, and a weak society, corruption has persisted amid the emergence of competitive multi-party elections, more autonomous state institutions, greater levels of pluralism, a network of anticorruption civil society organization, and a series of reforms. Democratization, in short, did little to curb the embedded nature of corruption in Mexico; instead, corruption has weakened and hollowed-out democracy. But while past reform efforts produced few results, there are some signs that the level of public outrage has brought the country to a turning point. With a much more engaged civil society leading the push to forge a rigorous system of accountability, some are cautiously optimistic that the new SNA, which clearly goes beyond the reforms of the past, will finally begin to address the deep-seated problem. Whether the system will enjoy any success and begin the process of rooting out corruption, strengthening the rule of law, or suffer the same fate as measures of the past, evoking once again a sense of dejá vú, remains to be seen (Morris 2016/17).

REFERENCES

Ackerman, J. 2017. Cómo para el fraude que viene. *Proceso* 2105 (March 11), 42–3,

Aguilar Zinser, A. 2000. El combate a la corrupción. *Reforma* (September 8).

Anderson, S. and P. Heywood. 2009. The politics of perception: use and abuse of Transparency International's approach to measuring corruption. *Political Studies* **57**, 746–67.

Arrellano Gault, D. and J. Guerrero Amparán. 2003. Stalled administrative reforms of the Mexican state. In Ben Ross Schneider and Blanca Heredia (eds), *Reinventing Leviathan: The Politics of Administrative Reform in Developing Countries*, pp. 151–79. Miami: North-South Center Press,

Bargent, J. 2016. Mexico impunity levels reach 99%: study. *Insight Crime* (February 4). http://www.insightcrime.org/news-briefs/mexico-impunity-levels-reach-99-study (accessed May 16, 2018).

Bohórquez, E. 2016. Anticorrupción en México: De los zares a los sistemas. *Ibero* (July), 4–7.

Bohórquez, E., I. Guzmán, and G. Petersen. 2016. Control de la corrupción en México: La impunidad neutraliza el efecto de los avances en transparencia. *Este Pais* (January).

Cantú, J. 2016. Y el presidente es todavía juez supremo. *Proceso* 2083 (October 2), 40.

Cártel de Juárez, proveedor del PRI y financiador en la campaña de Peña Nieto. 2017. Especial Aristegui Noticias (special report), March 12.

Casar, M. 2015. *México: Anatomía de la corrupción*. Mexico: IMCO and CIDE.

CIDAC. 2010. Encuesta Valores: Diagnóstico axiológico. CIDAC, Mexico. http://cidac.org/encuesta-de-valores-mexico-diagnostico-axiologico/ (accessed May 16, 2018).

Collins, R. 2011. Patrimonial alliances and failures of state penetration: a historical dynamic of crime, corruption, gangs, and mafias. *Annals of the American Academy of Political and Social Science* **636**(1), 16–31.

Dussauge Laguna, M. 2011. The challenges of implementing merit-based personnel policies in Latin America: Mexico's civil service reform experience. *Journal of Comparative Policy Analysis* **13**(1), 51–73.

Encuesta Nacional de Cultural Constitucional: Legalidad, Legitimidad de las instituciones y rediseño del Estado. 2011. México: IFE- IIJ, UNAM.

ENCUP (Encuesta Nacional sobre la Cultura Política y Prácticas Ciudadanas). 2012. México: Secretaría de Gobernación. http://www.encup.gob.mx/ (accessed May 16, 2018).

Ethos. 2017. *Reporte Ethos: Descifrando la corrupción*. Mexico: Ethos. http://ethos.org.mx/es/publicaciones/reporte-ethos/resumen-2/ (accessed May 16, 2018).

Grayson, G. 2010. *Mexico: Narco-violence and a Failed State?* New Brunswick, NJ: Transaction.

INEGI. 2015. *Encuesta Nacional de Calidad e Impacto Gubernamental*. Mexico: INEGI. http://www.beta.inegi.org.mx/proyectos/enchogares/regulares/encig/2015/ (accessed May 16, 2018).

Justice in Mexico News Report. 2011. https://justiceinmexico.org/news-rule-of-law/ (accessed May 16, 2018).

Latinobarometro. 2015. Opinión Pública Latinoamericano. http://www.latinobarometro.org/lat.jsp (accessed May 16, 2018).

Merino, M. 2014. Dos momentos. *El Universal. Mx Opinion* (November 5). http://www.eluniversalmas.com.mx/editoriales/2014/11/73172.php (accessed May 16, 2018).

Michener, Greg. 2005. Engendering political commitment: the *Grupo Oaxaca* – expertise, media projection – and the elaboration of Mexico's access to information law. Paper presented at the meeting of the Southern Political Science Association, New Orleans, LA: January 7–9.

Montes Mendoza, P. 2017. De dónde surge la impunidad de los gobernadores? *Letras Libres* **218** (February).

Morris, S. 1991. *Corruption and Politics in Contemporary Mexico*. Tuscaloosa, AL: University of Alabama Press.

Morris, S. 2009. *Political Corruption in Mexico: The Impact of Democratization*. Boulder, CO: Lynne Rienner.

Morris, S. 2013. Drug trafficking, corruption, and violence in Mexico: mapping the linkages. *Trends in Organized Crime* **16**(2), 195–220.

Morris, S. 2016/17. La corrupción en México a través de los años: continuidad y cambio. *Este País*. Part one, No. 308 (October 2016), 32–4; Part two, No. 313 (May 2017), 13–17.

Morris, S. and J. Klesner. 2010. Corruption and trust: theoretical considerations and evidence from Mexico. *Comparative Politics Studies* **43**(10), 1258–85.

Petersen, G. 2015. Como analizar y entender la lucha contra la corrupción en México. *Este País* (December).

Persson, A., B. Rothstein, and J. Teorell. 2013. Why anticorruption reforms fail – systemic corruption as a collective action problem. *Governance: An International Journal of Policy Administration and Institutions* **26**(3), 449–71.

Razafindrakoto, M. and F. Roubaud. 2010. Are international databases on corruption reliable? A comparison of expert opinion surveys and household surveys in sub-Saharan Africa. *World Development* **38**(8), 1057–69.

Rios, V. 2017. Los cuatro rounds que los ciudadanos han ganado a la corrupción. *Alto Nivel* (July 11).

Rock, M. 2009. Corruption and democracy. *Journal of Development Studies* **45**(1), 55–75.

Salas-Porras, A. 2014. Las elites neoliberales en México: Cómo se construye un campo de poder que transforma las prácticas sociales de las elites políticas. *Revista Mexicana de Ciencias Políticas y Sociales* **LIX**(222), 279–312.

Sandoval Ballesteros, I. 2009. Rentismo y opacidad en procesos de privatización y rescate. In Sandoval Ballesteros (ed.), *Corrupción y transparencia: debatiendo las fronteras entre Estado, Mercado y Sociedad*, pp. 121–35. Mexico: Instituto de Investigaciones Sociales, Siglo XXI.

Soreide, T. 2006. Is it wrong to rank? A critical assessment of corruption indices. CMI Working Paper 2006:1, Chr. Michelsen Institute.

SourceMex, 2017. Former Veracruz Governor goes underground to avoid corruption charges. March 1. http://ladb.unm.edu/sourcemex (accessed May 16, 2018).

Transparency International. Global Corruption Barometer. https://www.transparency.org/ (accessed May 16, 2018).

Transparency International. Corruption Perception Index. https://www.transparency.org/ (accessed May 16, 2018).

Transparencia Mexicana. 2001, 2003, 2005, 2007, 2010. Encuesta nacional de corrupción y buen gobierno. Mexico. http://www.tm.org.mx/ (accessed May 16, 2018).

Triesman, D. 2007. What have we learned about the causes of corruption from ten years of cross-national empirical research? *Annual Review of Political Science* **10**, 211–44.

Ugalde, L. 2015. Por qué más democracia significa más corrupción? *Nexos* (February). http://www.nexos.com.mx/?p=24049 (accessed May 16, 2018).

Uslaner, E. 2008. *Corruption, Inequality, and the Rule of Law*. New York: Cambridge University Press.

World Bank. 2007. *Democratic Governance in Mexico: Beyond State Capture and Social Polarization*. World Bank Report No. 37293-MX. Washington, DC.

World Bank Group. Worldwide Governance Indicators. http://info.worldbank.org/governance/wgi/#home (accessed May 16, 2018).

World Values Survey. 2012. http://www.worldvaluessurvey.org/wvs.jsp (accessed May 16, 2018).

Zepeda Lecuona, G. 2004. *Crimen sin castigo: Procuración de justicia penal y Ministerio Público en México*. Mexico City: Centro de Investigación para el Desarrollo and Comisión Federal de Electricidad.

9. Persistent malfeasance despite institutional innovations and public outcry: a survey of corruption in Brazil

Kelly Senters and Matthew S. Winters

In the first Transparency International Corruption Perceptions Index, in 1995, Brazil was ranked the fifth most corrupt – after Indonesia, China, Pakistan, and Venezuela – of the 41 countries surveyed. As more countries were added to the index, Brazil consistently ranged between the 45th and 55th percentiles (where higher percentiles indicate better control of corruption) in the rankings in the late 1990s and early 2000s. From 2003 to 2014, the country placed around the 60th percentile, ranking as the 72nd best out of 180 countries in 2007, for instance. Recently, however, Brazil has fallen several places in the rankings, placing 79th out of 176 countries in the 2016 rankings, leaving it back at the 45th percentile.

The most recent years of Brazilian history have revealed the persistence and pervasiveness of large-scale political corruption in the country, despite high-profile anti-corruption initiatives undertaken over the past 20 years. In this chapter, we first describe the persistence of corruption in Brazil with reference to major scandals, statistics about the criminal accusations made against politicians at all levels of government, and information about the variation in state- and local-level corruption across various regions of the country. In the second section, we describe the institutional innovations that the country has introduced in its attempts to curb corruption. We then describe patterns in Brazilian public opinion toward corruption before concluding the chapter with some informed speculation about what might happen in light of the latest political scandal that has already led to the impeachment of one president and a court-of-first-incidence conviction of a former president.

PERSISTENT CORRUPTION IN BRAZIL SINCE THE 1985 TRANSITION TO DEMOCRACY

Contemporary Brazil is no stranger to political corruption. High-profile corruption scandals have regularly permeated the political environment in Brazil since its 1985 transition to democracy. Every presidential administration in the present democratic era has witnessed the eruption of at least one major corruption scandal during its tenure (Power and Taylor 2011). At the level of the highest political office, these scandals have led to the impeachment of two Brazilian presidents. Fernando Collor do Melo was impeached in 1992 following the public revelation of fraudulent business dealings, and Dilma Rousseff was removed from office in 2016 for the inappropriate presentation of government accounts.

The depth of political corruption in Brazil can be witnessed in the extent to which the major scandals of the last three decades have affected not only presidents but also large numbers of federal legislators. These scandals include the Anões do Orçamento ("budget dwarves") scandal of 1993–94, the Sanguessuga ("bloodsucker") scandal of the mid-2000s, the Mensalão ("big monthly payment") scandal of 2005, and the Operação Lava Jato ("car wash") scandal that emerged in 2014 and continues to reveal more information about nefarious financial relations within the Brazilian national government at the time of this writing (Power and Taylor 2011; Carson and Prado 2014). The allegations of corruption within these scandals range from the allocation of funds to phantom non-profit institutions by budget committee members in the Anões do Orçamento scandal to the funneling of public and private funds to members of the Worker's Party in an effort to drum up support for legislative initiatives in the Mensalão scandal. Between two prominent impeachments and four wide-reaching scandals, political corruption has proved a regular grievance and has impeded economic development in Brazil's present democratic period.

While these high-profile scandals are notable for both the political reaches to which they soar and the brazenness of some of the financial exchanges involved, Brazil's broader experience with corruption also involves the pervasive permeation of malfeasance in local politics. Brazilian state and local governments are commonly believed to be rife with corruption. In local contexts, corruption is most visible in the areas of procurement and public works (Melo 2014). Local politicians have frequently been found to use fake receipts and to create phantom firms in an attempt to falsely signal legal compliance while diverting funds allocated for public goods provision (Ferraz and Finan 2011). Additionally, mayors and councilmen have been found to use non-competitive procurement processes that privilege politically allied business partners (Ferraz and Finan 2011). In addition to these more discreet strategies, local-level bureaucratic corruption commonly manifests itself through the public's interactions with government officials and, most notably, bribe-soliciting police (Melo 2014). Across AmericasBarometer surveys conducted every other year between 2006 and 2014, the proportion of Brazilian respondents saying that they have had to pay a bribe in their interactions with public officials has ranged between 9 and 21 percent (Senters et al. forthcoming). Perhaps, in part, a function of bribe solicitations, public opinion surveys reveal that the Brazilian public perceives the police, political parties, and the legislature to be widely corrupt (Melo 2014). The average response given by Brazilians when asked to evaluate the commonality of corruption among public officials on a scale from one to four (where one indicates "very common") has been around 2.0 in the 2008, 2010, and 2012 rounds of the AmericasBarometer survey. This evaluation is actually slightly better than the region-wide average of 1.8 during those same years but not at the level of 2.3 obtained by Canada, for example.

Although corruption in Brazil remains widespread by objective measures and prominent in the national discourse, the country has seen a number of efforts to identify and sanction malfeasance in recent years. Such initiatives were rare in the "reign of impunity" in the early years of the present democratic period, but with the progression of time, Brazilian politicians at all levels of government have been convicted of or are currently facing corruption charges (Pereira et al. 2011). In addition to the impeachments of Collor and Rousseff, former president Luiz Inácio Lula da Silva is presently on trial as a defendant in six criminal cases for his alleged involvement in corrupt political practices; in July 2017, he was convicted in one of these cases, although he remains free on appeal as of this writing.[1] In 2008, over one-third of federal deputies faced corruption charges (Melo 2014). With the eruption of the Lava Jato scandal, this percentage has ballooned to almost 60 percent of sitting lawmakers who have been confronted with legal challenges due to alleged malfeasance. The number of state legislators identified as engaging in fraudulent practices in 2008 was, similar to the number of federal deputies, approximately one-third (Melo 2014). It is unclear whether or not the number of implicated state officials has been increasing in more recent years. At the most local level, municipal audits from the mid-2000s revealed evidence of corruption in 80 percent of the municipalities audited by the federal Comptroller General's Office (Ferraz and Finan 2011).

These statistics conceal important inter-regional and inter-state variations in politicians' involvement in corruption. According to Melo (2014), corruption has traditionally been more prevalent in states in the north and central-west regions of the country. For example, in 2008, 75 and 63 percent of federal deputies faced criminal charges in the northern states of Tocantins and Rondônia, respectively, and 73 percent of politicians

---

[1]   Lula, who remained widely popular despite the Mensalão scandal that erupted at the end of his first term in office, had been expected to run again for the presidency in 2018.

from Goiás in the central-west region faced charges of malfeasance (Melo 2014). These proportions exceed the national average.

Variation in subnational corruption patterns might be attributable to disparities in the proper functioning of checks and balances, institutional quality, and the degree of political pluralism across states. Alston et al. (2008) find that a decrease in checks and balances within a state – measured using an index that captures the existence, effectiveness, and independence of seven state institutions[2] – significantly increases the probability of corruption, as observed using a measure of the average variation in wealth of state deputies. Biddle (2013) looks across municipalities and finds that the existence of local radio stations contributes to explaining inter-municipal variation in corruption (as revealed in municipal audits). Perhaps unsurprisingly, Biddle (2013) makes the broad argument that corruption is most prevalent in areas where resources funneled into public accountability are low.

Aside from institutional quality, the degree of political competition and fragmentation within subnational polities influences levels of malfeasance (Biddle 2013; Melo 2014). Melo (2014, p. 34) finds that "a 50 percent margin [of victory of mayoral candidates over their rivals] is predicted to cause a very substantial increase in the number of [state-level] corruption irregularities, from 6.2 to 9.8." Biddle (2013) substantiates this position and observes that fragmentation in local councils increases the likelihood of corruption in municipalities.

Widespread corruption at both national and subnational levels of government has been controlled in recent years with the increased number and proper functioning of accountability-enhancing institutions in Brazil (Michener and Pereira 2016). The prominent and highly visible nature of national corruption scandals has served as an important periodic impetus to develop and empower new institutions designed to discourage politicians from engaging in corruption. For example, the congressional budgetary malfeasance revealed in the Anões do Orçamento scandal led to the adoption of budgeting reforms that tightened procedural and reporting requirements and more evenly distributed political power that was previously concentrated in the congressional reporter (Melo 2014; Michener and Pereira 2016). A scandal involving the dramatic diversion of funds allocated to the construction of the São Paulo Regional Labor Court (TRT) in 1998 resulted in the empowerment of the Comptroller General's Office in an effort to improve internal control mechanisms and

---

[2] These institutions include the judiciary, public prosecutors, state audit offices, and the media.

enhanced coordination in matters involving the oversight of federal funds (Carson and Prado 2014). Finally, the funneling of private funds to members of the Worker's Party (PT) in the Mensalão scandal triggered reforms in campaign finance (Carson and Prado 2014). These initiatives represent a mere sample of the efforts initiated to enhance accountability in Brazil.

In the next section, we provide a historical introduction to these and other prominent accountability-enhancing institutions in Brazil. These institutions are similarly charged with the related tasks of investigating, overseeing, and sanctioning political malfeasance. We then proceed to document the consequences of corruption as uncovered by these institutions.

BRAZIL FIGHTING BACK: INSTITUTIONAL INNOVATIONS IN THE FIGHT AGAINST CORRUPTION

Over the past 30 years, there have been numerous initiatives undertaken to curb the persistent and widespread corruption at all levels of government in Brazil. The country's investment in combatting corruption in the current democratic period commenced with the installation of the 1988 constitution. Provisions in that document guided the federal bureaucracy in Brazil toward universal procedures and merit-based recruitment (Pereira 2016), vested increased power in the Public Prosecutors Office and the audit courts, and guaranteed full press freedom (Carson and Prado 2014; Melo 2014). In the present democratic period, the powers of the judiciary in the country have also been expanded to allow for an increasingly autonomous and professional federal judicial system to investigate elected and appointed officials, and the legislation that regulates the punishment of individuals and companies implicated in corruption has become stricter. Finally, the number and power of monitoring institutions has grown substantially, represented most notably by an ambitious program of regular municipal audits through which cities are randomly selected for inspection.

The institutions primarily responsible for the investigation and sanctioning of corruption are, or have been, the Federal Court of Accounts (*Tribunal de Contas da União*, TCU), the Federal Public Prosecutor (*Ministério Público Federal*, MPF), and the Comptroller General's Office (*Controladoria-Geral da União*, CGU). The TCU is the body of Brazil's federal legislature responsible for preventing, investigating, and sanctioning corruption and malfeasance. The MPF exists outside the three formal branches of governments and consists of independent public prosecutors responsible for conducting criminal investigations. The CGU was created as part of a transformation of the Federal Control Secretary (*Secretaria Federal de Controle*) in 2001 and was the comptroller and anti-corruption agency of the federal government; it played the lead role in auditing small- and medium-sized municipalities across the country (Melo 2014). The creation of the CGU led to the introduction of new initiatives, laws, and rules aimed to make use of the institution's independent powers (Praça and Taylor 2014). As a part of government reorganization under President Temer, the CGU was folded into the new Ministry of Transparency, Inspection, and Control (MTFC) in May 2016. In general, these bodies have exhibited broad mandates, high levels of independence and professionalism, adequate funding, and meritocratic recruitment (Speck 2011; Carson and Prado 2014; Melo 2014). According to data on Brazilian institutions, the CGU, especially, has been viewed as an extremely autonomous and relatively high capacity institution that is free from political influence and staffed by career civil servants drawn from the Planning and Finance Ministries (Bersch et al. 2016, 2017).

Perhaps the most prominent and widely studied anti-corruption program in Brazil is the randomized auditing of Brazilian municipal expenditures under the CGU and MTFC. Beginning in 2002, the new government under Lula empowered the CGU to randomly select Brazilian municipal governments for the auditing of their expenditures. The program initially investigated the accounts of 26 randomly selected municipalities (one in each Brazilian state) (Ferraz and Finan 2011). However, it quickly expanded to investigate the spending of 50 and, eventually, 60 randomly selected Brazilian municipalities per month.[3] Municipalities are randomly selected for auditing through a lottery at the *Caixa Econômica Federal* in Brasília (Ferraz and Finan 2011). Under the program, auditors investigate irregularities in municipal expenditures by assessing accounts and documents, examining public works and services, and interviewing community members. They, then, report malfeasance to the TCU, public prosecutors, and local legislatures

The body of data provided by these municipal audits has been used to study the effects of corruption and corruption revelation.[4] The most prominent among the findings produced from audit studies is that voters punish incumbent mayors seeking reelection who have been revealed to be

---

[3] Only municipalities with populations under 450,000 are included in the lottery (Ferraz and Finan 2011).

[4] The random selection of municipalities for auditing helps scholars create plausible comparison groups that they use to make compelling inferences about the effects of corruption and (threatened) corruption revelation.

corrupt by the audits (Ferraz and Finan 2008; Brollo 2009). Ferraz and Finan (2008) pioneered the use of the audits to study the effects of political corruption revelation. They estimated a drop of seven percentage points in the probability of reelection for incumbent mayors in municipalities where at least two audit violations associated with corruption had been revealed before the election. Brollo (2009) confirmed that corruption revealed in the audits reduced the probability of reelection but clarified that the information needed to be released within several months of the elections in order to have this effect.

Using data from the State Audit Court of Pernambuco, Pereira et al. (2009) and Pereira and Melo (2015) find somewhat different results. Pereira et al. (2009) find that revealed municipal corruption is not as damaging to incumbent politicians as irregularities revealed in electoral campaigns, while Pereira and Melo (2015) find that revealed corruption harms incumbents but less so if they have been engaging in high levels of public spending in the city.[5] These works suggest plausible variation in cross-state experiences with corruption.

Several other studies from Brazil substantiate the findings from the municipal audit studies by providing more direct evidence that voters will electorally sanction politicians who are revealed to have been involved in corruption. De Figueiredo et al. (2010) undertook a field experiment in São Paulo in which they distributed flyers describing corruption convictions associated with each of the two candidates in the 2008 mayoral election. They find that turnout and the vote share for one of the two candidates dropped in areas where the flyers were distributed. Winters and Weitz-Shapiro (2013) provide information about corruption and service delivery in the context of a hypothetical mayoral election to a subset of survey respondents: those respondents who are exposed to information about corruption say that they would vote against the corrupt candidate, and this negative reaction is not substantially moderated by positive information about service delivery. In subsequent work, Winters and Weitz-Shapiro (2016) show that voters are more likely to sanction mayors who are personally involved in corruption as compared to mayors overseeing corrupt municipal administrations.

While the existing evidence suggests that politicians have something to fear from having corrupt behavior revealed, a question remains about whether they fear the revelation of corruption enough to reduce their involvement in it. Ferraz and Finan (2011) argue that fear of the revelation

---

[5]   The audits that these authors study are not implemented at random and, therefore, lack some of the analytical advantages of the CGU audits for making plausible causal inferences.

of corruption influences mayors' behaviors and, specifically, their calculations on the advantages and disadvantages of engaging in malfeasance. Comparing mayors who are eligible for reelection to those who are not, Ferraz and Finan (2011) use CGU audits to show that mayors who are eligible for reelection misappropriate 27 percent fewer resources than mayors who are term-limited. Likewise, Zamboni and Litschig (2013) show that increasing the propensity of being audited reduces the likelihood of corruption and changes the willingness of municipal officials to use certain procurement modalities.

In contrast with Ferraz and Finan (2011), Pereira et al. (2009) find that politicians running for reelection are actually more likely to engage in corrupt behavior. They justify this seemingly counterintuitive finding by claiming that high rent extraction obtained from being in office offsets the cost of engaging in corruption. Similarly, Mondo (2016) fails to find statistically significant evidence that electoral accountability – measured by voters removing corrupt incumbent mayors from office – reduces future misdeeds in corruption-inflicted municipalities.

Perhaps these opposing camps are reconcilable when one accounts for supplementary strategies corrupt politicians can invoke to counteract popular disfavor stemming from malfeasance. Jucá et al. (2016) suggest that corrupt incumbents can effectively offset some electoral penalization following corruption revelation with increased campaign spending. They calculate that in order to avoid penalty for malfeasance, implicated federal deputies must spend approximately BRL 1,500,000 (approximately USD 500,000 in 2017) on their campaigns (Jucá et al. 2016). They suggest that this amount is unfeasible for the average federal deputy engaged in malfeasance, who spent approximately BRL 840,000 (USD 325,000) on campaigning over the 1995–2010 period (Jucá et al. 2016). Pereira and Melo (2015) similarly argue that public expenditure mitigates the negative electoral consequences of political malfeasance. These findings correspond with conventional wisdom in Brazil that voters overlook corruption when politicians "*rouba, mas faz*" ("rob but get things done") (Winters and Weitz-Shapiro 2013; Pereira and Melo 2015).[6]

The most recent studies in Brazil have started to look at how the revelation of corruption has longer-term consequences on municipal politics. Timmons and Garfias (2015) show that in municipalities where the CGU

---

[6]   The actual findings in Winters and Weitz-Shapiro (2013), however, are in the contrary direction. In their survey experiment involving a hypothetical mayoral election, they do not find evidence that voters are willing to condone corruption in the presence of public works provision.

audits revealed corruption, property tax revenue has fallen and participatory budgeting has become more likely.

Finally, the audit data has been helpful for scholars trying to understand the origins of local-level corruption in Brazil. In a study that examines the relationship between per capita transfer award amounts and corruption incidence, Brollo et al. (2013) found that larger transfers per capita correspond not only with increased corruption but also with less capable candidates running for municipal office. Avelino et al. (2013) showed that more well-established municipal health councils lead to lower levels of corruption in municipal health expenditures, providing important evidence that decentralized governance institutions can improve governance outcomes.

Current research is building on the project of Avelino et al. (2013) and works to understand which Brazilian institutions are likely to be most effective in combatting corruption. Bersch et al. (2016) show how their agency-specific measures of capacity (based on the extent of meritocratic recruitment, the longevity of civil servant careers, and average salaries) and autonomy (based on political party membership among civil servants) predict the number of corruption scandals associated with individual Brazilian government institutions. In addition, they provide evidence that institutions associated with a political party are likely to exhibit lower capacity (and, therefore, to be more susceptible to corruption). Carson and Prado (2014) study the 33 subnational audit courts in the country (27 state courts and six municipal courts) and show how their structure varies with political competition and their integration with Public State Ministries (*Ministérios Públicos Estaduais*, MPEs). Melo et al. (2009) find that states with more frequent alternation in political power espouse more activism in these state audit courts than their counterparts with less political competition.

In sum, a host of accountability-enhancing institutions have been created and developed since Brazil's 1985 democratic transition. Scholars have leveraged the adoption of these institutions – especially the system of randomized municipal audits – to study the causal effects of corruption and corruption revelation. Based on this research, scholars have affirmed that corruption impedes the likelihood of incumbent mayoral reelection. Moreover, they have shown that citizens may react to revealed corruption by changing the ways in which they interact with their local governments and, specifically, in their electoral choices. While the federal audits have been undertaken by a highly autonomous and capable agency, research has shown that there is variation in the susceptibility of different Brazilian institutions to corruption and even that there is variation in the level of politicization of some of the anti-corruption institutions.

That Brazilian voters are willing to take action against corruption suggests that the latest corruption scandal may lead to major political turnover in Brazil. In the next section, we explore how the Brazilian public thinks about and reacts to corruption in greater depth and provide some evidence that if major change does not come soon, then it may need to wait for yet another round of grand corruption and public reaction to it.

"*VAMOS PARA RUA*": PUBLIC PERCEPTIONS OF AND REACTIONS TO CORRUPTION

The fact that Brazilian voters publicly express their exasperation with corruption represents one part of a larger paradox when it comes to the persistence of corruption in Brazil: citizens in the country have frequently mobilized against corruption and yet corruption remains pervasive. We make a tentative argument here that this may have to do with the lack of sustained public attention to corruption at moments in which there is no major scandal in the news.

The most visible display of popular discontent with political malfeasance comes – with the call of "*vamos para rua*" ("we're going to the streets") – in the form of mass protests. Allegations of President Fernando Collor de Melo's corruption and self-enrichment in 1992, for example, prompted a series of large-scale protests in August and September of that year in which the Brazilian public demanded his impeachment. Fast-forwarding to 2013, protests that were initially sparked by public transport price increases grew into large demonstrations across the country motivated by perceived failings of the political elite; corruption was foremost among the public's grievances (Saad-Filho 2013; Winters and Weitz-Shapiro 2014). Although protests waned in the following months, they powerfully reemerged in the lead up to Dilma's impeachment. In the spring of 2016, at least three million Brazilians rallied to express their discontent with the Rousseff government, with tens of thousands calling for her impeachment while many thousands argued against impeachment (Watts 2016). As recently as June 30, 2017, nationwide protests deadlocked Brazilian cities over the unpopular labor and pension legislation proposed by the scandal-plagued Temer administration (Fonseca 2017).

Linked to these protests have been other changes in the way that citizens position themselves in the political landscape. Winters and Weitz-Shapiro (2014), for instance, show that Brazilians became less willing to express a partisan identity (and particularly identification with the ruling Workers' Party, or PT) at the time of the 2013 protests. The rate at which respondents declined to express a partisan identification climbed five percentage points from the month before the protests to the month after.[7] Interviews and assessments conducted as a part of Alonso and Mische's (2017) recent work further substantiate these findings; they argue that the 2013 protests signaled a strong rejection of all political parties due either to a craving for national unity or disenchantment with the PT regime.

Despite these very prominent manifestations of anti-corruption sentiment and disassociation with political parties, it is unclear whether the Brazilian public is willing to consistently pressure politicians to reduce corruption. Senters et al. (forthcoming) study the long-term trends in Brazilian attitudes toward corruption using data from about 50 public opinion surveys. From the period 1988–2005, they found that a relatively consistent 5 percent of the population described corruption as the "most important problem" facing Brazil. In the wake of the Mensalão scandal in 2005, this number shot up to 20 percent. However, it quickly dropped back down to about half that level (which is still higher than the 1990s average) over the period 2006–14, only to shoot up again in 2015 as evidence began to emerge from Operação Lava Jato. Looking only at Latinobarometer data, the authors show how the proportion of Brazilians concerned about corruption was four times the Latin American average at the time of Mensalão but then dropped back down to the regional average over the period 2008–13.

Although it is difficult to assert that individuals should remain doggedly committed to pursuing particular public policy reforms, it also does not seem unfair to speculate that this ebb-and-flow in the attention of the Brazilian public to corruption may help explain its persistence over the past 30 years. Despite moments of impressive mobilization, the crowds have dissipated afterward, thereby relieving the pressure on politicians to take additional action. For all of the institutional innovation and reform that Brazil has seen, at the highest levels, it seems not to have been enough.

CONCLUSION

The revelations of Operação Lava Jato beginning in 2014 suggest that corruption in Brazil is as rampant as ever. This is despite the fact that scholars generally concede that the country has undertaken major initiatives to create institutions that can help citizens hold corrupt politicians accountable (Praça and Taylor 2014; Michener and Pereira 2016). Praça

---

[7]   Winters and Weitz-Shapiro (2015), on the other hand, provide some evidence that highly educated respondents may be more likely to describe themselves as partisan in the wake of information about corruption.

and Taylor (2014) interpret the enhanced transparency of public accounts and availability of government data, the increased number of civil servants removed from their posts as a consequence of malfeasance, and the rise in the number of actions taken by accountability institutions as tangible evidence of institutional progress. Moreover, they view a series of laws enacted to preemptively deter corruption – such as the Clean Record Law (*Lei de Ficha Limpa*), which allowed electoral courts to ban politicians from running for office based on criminal convictions, and the 2013 anti-corruption law that extended criminal liability to corporations[8] (Richard 2014) – as important advancements for proper democratic functioning.

In spite of these advancements, there is still much room for improvement in accountability and for more cohesive integration between institutions involved in the broad network of accountability institutions in Brazil. In the words of Taylor and Buranelli (2007, p. 79), "the weakness of the 'web of accountability' in Brazil is not due solely, or even primarily, to the weakness of individual institutions in that web . . . [but rather] the larger problem is the imperfect 'orthodontia' of the institutions that make of the web of accountability in Brazil." Carson and Prado (2014) also concede that the way forward in the fight for accountability depends integrally on improving coordination between different types of accountability institutions; they specifically emphasize strengthening institutions charged with sanctioning.

In addition to improvements in oversight and sanctioning, scholars, anti-corruption agencies, and others have identified a number of places where the anti-corruption regulatory framework in Brazil remains stunted. There remain possibilities for fighting illegal campaign contributions to politicians who are expected to compensate donors with political contracts if elected; for speeding up criminal proceedings, protecting the identities of whistleblowers, and extending the statute of limitations on corruption-related crimes; for improving asset retraction capabilities; and for adopting more stringent requirements for election to the legislature (Lagunes and Rose-Ackerman 2017). In this chapter, we put forth the preliminary argument that public mobilization and involvement can help to bring about desired changes in these areas.

The third section in this chapter highlighted the intense anti-corruption mobilization of the Brazilian public in the wake of high-profile corruption scandals and situated these periods of heightened activism among longer periods of tranquility. The observed ebb-and-flow in anti-corruption

---

[8]   This law earned comparisons to the US Foreign Corrupt Practices Act and the UK Bribery Act (Melo 2014). However, implementation of this law lagged. As of 2016, no company had yet been punished under the law (Vasconcellos de Figueiredo 2016).

mobilization suggests that the public loses interest in corruption after the initial outbreak and, consequently, that the extent to which citizens hold politicians accountable fluctuates in accordance with the eruption and waning of scandals. On the one hand, the Brazilian experience provides concrete evidence of citizens mobilizing around corruption by coordinating to remove corrupt public officials from office or by taking to the streets to demand legal changes. However, in between major scandals there is evidence that the public's interest wanes, and it seems possible that this faltering of attention to corruption allows corruption-inclined politicians to embark on the road to robbing the public coffers once more.

Corruption is a remarkable constant in and long-term staple of Brazilian politics. Through the adoption and advancement of accountability-enhancing institutions, both informed citizens and scholars have learned about the scope of malfeasance in the country and about the broader effects of its revelation, although our understanding of subnational variation in these experiences remains limited. We hope that continued research on corruption in Brazil will lead to useful refinements to the institutions that seek to combat corruption in the country.

REFERENCES

Alonso, A. and A. Mische. 2017. Changing repertoires and partisan ambivalence in the new Brazilian protests. *Bulletin of Latin American Research* **36**(2), 144–59.

Alston, L., M. Melo, B. Mueller and C. Pereira. 2008. The choices governors make: the roles of checks and balances and political competition. In *Anais do XXXVI Encontro Nacional de Economia* (Proceedings of the 36th Brazilian Economics Meeting). https://www.colorado.edu/ibs/es/alston/papers/The_Choices_Governors_Make.pdf (accessed May 18, 2018).

Avelino, G., L. Barberia and C. Biderman. 2013. Governance in managing public health resources in Brazilian municipalities. *Health Policy and Planning* **29**(6), 694–702.

Bersch, K., S. Praça and M. Taylor. 2016. Bureaucratic capacity and political autonomy within national states: mapping the archipelago of excellence in Brazil. In A. Kohli and D. Yashar (eds), *States in the Developing World*, pp. 157–83. New York: Cambridge University Press.

Bersch, K., S. Praça and M. Taylor. 2017. State capacity, bureaucratic politicization, and corruption in the Brazilian state. *Governance* **30**(1), 105–24.

Biddle, L. 2013. Corruption in democratic Brazil. PhD dissertation. University of North Carolina at Chapel Hill, NC.

Brollo, F. 2009. Who is punishing corrupt politicians – voters or the central government: evidence from the Brazilian anti-corruption program. Boston University – Department of Economics – The Institute for Economic Development Working Papers Series. http://www.bu.edu/econ/files/2012/11/dp168.pdf (accessed July 16, 2017).

Brollo, F., T. Nannicini, R. Perotti and G. Tabellini. 2013. The political resource curse. *American Economic Review* **103**(5), 1759–96.

Carson, L. and M. Prado. 2014. Mapping corruption & its institutional determinants in Brazil. IRIBIA Working Paper No. 8.

De Figueiredo, M., F. Hidalgo and Y. Kasahara. 2010. When do voters punish corrupt politicians? Experimental evidence from Brazil. Yale University. http://cega.berkeley.edu/

assets/cega_events/44/CEGA_ResearchRetreat2012_deFigueiredo_Paper.pdf (accessed May 18, 2018).

Ferraz, C. and F. Finan. 2008. Exposing corrupt politicians: the effects of Brazil's publicly released audits on electoral outcomes. *Quarterly Journal of Economics* **123**(2), 703–45.

Ferraz, C. and F. Finan. 2011. Electoral accountability and corruption: evidence from the audits of local governments. *American Economic Review* **101**(4), 1274–311.

Fonesca, P. 2017. Brazil unions protest Temer's reforms amid political crisis. https://www.reuters.com/article/us-brazil-politics-protests- idUSKBN19L270 (accessed May 18, 2018).

Jucá, I., M. Melo and L. Rennó. 2016. The political cost of corruption: scandals, campaign finance, and reelection in the Brazilian Chamber of Deputies. *Journal of Politics in Latin America* **8**(2), 3–36.

Lagunes, P. and S. Rose-Ackerman. 2017. Brazil's struggle against corruption. https://www.csmonitor.com/World/Americas/Latin-America-Monitor/2017/0214/Brazil-s-struggle-against-corruption (accessed May 18, 2018).

Melo, M. 2014. Brazil: democracy and corruption. Democracy Works Conference Paper Series. Centre for Development and Enterprise and Legatum Institute.

Melo, M., C. Pereira and C. Figueiredo. 2009. Political and institutional checks on corruption: explaining the performance of Brazilian audit institutions. *Comparative Political Studies* **42**(9), 1217–44.

Michener, G. and C. Pereira. 2016. A great leap forward for democracy and the rule of law? Brazil's Mensalão trial. *Journal of Latin American Studies* **48**(3), 477–507.

Mondo, B. 2016. Measuring political corruption from audit results: a new panel of Brazilian municipalities. *European Journal on Criminal Policy and Research* **22**(3), 477–98.

Pereira, A. 2016. Is the Brazilian state "patrimonial"? *Latin American Perspectives* **43**(2), 135–52.

Pereira, C. and M. Melo. 2015. Reelecting corrupt incumbents in exchange for public goods: *Rouba mas faz* in Brazil. *Latin American Research Review* **50**(4), 88–115.

Pereira, C., M. Melo and C. Figueiredo. 2009. The corruption-enhancing role of re-election incentives? Counterintuitive evidence from Brazil's audit reports. *Political Research Quarterly* **62**(4), 731–44.

Pereira, C., L. Rennó, and D. Samuels (eds). 2011. Corruption, campaign finance, and re-election. In T. Power and M. Taylor (eds), *Corruption and Democracy in Brazil: The Struggle for Accountability*, pp. 80–102. South Bend, IN: Notre Dame University Press.

Power, T. and M. Taylor. 2011. *Corruption and Democracy in Brazil: The Struggle for Accountability*. South Bend, IN: University of Notre Dame Press.

Praça, S. and M. Taylor. 2014. Inching toward accountability: the evolution of Brazil's anticorruption institutions, 1985–2010. *Latin American Politics and Society* **56**(2), 27–48.

Richard, M. 2014. Brazil's landmark anti-corruption law. *Law and Business Review of the Americas* **20**, 141–7.

Saad-Filho, A. 2013. Mass protests under "left neoliberalism": Brazil, June–July 2013. *Critical Sociology* **39**(5), 657–69.

Senters, K., R. Weitz-Shapiro and M. Winters. Forthcoming. Continuity and change in public attitudes toward corruption. In B. Ames (ed.), *Handbook of Brazilian Politics*. London: Routledge.

Speck, B. 2011. Auditing institutions. In T. Power and M. Taylor (eds), *Corruption and Democracy in Brazil: The Struggle for Accountability*, pp. 127–61. South Bend, IN: Notre Dame University Press.

Taylor, M. and V. Buranelli. 2007. Ending up in pizza: accountability as a problem of institutional arrangement in Brazil. *Latin American Politics and Society* **49**(1), 59–87.

Timmons, J. and F. Garfias. 2015. Revealed corruption, taxation, and fiscal accountability: evidence from Brazil. *World Development* **70**, 13–27.

Vasconcellos de Figueiredo, F. 2016. Building up a convenient accountability: how the "anti-corruption" law in Brazil was put into force. *Brasiliana: Journal for Brazilian Studies*, **4**(2), 550–78.

Watts, J. 2016. Dilma Rousseff impeachment: what you need to know – the Guardian briefing. https://www.theguardian.com/news/2016/aug/31/dilma-rousseff-impeachment-brazil-what-you-need-to-know (accessed May 18, 2018).

Winters, M. and R. Weitz-Shapiro. 2013. Lacking information or condoning corruption: when do voters support corrupt politicians? *Comparative Politics* **45**(4), 418–36.

Winters, M. and R. Weitz-Shapiro. 2014. Partisan protesters and nonpartisan protests in Brazil. *Journal of Politics in Latin America* **6**(1), 137–50.

Winters, M. and R. Weitz-Shapiro. 2015. Political corruption and partisan engagement: evidence from Brazil. *Journal of Politics in Latin America* **7**(1), 45–81.

Winters, M. and R. Weitz-Shapiro. 2016. Who's in charge here? Direct and indirect accusations and voter punishment of corruption. *Political Research Quarterly* **69**(2), 207–19.

Zamboni, Y. and S. Litschig. 2013. Audit risk and rent extraction: evidence from a randomized evaluation in Brazil. Unpublished manuscript. Universitat Pompeu Fabra, Barcelona.

10. Corruption in East Central Europe: has EU membership helped?

Agnes Batory

Corruption is rarely absent from the front pages of newspapers in East Central Europe (ECE). In 2012, Ivo Sanader, a former prime minister of Croatia, was sentenced to ten years in prison for taking bribes. In 2017, Adrian Nastase, a Romanian former premier, was given four years for the same, having been released after a previous sentence only a few months earlier. Andrej Babis, whose "Ano" party won the fall 2017 election in the Czech Republic, had been indicted on charges of misusing European Union (EU) subsidies just before the election. The list could be continued almost endlessly, although with somewhat less high-profile cases. The good news is that, as at least the first two politicians' examples show, those abusing their positions can no longer be sure of being let off the hook. But in many ways the story remains depressingly familiar: after nearly three decades since regime change in 1989/1990, corruption, especially high-level political corruption, remains deeply ingrained within the political life of the EU's newest member states. For ordinary citizens in the East Central European region, perhaps equally or more importantly, petty bribery to secure acceptable medical care, for instance, has also remained a fact of life. Practically everyone believes that corruption is widespread in their country and about one-third of East Europeans claim that they are personally affected by corruption in daily life (European Commission 2013).

The post-communist member states of the EU – the Baltic republics, Poland, the Czech Republic, Slovakia, Hungary, Romania, Bulgaria, Slovenia, and the latest member of the club, Croatia – offer an interesting insight into the dynamics of corruption and corruption control in new democracies. They share historical legacies and similar levels of economic development, as well as the recent experience of joining the world's most developed regional integration, with supranational institutions able to monitor and to some extent enforce anti-corruption standards. Yet, these countries also display variation in terms of corruption control efforts and changing patterns and forms of corruption. This allows investigating the impacts of EU accession over a decade of membership (in most cases), and probing into the causes of the resilience of corruption as a seemingly permanent feature of political life. Has EU membership helped, and if so in what way? Have domestic anti-corruption efforts made a difference, and to what extent? This chapter investigates these main questions with respect to the ECE countries.

The next section maps the region's historical legacies and what existing scholarship has to say about a specifically post-communist blend of corruption. It then moves to a more recent event, namely, the East Central European countries' accession to the EU in 2004 (2007 for Romania and Bulgaria) and discusses the EU's expected impact on corruption control efforts, in terms of normative commitments and hard law. The following section traces trends in corruption in the region while acknowledging the limitations of available data (and the well-known general difficulties involved in measuring corruption). This section also explores whether new forms of corruption might be in evidence in the post-accession era. The third section deals with anti-corruption measures, with a special focus on anti-corruption institutions. The final section provides a brief conclusion.

HISTORICAL LEGACIES AND EU ACCESSION

Writing in 2002, legal scholar Andras Sajo (2002, p. 2) commented that "[t]he public's belief in, and allegiance to, the rule of law is fragile, for the whole concept was parachuted into Eastern Europe. It is alien to most of the local political cultures, acquainted as they are with only the primacy of surviving by mutual social favours." This view is characteristic of a rich body of literature discussing the "post-communist condition" and the legacies of four decades of state socialism, essentially arguing that while sharing characteristics with any case of radical change, Central and Eastern Europe exhibits a distinctive environment for corruption. Legacies matter since post-communist countries "are regimes in flux in which traits of the old systems coexist with those that are already truly democratic and others that are transitional" (Karlkins 2005, p. 16).

Common observations in these studies point first to clientelistic structures related to the former regimes' *nomenclatura* which did not fade away with regime change but rather, as often argued, preserved the influence of largely hidden social networks which in turn acted to obstruct reform and/ or undermine the formal institutions of the new democracies. A second legacy commonly referred to is excessive bureaucracy, with Kafka-esque inefficiency, indifference to citizens' rights or needs, and the day-to-day coping strategies this has forced citizens into. Under communism, "[c]itizens could expect neither serious consideration nor fair treatment without some means of 'interesting' the official in their case," which has led, in some countries, to a culture of corruption, not just relative willingness to submit to extortion by officials, but to some extent a preference for a system where it is possible to bend the rules, in sharp contrast with the ideal-type objective Weberian bureaucracy (Miller et al. 2001, p. 17).

Communism is also argued to have given rise to attitudes carried over to the era after formal democratization, which hold that because the old regime and institutions were not seen to be legitimate, it was acceptable not to observe the laws and to subvert state authority (Karlkins 2005). One aspect of this system was citizens' disregard for public property. In the people's republics what belonged to the state was supposed to belong to "the people" (everyone) which was widely interpreted as belonging to no one (or, in practice, to the ruling party elite) – leading to the widespread sentiment that "he who doesn't steal from the factory steals from his family." In the same vein, it was common, and if not liked certainly tolerated, for officials to supplement their meager income from selling or exchanging their services to or with others in similar positions. This blurring of the division between public and private is clearly very problematic for corruption defined as the abuse of public position for private gain.

In addition to these post-communist specificities, ECE countries were also subject to the influences the literature identifies with respect to conditions of rapid, radical change, and particularly democratization, in general. These include supply-side factors, that is, increased opportunities for corruption, brought by large-scale privatization campaigns and the inflow of foreign aid and foreign direct investment – the former in particular was widely seen as a hotbed of corrupt practices (Moran 2001). Open borders also had the unintended consequence of allowing international criminal networks to form or flourish, and for the proceeds of corruption to be ferreted away in foreign bank accounts. In conjunction with this, democratization is often accompanied by a (temporary) weakening of state capacities, including coercive powers (law enforcement) and judicial powers: "control from above" struggles to keep up with the changes while "control from below," from civil society actors, is not yet able assert its influence (e.g., Back and Hadenius 2008). Moreover, as Huntington (1968) pointed out nearly half a century ago, the period directly after democratization – in his terms, "modernisation" – is by default about building up the machinery of political competition, in turn implying a need for funds for party organizations and election campaigning, which tend to come from converting public funds to party assets. Finally, democratization almost immediately increases the likelihood that corruption is exposed, while the acceptance of corruption as a "given" decreases, and corruption scandals become a salient feature of political life. All of these factors combined mean that during transition "corruption explodes" (Mungiu-Pippidi 2006, p. 89).

For East and Central Europe, this is an accurate characterization of the transition years in the 1990s. Around the turn of the twenty-first century, however, a gradual shift took place. On the one hand, the structural factors that almost inevitably create ideal conditions for corruption, notably large-scale privatization and the creation and adaptation of political parties to electoral competition, slowly phased out. On the other hand, the race for EU membership gained momentum, which both supported reformers within the countries and gave enormous leverage to the EU, and the European Commission in particular, to push structural changes in the candidate countries. As is well known, the Copenhagen criteria of 1993 set out the conditions of membership, requiring that in order to be admitted, countries display

> stability of institutions guaranteeing democracy, the rule of law, human rights and respect for and protection of minorities; [have] a functioning market economy and the ability to cope with competitive pressure and market forces within the EU; [and have the] ability to take on the obligations of membership, including the capacity to effectively implement the rules, standards and policies that make up the body of EU law (the *acquis*), and adherence to the aims of political, economic and monetary union. (European Council 1993)

Although the criteria do not explicitly mention corruption, during the accession negotiations the Commission clearly treated it as being covered by conditionality and monitored candidate countries' progress (or lack thereof) in bringing it under control. In the case of Bulgaria and Romania, conditionality in this respect survived EU accession in 2007 in the form of the Cooperation and Verification Mechanism. These is little doubt that this kind of membership conditionality, coupled with rigorous monitoring, constituted a powerful driver of reform in the candidate countries (Vachudova 2009). It can also be credited with creating strong momentum for the establishment of anti-corruption institutions and mechanisms – albeit perhaps more with the intention of pleasing the Commission than actually combating corruption (Batory 2012).

When most of the East and Central European countries finally joined the EU in 2004, there was much speculation about whether they would keep up the good work – not just in terms of corruption control efforts but in terms of meeting the obligations of membership generally, for instance, in the form of transposing EU legislation. With respect to the latter, while there is no evidence that the new member states were more lax than the longer-standing EU members (the EU 15) (Sedelmeier 2008; Knill and Tosun 2009), there is considerable evidence that the post-communist member states often adopted legal rules from the EU without commitment to their actual implementation on the ground, turning them into "dead letters" or "empty shells" (Falkner and Treib 2008; Dimitrova 2010). Much of the cross-country variation among the ECE countries seemed to be a function of whether domestic actors ("veto players") managed to lock in pre-accession institutional change, thereby preserving some of the force of membership conditionality (Sedelmeier 2012). With respect to anti-corruption efforts too, whether the pre-accession momentum was kept up seemed to depend largely on the presence of domestic champions inside or outside government (Batory 2010; Moroff and Schmidt-Pfister 2010). Thus, while, for instance, in Slovakia after "EU entry a lack of urgency [was] . . . evident and most ministries have reduced their [anti-corruption] activity," in Latvia work continued due to a combination of domestic factors (Pridham 2008, p. 378).

Putting aside conditionality and its after-life, it is worth briefly summarizing the actual obligations arising from EU membership, since this should be a stable and long-lasting source of influence on corruption control efforts in the ECE countries. First, EU membership certainly creates a normative commitment in this respect – expressed not only in terms of the Copenhagen criteria that the new member states accepted, but also in terms of the foundational values of the Union, embodied in Article 2 of the Treaty on European Union, which refers to the rule of law, respect for human dignity, freedom, democracy, equality, and human rights. It would be difficult to argue that the rule of law can be upheld in highly corrupt environments.

Second, there is a body of EU law (secondary law) that creates specific legal obligations for member states which are enforceable by the European Commission and, ultimately, the European Court of Justice. These include, most notably, conventions on the protection of the EU's financial interests, directives on public procurement, money laundering or the confiscation of criminal assets (EPRS-Rand Research 2016). There are also EU institutions, such as the EU Anti-Fraud Office (OLAF), with competences in corruption control. These existing legal instruments and competences have so far failed to coalesce into a comprehensive anti-corruption policy on the EU level (despite the Commission's efforts to forge such a policy under the 2010 Stockholm Programme). Nonetheless, they are strong tools in the limited areas where they apply. By anchoring new democratic institutions, EU membership more broadly can clearly be expected to have a noticeable effect on corruption in ECE countries. After more than a decade in the club, this should be manifested in lower levels of corruption. Whether this is indeed the case is the subject of the next section.

TRENDS IN LEVELS AND FORMS OF CORRUPTION

Although so far East and Central European countries have been treated as one rather homogeneous bloc, it is important to clarify that there is in fact a great deal of variation among formerly communist EU members. Some, notably Estonia, are relatively clean, even in wider comparison to the EU, while Bulgaria and Romania feature among the most corrupt EU countries. Is should also be noted that Italy and Greece, two long-standing EU member states, are in fact ranked lower than the bulk of the "new" member states – at least according to Transparency International's 2016 Corruption Perception Index (CPI). The CPI, alongside other commonly used quantitative indicators such as the World Bank Governance Indicators/control of corruption (WGI) index, suffers from major methodological limitations that cannot be discussed here. Suffice to say that, as recent scholarly analysis concludes, "the study of changes in levels of corruption is still in its infancy" (Escresa and Picci 2016).

Acknowledging these methodological concerns, on the CPI's 1 (most) to 100 (least) corrupt scale, in 2016 the cleanest European country (Norway) scored 90, and the cleanest post-communist country (Estonia) 70; Bulgaria was at the bottom, with a score of 41, admittedly barely behind Italy and Greece. Romania, Hungary, Croatia, and Slovakia also scored below 50; the Czech Republic, Latvia, and Lithuania below 60; and Slovenia and Poland just above 60 thereby ranking as relatively clean in ECE regional perspective. This divergence was by-and-large confirmed by mass surveys of actual experiences with corruption. Transparency International's Global Corruption Barometer showed that, in 2016, only 3 percent, 5 percent, and 7 percent of Slovenian, Estonian, and Polish households, respectively, paid a bribe when using public services, whereas the corresponding figure was 29 percent, 22 percent, and 17 percent for Romania, Hungary, and Bulgaria (Transparency International 2016a).

Considering the control of corruption, rather than corruption per se, a similar picture emerges. The WGI data show Estonia's, again the high performer, percentile rank at 85, Slovenia's at 77, and Poland's at 76, whereas the laggards, Bulgaria, Romania, and Hungary scored 51, 58, and 61, respectively (World Bank various dates). A more complex composite indicator of factors normally expected to act against corruption (in this case, performance on administrative burden, trade barriers, transparency/e-government, auditing standards, judicial independence, and civic engagement), labeled the Public Integrity Index again shows Slovenia as the leader of the pack and Romania and Bulgaria as the weakest performers, with a somewhat different ordering in between these extremes (Mungiu-Pippidi 2013). In sum, a wide range of indicators confirms relatively large differences in levels of corruption among the ECE countries. Moreover, a quick glance at the map strongly suggests – especially when the EU as a whole is considered – that in addition to an East/West divide between post-communist new democracies and more established democracies, a north/south dimension is also salient (Pujas and Rhodes 1999). The EU's corruption problem is concentrated in its southeastern corner.

When considering trends discernable from the most widely used quantitative indicators, the most salient observation is – in the words of a recent study by Rand Europe for the European Parliament – that "[there is no] strong empirical evidence of movement over time in the corruption rankings across Member States. Member States that had high levels of corruption in 1995 continue to do so in 2014 [according to CPI and WGI data]" (EPRS-Rand Europe 2016). Thus, with very few exceptions, corruption is not evidently less prevalent in the ECE countries post-accession than it was prior to EU membership. To some extent, this pattern seems to apply even to the top performer, Estonia. In the 1990s and early in this century, the country made impressive progress in reducing corruption, far outpacing its regional competitors. The explanation for the Estonian success story involves a mixture of cultural factors (traditional cultural links and affinity with the Nordic countries), regime change specificities (a radical rapture with the past resulting in the displacement of the old *nomenclatura*), and economic factors, including radical neoliberal market reforms (Karlnins 2014, p. 14). Since then, however, corruption control gains have plateaued, with levels of corruption stable in the last decade (Karlnins 2014).

Hungary is an outlier in another sense: this is the country where rather than lack of visible progress, proxies show clear deterioration in recent years. The country's CPI score went from 55 in 2012 to 48 in 2016, and similar tendencies can be observed in many other measures. However, in the case of Hungary, the worsening corruption situation is clearly part of a broader tendency of backsliding in democratic quality since 2010 – observable in freedom of the media, electoral process, and civil society ratings (e.g., *Freedom House Nations in Transit*; Hegedus 2017). The Hungarian case is also instructive with respect to forms of corruption. The country shows relatively high levels of street-level corruption (22 percent bribery incidence according to the 2016 Global Corruption Barometer), where the health care sector is the most affected. However, the country also displays an arguably more damaging, and certainly more intractable, pattern of high-level corruption, characterized by several sources as state capture – or even as a "mafia state" (Magyar 2016).

State capture is a particular form of extractive regime (Gryzmalla-Busse 2008), one where institutions are shaped by a small business and/or political elite with the purpose of facilitating the syphoning off of public resources. In a captured state, rent extraction is institutionalized through legislation, with the captors shaping the rules of the game to their own economic and/or political advantage (Hellman and Kauffman 2001). State capture may involve powerful economic actors (oligarchs) penetrating the state, or high-level policy-makers cultivating corruption networks and loyal oligarchs for diverting public resources from their intended use. In either case, the detection and prosecution of high-level corruption becomes difficult if not impossible: sometimes even the most damaging examples of rent extraction are "legal," and the public bodies that would normally act against corruption are often captured themselves. A key question is whether the judiciary has remained independent or not.

Studying state capture empirically is fraught with difficulty. The World Bank's Business Environment and Enterprise Performance Surveys (BEEPS) are used for this purpose, for instance, by Knack (2007). The BEEPS tap into firms' experiences with public officials (administrative corruption) as well as bias in public procurement (government contracts having to be secured by providing gifts) as something going beyond petty administrative corruption. This latter measure is, however, relatively limited as a proxy for state capture, being associated only with firms' efforts to secure government favors, rather than with the organized effort for the massive and systematic expropriation of public resources the term implies. Having said that, BEEPS' responses do indicate that corruption in government contracting increased in a number of new EU countries following accession.

A more promising approach is followed by Fazekas and Toth (2016b, p. 320), who utilize "a distinct network structure [associated with state capture] in which corrupt actors cluster around parts of the state allowing them to act collectively in pursuance of their private goals to the detriment of the public good." Analysing micro-level contractual networks in public procurement in Hungary, they conclude that "state capture is established daily practice in approximately 60 percent of public sector organizations conducting public procurement" in that country (Fazekas and Toth 2016b, p. 332). To take another example, in Zagreb, Croatia, there was no competition in over a quarter of public procedures between 2011 and 2016, representing a very large volume of public money at least potentially used to purchase goods and services from friendly contractors (Toth et al. 2017). Another study found considerable evidence of systematically distorted, manipulated public procurement processes in the Czech Republic, Hungary, and Slovakia, which suggests that "institutionalized grand corruption" is a feature of several political systems in the region (Fazekas et al. 2014).

Comparative analysis of procurement data in Bulgaria, Croatia, Hungary, and Romania also suggests that public contracting is the main tool for the allocation of public resources to private actors who then channel funds back to the decision-makers, often through long chains of intermediaries. This pattern is seen in major changes following elections in the range of companies that tend to profit from public contracts: politically connected companies – and the oligarchs who own them – amass spectacular fortunes while their competitors fall out of favor (Mungiu-Pippidi 2015). One high-profile example for this can be cited from Hungary, where the mayor of Prime Minister Viktor Orban's hometown (a plumber by training) has become one of the richest men in the country – regularly featuring in *Forbes* magazine's top 30 richest list – since his friend has taken office, while the business empire of the former treasurer of the main governing party, who had fallen out with the prime minister, suffered major losses in revenue. Incidentally, by 2015 the Orban family was estimated to have amassed assets worth over 20 million euros (Puhl 2017).

The impact of EU membership on this kind of high-level political corruption is mixed in the post-communist member states. On the one hand, there is considerable evidence that EU structural and cohesion funds fuel high-level corruption. Cohesion funds represent very large financial resources, provided by the Union to the poorest regions in the member states, of which the ECE countries are major recipients, with allocations representing 2–4 percent of gross domestic product (GDP) annually – a windfall for governments in the region. As with development aid, the influx of resources is associated with high corruption risks in the ECE countries (Fazekas et al. 2014). In fact, corruption seems to be even more prevalent when EU funds are spent in nationally funded projects, as seen, for instance, in higher contracted prices than in other procurement processes (Fazekas and Toth 2016a). Indeed, overpricing is widespread in EU-funded projects in Hungary, for instance, where often unsuitable or unnecessary development goals are realized and where all participants accept the practice of "rents" built into contracts (Kallay 2015). In Slovakia, for a while "the Structural Funds gradually became a symbol of corruption in the distribution of government subsidies" (Beblavy and Sičáková-Beblavá 2014, p. 550). This is despite the fact that the EU institutions – notably the Commission and OLAF – can, and in high-profile cases do, use their competencies provided under EU law (as discussed above), for instance, legislation on public procurement or state aid (Beblavy and Sičáková-Beblavá 2014).

On the other hand, political pressure from the EU can still help to spur, or support, domestic actors in the fight against high-level political corruption – which is the subject of the next section.

ANTI-CORRUPTION EFFORTS

The case in point – a member state where EU backing seems to have had a major if indirect impact on high-level corruption – is Romania. More precisely, the EU's consistent support for the country's high-level National Anti-Corruption Directorate (*Directia Nationala Anticoruptie*; DNA), a specialized prosecution service, was crucial in the latter managing to maintain its independence, often in the face of strong opposition from the government of the day. The DNA has proven particularly zealous in investigating and prosecuting high-level politicians and oligarchs on corruption charges, with a high rate of convictions resulting in mayors, ministers, and even a former prime minister finding themselves behind bars. Opinion is divided on the effectiveness of this approach. As one Romanian expert comments, "[it] is the country where generalized corruption and the toughest anticorruption in Europe have been co-existing for the past ten years. The result is not less corruption, but crowded jails" (Mungiu-Pippidi 2017, p. 5). However, even if one has a more positive assessment of DNA's performance, what we see in Romania is more likely the residual effect of pre-membership conditionality aided by continued strict monitoring under the Cooperation and Verification Mechanism (which was specific to Romania and Bulgaria) than the impact of membership per se.

Looking at East and Central European countries more broadly with respect to corruption control, findings are mixed. Kartal's study (2014, p. 945), relying on *Freedom House Nations in Transit* data, concludes that "the average level of corruption control increases prior to [EU] membership but it decreases noticeably afterwards." Relying largely on qualitative methods, Transparency International's comparative National Integrity System study similarly points out that "Apart from Bulgaria and Romania, which continue to have serious integrity deficits, the majority of the newly acceded countries can be classified as exhibiting mixed progress in the fight against corruption," but also that "'The evidence suggests that since accession to the EU in 2004, there has been a rolling back on progress made in the fight against corruption in the Czech Republic, Hungary and Slovakia" (Transparency International 2012, pp. 16–17). In other words, several East European governments appear to have relaxed their efforts to roll back corruption once conditionality was no longer effective.

This conclusion is by and large supported by the European Commission's Anti-corruption Reports, published in 2014. As fitting the rather political nature of the exercise, the reports provided a careful mix of praise and criticism, but in most cases conveyed a message that a lot remains to be done. There is also a recurring theme suggesting that anti-corruption agencies or commissions are often subject to, at least attempts of, political interference. To sample the reports, the Commission pointed out that "corruption remains widespread" and anti-corruption institutions should be shielded from political influence in Bulgaria; Latvia should build on the achievements of its anti-corruption bureau by "strengthening its independence"; Poland is recommended to "strengthen safeguards against the potential politicization" of its agency; and Slovenia "should safeguard the operational independence and resources of anti-corruption bodies" (European Commission 2014). Romania was recommended to ensure that "all necessary guarantees remain in place for the independence and continuation of non-partisan investigations into high profile cases," a not too veiled request to leave the National Anti-Corruption Directorate well alone.

CONCLUSION

Why have gains from EU membership failed to materialize for (continued) progress in curbing corruption? As already indicated, part of the answer lies in the new member states having relaxed their efforts to control corruption once EU membership conditionality was now longer applicable. Although EU law provides the EU with a limited set of enforceable standards, it simply does not have comparably strong and comprehensive mechanisms for putting countries under pressure once admitted. This is not to suggest that nothing has changed; indeed, as discussed, some of the East European countries made good progress in specific areas of corruption control, although mainly prior to, rather than following, EU accession. For the citizens of the "cleaner" countries of the region, petty bribery in basic public services is, if not quite a thing of the past, far less prevalent than a couple of decades ago. Fewer than one out of ten Estonian, Polish or Czech households report having had to make an informal payment when interacting with public officials, and this is clearly an important achievement (Transparency International 2016a).

However, contrary to expectations that EU membership would set these countries on a sure path to good (or at least better) governance, rather than overall positive tendencies a more nuanced picture emerges, one involving success stories but also setbacks, and even reversals and backsliding. One observation here is continuing, widespread administrative corruption in the less fortunate countries of the region – including, as of yet, Romania, which has shown one of the world's most active, if not aggressive, approaches to the prosecution of corruption crimes among high-level political figures. The other, perhaps even more pessimistic, conclusion is that what we witness in the post-communist countries of the

EU is not so much a reduction in corruption than changing forms and methods for the illicit diversion of public resources to private hands. As Transparency International (2016b) put it, "the new face of corruption in Europe ... [is] not the lawless, 'anything goes' environment of the immediate post-Soviet period, but the deliberate shaping of the laws and institutions to favour a ruling party and its cronies – all under the guise of a nationalist, 'illiberal' agenda."

REFERENCES

Websites accessed 29 November 2017.

Back, H. and A. Hadenius. 2008. Democracy and state capacity: exploring a J-shaped relationship. *Governance* **21**, 1–24.

Batory, A. 2010. Post-accession malaise? EU conditionality, domestic politics and anti-corruption policy in Hungary. *Global Crime* **11**(2), 164–77.

Batory, A. 2012. Political cycles and organisational life cycles: delegation to anti-corruption agencies in Central Europe. *Governance* **25**(4), 639–60.

Beblavy, M. and E. Sičáková-Beblavá. 2014. The changing faces of Europeanisation: how did the European Union influence corruption in Slovakia before and after accession? *Europe-Asia Studies* **66**(4), 536–56.

Dimitrova, A. 2010. The new member states of the EU in the aftermath of enlargement: do new European rules remain empty shells? *Journal of European Public Policy* **17**(1), 137–48.

Escresa, L. and L. Picci. 2016. Trends in corruption around the world. *European Journal on Criminal Policy and Research* **22**(3), 543–64.

European Commission. 2013. Special module of the Eurobarometer survey on corruption. Eurobarometer 367. http://ec.europa.eu/public_opinion/archives/ebs/ebs_397_en.pdf

European Commission. 2014. EU anti-corruption reports (various country reports). https://ec.europa.eu/home-affairs/what-we-do/policies/organized-crime-and-human-trafficking/corruption/anti-corruption-report_en

EPRS (European Parliamentary Research Service)-Rand Europe. 2016. The cost of non-Europe in the area of organised crime and corruption. http://www.europarl.europa.eu/thinktank/en/document.html?reference=EPRS_STU(2016)579319

European Council. 1993. Conclusions of the Presidency –- Copenhagen, June 21–22. https://www.consilium.europa.eu/media/21225/72921.pdf

Falkner, G. and O. Treib. 2008. Three worlds of compliance or four? The EU-15 compared to new member states. *Journal of Common Market Studies* **46**(2), 293–313.

Fazekas, M. and I. Tóth. 2016a. Corruption in EU funds? Europe-wide evidence on the corruption effect of EU funded public contracting. In J. Bachtler, P. Berkowitz, S. Hardy, and T. Muravska (eds), *EU Cohesion Policy*, pp. 186–202. London: Routledge.

Fazekas, M. and I. Tóth. 2016b. From corruption to state capture: a new analytical framework with empirical applications from Hungary. *Political Research Quarterly* **69**(2), 320–34.

Fazekas, M., J. Chvalkovska, J. Skuhrovec, I. Tóth, and L. King. 2014. Are EU funds a corruption risk? The impact of EU funds on grand corruption in Central and Eastern Europe. In A. Mungiu-Pippidi (ed.), *The Anticorruption Frontline. The Anticorrp Project*, vol. 2. pp. 68–89. Opladen: Barbara Budrich.

Grzymala-Busse, A. 2008. Beyond clientelism: incumbent state capture and state formation. *Comparative Political Studies* **41**(4–5), 638–73.

Hegedus, D. 2017. *Freedom House Nations in Transit*. Country report on Hungary 2017. https://freedomhouse.org/sites/default/files/NIT2017_Hungary.pdf

Hellman, J. and D. Kauffman. 2001. Confronting the challenge of state capture in transition economies. *Finance and Development* **38**(3), 1–8.

Huntington, S. 1968. *Political Order in Changing Societies*. New Haven, CT: Yale University Press.

Kallay, L. 2015. The corruption risks of EU funds in Hungary. Transparency International Hungary. https://transparency.hu/wp-content/uploads/2016/05/The-Corruption-Risks-of-EU-Funds.pdf

Karlkins, P. 2005. *The System Made Me Do It: Corruption in Post-communist Societies*. London: Routledge.

Karlnins, V. 2014. Estonia: almost there. Background Paper on Estonia. Anticorrp project. http://anticorrp.eu/wp-content/uploads/2014/03/Estonia-Background-Report_final.pdf

Kartal, M. 2014. Accounting for the bad apples: the EU's impact on national corruption before and after accession. *Journal of European Public Policy* **21**(6), 941–59.

Knack, S. 2007. Measuring corruption: a critique of indicators in Eastern Europe and Central Asia. *Journal of Public Policy* **27**(3), 255–91.

Knill, C. and J. Tosun. 2009. Post-accession transposition of EU law in the new member states: across-country comparison. In F. Schimmelfennig and F. Trauner (eds), *Post-accession compliance in the EU's new member states, European Integration online Papers* (EIoP), Special Issue **2**(13), Art. 18. http://eiop.or.at/eiop/texte/2009-018a.htm

Magyar, B. 2016. *Post-communist Mafia State: The Case of Hungary*. Budapest: Central European University Press.

Miller, W., A. Grodeland, and T. Koshechkina. 2001. *A Culture of Corruption: Coping with Government in Post-communist Europe*. Budapest and New York: Central European University Press.

Moran, J. 2001. Democratic transitions and forms of corruption. *Crime, Law and Social Change* **36**(4), 379–93.

Moroff, H. and D. Schmidt-Pfister (eds) 2010. Anti-corruption for Eastern Europe. Special Issue of *Global Crime* **11**(2).

Mungiu-Pippidi, A. 2006. Corruption: diagnosis and treatment. *Journal of Democracy* **17**(3), 86–99.

Mungiu-Pippidi, A. 2013. The good, the bad and the ugly: controlling corruption in the European Union. Advanced Policy Paper for Discussion in the European Parliament. Hertie School of Governance. http://www.bertelsmann-stiftung.de/fileadmin/files/BSt/Presse/imported/downloads/xcms_bst_dms_37612_37613_2.pdf

Mungiu-Pippidi, A. (ed.) 2015. *Government Favouritism in Europe: Anti-corruption Report* 3. Opladen: Barbara Budrich.

Mungiu-Pippidi, A. 2017. Romania's anti-corruption explosion. *Euronews*, February, 5. http://www.euronews.com/2017/02/05/view-romania-s-anticorruption-implosion

Pridham, P. 2008. The EU's political conditionality and post-accession tendencies: comparisons from Slovakia and Latvia. *Journal of Common Market Studies* **46**(2), 365–87.

Puhl, J. 2017. Web of dependencies: a whiff of corruption on Orban's Hungary. *Spiegel* online (January 17). http://www.spiegel.de/international/europe/a-whiff-of-corruption-in-orban-s-hungary-a-1129713.html

Pujas, V. and M. Rhodes. 1999. A clash of cultures? Corruption and the ethics of administration in Western Europe. *Parliamentary Affairs* **52**(4), 688–702.

Sajo, A. 2002. Clientelism and extortion: corruption in transition. In S. Kotkin and A. Sajo (eds), *Political Corruption in Transition*, pp. 1–23. Budapest: Central European University Press.

Sedelmeier, U. 2008. After conditionality: post-accession compliance with EU law in East Central Europe. *Journal of European Public Policy* **15**(6), 806–25.

Sedelmeier, U. 2012. Is Europeanisation through conditionality sustainable? Lock-in of institutional change after EU accession. *West European Politics* **35**(1), 20–38.

Tóth, I., M. Hajdu, and E, Purczeld. 2017. Corruption risks, intensity of competition and estimated direct social loss in public procurement of Zagreb 2011–2016. Corruption Research Center Budapest. http://www.crcb.eu/wp-content/uploads/2017/10/cms_2016_final_report_171010_.pdf

Transparency International. 2012. Corruption risks in the Visegrad countries: Visegrad

Integrity System Study. https://transparency.hu/wp-content/uploads/2016/05/Corruption-risks-in-the-Visegrad-Countries-Visegrad-Integrity-Study-ENG.pdf

Transparency International. 2016a. Global Corruption Barometer 2016: Europe and Central Asia. https://www.transparency.org/whatwedo/publication/people_and_corruption_europe_and_central_asia_2016

Transparency International. 2016b. Crisis, what crisis? https://transparency.eu/cpi16/

Vachudova, M. 2009. Corruption and compliance in the EU's post-communist members and candidates. *Journal of Common Market Studies* **47**(s1), 43–62.

World Bank. various dates. Worldwide Governance Indicators. http://info.worldbank.org/governance/wgi/#reports

11. Corruption in Ukraine: Soviet legacy, failed reforms and political risks

Johannes Leitner and Hannes Meissner

Transparency International, the European Bank for Reconstruction and Development, the World Bank, and numerous local rankings confirm the widespread experience of corruption in Ukraine (Dollbaum 2017). Even in 2015, with the armed conflict in Eastern Ukraine going on and the occupation of Crimea through Russian forces covering the headlines, a national survey in different Ukrainian regions revealed that "corruption in general" was the number three problem when asked about the most serious problems in Ukraine today. The most pressing problem was the military action in the Donetsk and Lugansk regions, followed by the high cost of living (Kiev International Institute of Sociology 2016). The latter was a result of the economic recession which shook Ukraine between 2014 and 2016.

At the same time, for decades politicians from different parties in power and of all levels of governance have declared the fight against corruption as a top priority of their political agenda. Nevertheless, social surveys and international rankings show a quite stable picture with little changes in the level of perception of corruption in Ukraine.

One of the major consequences of the prevailing systemic corruption in Ukraine is its lagging economic development. While most successor states of the Soviet Union experienced dynamic economic development after their independence in 1991, Ukraine has been stumbling, despite its agricultural and industrial heritage that promised a bright future for the country and its citizens (Anders 2015; Denisova-Schmidt et al. 2017). Not only did economic growth lag behind its potential over the last two decades, but today Ukraine, along with Russia, is recovering at a much lower rate from the recent economic recession than are its Eastern European peers. While the Central and East European states, and in particular the member states of the European Union, show gross domestic product (GDP) growth rates of around 4 percent, Ukraine only grew at 2 percent (Gligorov et al. 2017). Certainly, the sluggish economic recovery is not exclusively attributable to the omnipresence of corruption in the country, but correlations between the level of corruption and economic growth have repeatedly been shown (Mauro 1995; Rose-Ackermann 1999; Mo 2001), even though empirical evidence is not unambiguous (Huang 2016).

Apart from its impact on a country's economic performance, another aspect is how individual citizens are influenced through corruption. In 2015, the Kiev International Institute of Sociology carried out the fourth wave (after 2007, 2009 and 2011) of a survey on corruption in Ukraine. In spite of the very recent annexation of Crimea by Russian forces, and the military conflict in Eastern Ukraine's Donbass region, Ukrainians showed to be highly concerned with "corruption in general": 94.4 percent of respondents rated this problem as very serious, and "corruption in government" was rated by 93.8 percent as being a very serious problem (Kiev International Institute of Sociology 2016). Both concerns ranked among the top five issues in a time of military conflicts, economic decline and rising cost of living.

Corruption typically occurs on three different levels: (1) the macro, or political level; (2) the organizational, or firm level; and (3) the micro, or individual level. Firm-level analyses on corruption have often been neglected, which is why we shed light on the issue of firm level because apart from the political sphere, individual firms are crucial economic and societal actors in a country and therefore substantially shape the appearance and prevalence of corruption. In Ukraine and other post-Soviet states, it is not unusual that firms tend to become members of associations with limited access to avoid paying taxes and protect themselves from legal authorities. At the same time, members of these associations obtain preferred market access, which in some cases makes them profit from monopolistic market forces (Denisova-Schmidt and Prytula 2017). Member firms of those associations take advantage of favoritism, a specific form of corruption characterized by non-monetary transactions. In contrast, corruption entails monetary aspects in its transactions (Meissner 2017b). To better understand the patterns that shape corruption in Ukraine, the next section sheds light on the legacies that Ukraine inherited from its history.

(PRE-)SOVIET LEGACY AND CONSEQUENCES OF SOVIET UNION DISSOLUTION

Although bribery and corruption in Ukraine take place on an occasional basis too, it is important to note that the phenomenon is deeply rooted in its politics, economics and the society as a whole. Thus, the practices observable to outsiders are comparable to the visible parts of an iceberg. In order to understand the full picture, it is necessary to look beneath the surface at informal structures and practices, including connections to patronage and clientelism networks. Clientelism is a mutual relationship between a person or group of persons higher ranked in the societal or political order and an entourage seeking protection and particular advantages (Heinritz 2008; Meissner 2011).

In the country-specific context, clientelism (and corruption) are a part of the Soviet and pre-Soviet legacy (Meissner 2010). As Fairbanks (1996, p. 347) highlights, "clientelism is one of the most important aspects of politics in the Soviet Union, and the Soviet Union is one of the most important cases of clientelist politics in the modern world." This has to do with the fact that the Soviet system conserved patterns of pre-modern forms of interaction, based on trust, personal interactions and exchange. At the same time, the centralist, top-down structure of institutions and procedures as well as the lack of check and balances proved to be advantageous for patronage. Consequently, high-ranked members of the *nomenklatura* privatized official positions and material goods in order to hand them down to their clientele in exchange for loyalty. As a result, the distribution of power focused on promoting the specific interests of a personal network. These informal networks often had a pyramid structure because subordinates set up their own clientelist networks at lower levels again (Meissner 2011, p. 6).

Against the background of the importance of informal networks, corruption is deeply rooted in post-Soviet societies. Since everybody is engaged in similar practices, corruption in general, and the fact that some are benefiting more than others in particular, is accepted as something normal, which makes it difficult to tackle. Such mentalities are a breeding ground for collusion with political elites who siphon off public revenues (Meissner 2010).

Even if the dissolution of the Soviet Union threatened many old Soviet networks, it did not shatter the underlying logic of clientelism (and corruption). As Hale (2015) stresses, "the most important actors in post-Soviet politics" are "extended and loosely connected hierarchical networks led informally by powerful patrons" (Hale 2015, p. 95). The disintegration of the Soviet Union turned out to be "a critical juncture that set . . . dynamics in motion" (Hale 2015, p. 96). Eventually, "creative entrepreneurs recognized opportunities to build something new out of the shaken networks, resources, and relationships," stringing together "new networks or mobilizing old networks for new purposes." As Hale (2015) further highlights, "some of these networks rose through business, others through state authority" (Hale 2015, p. 98).

In Ukraine, informal politics are traditionally characterized by the strong influence of competing regional networks. It was former President Leonid Kuchma's strategy to promote representatives of regional elites. Having received influential positions in national offices in Kiev, they then patronized their own clients, among them entrepreneurs (Pleines 2012, p. 127). Pleines (2012) describes the constellation as an informal win-win constellation: "The regional politicians reaped influential positions on the national level as well as bribery payments from the business world. The entrepreneurs received preferential treatment via policies that brought them immense profits" (Pleines 2012, p. 127). In this context, so-called oligarchs became a key feature of Ukraine's political system. Initially, connections to political elites were a precondition for their business success before they became politically active themselves in the second half of the 1990s in order to cement their positions (Pleines 2012; Robinson 2013). Pleines (2012) describes three basic strategies for how this took place. They first acquired mass media in order to manipulate public opinion. Second, they developed informal networks, incorporating political elites. Third, they gained public office.

However, this must not imply that oligarchs are the only stakeholders in this system. Apart from the fact that the term "oligarch" is ill-defined, regional politicians of networks, entrepreneurs and wealthy businesspeople are major players as well (Pleines 2012). According to Fisun (2003), the characteristic features of this informal system are the "wide strata of . . . rent-seeking actors, acting together with/or in place of governmental institutions via clientelistic networks of patronage and pork barrel rewards," as well as a high level of competition (Fisun 2003, p. 6). The central broker is the president. He maintains "a system of personal ties . . . based first and foremost on regional . . . unity, as well as on present-day rent-seeking interests." He maintains power by capturing state resources and redistributing them to his own clientelistic network (Fisun 2012, p. 3). In the beginning, the main source of rents accrued from energy transit from Russia to Europe. These rents were supplemented from foreign borrowing from the late 1990s (Robinson 2013).

Against this backdrop, the rise of the Donetsk network and later President Viktor Yanukovych started in 2002, when he was appointed prime minister. Under his rule as president of Ukraine (2010–14), clientelism and corruption assumed huge dimensions. However, this must not imply that other actors were less prone to corruption. His predecessor Yulia Tymoshenko, for example, had no better reputation. As she became head of Ukraine's United Energy Systems, she received the unofficial title of the country's "gas princess" (Hale 2015, p. 100).

The constellation of how formal and informal power mixed under Yanukovych's reign gained a lot of attention by investigative journalists as well as local researchers. Meanwhile, this information is well established in the international scientific community dealing with Ukraine. Menon and Rumer (2015) provide an in-depth analysis of the functioning of this system, including its large-scale corruption schemes, centered on the Yanukovych network. As the authors highlight, such practices were treated as "no secret" by President Yanukovych (Menon and Rumer 2015). Marples (2015) describes this system as "a Donetsk-based regime of apparatchiks and gangsters with their own private mansions and assets abroad" (Marples 2015, p.16). According to Schneider-Deters (2013), Yanukovych started to strengthen his system of personal ties centered on his family and rooted in the region of Donetsk straight after winning the presidential election in 2010 (Schneider-Deters 2013). In the period that followed, he used his presidential power for "a zero-sum 'winner-take-all'" game (Fisun 2012, p.2). He reduced cooperation with other networks to the necessary minimum and dedicated himself to promote the business activities of his closest family circle (Fisun 2012). By doing so, he accumulated ever more wealth at a high pace and was perceived as finally becoming an oligarch himself. Growing authoritarianism, this "winner-takes-all" behavior, extreme forms of rent-seeking, massive wealth inequalities and fierce elite competition contributed to increasing instability of this system (Fisun 2012). In March 2014, Yanukovych was overthrown as a result of mass protests triggered by the announcement of the Ukrainian government that it would not sign the Association Agreement with the European Union and seeking closer economic relations with Russia instead.

STATE CAPTURE, CORRUPTION AND POLITICAL RISK

In the search to operationalize the aforementioned constellation of power, the literature on "state capture" provides helpful links (Sindzingre 2010; Laruelle 2012). State capture means that private elites, such as family and friendship networks, clans and businesspeople acquire political power and then misuse their positions to realize their very own particularistic interests. The general interest, in turn, does not play any decisive role anymore when the ruling elite's acts are in sharp contrast to the needs of the society. In this context, the members of the ruling elite first and foremost strive to enrich themselves (although prestige can also be a driver). In order to secure their position in the state, the economy and society permanently, they often draw on authoritarian power strategies. At the same time, clientelism is a helpful tool to stabilize their power informally. As they use their access to public goods (such as state funds, the budget, state companies and the public service) in order to hand down the resources to their clientele (by state tenders, distribution of lucrative jobs), they gain political loyalty from these people in turn (Meissner 2017b).

Corruption is a core feature of state capture. It can be operationalized in three different ways, all of which pose a political risk to international operating companies. Leitner (2017) defines political risks as "any occurrence in the international business context where public actions or non-state actors that are active in the host country of the international activities interfere with private international businesses and adversely impact the performance of the international operation" (Leitner 2017, p. 29).

One of these factors is monetary corruption, which we call "systemic corruption" (Meissner 2017b, p. 21). It means that members of the ruling elite systematically misuse public office to gain immediate monetary advantages. This can take place in different ways, for example, by the sale of public positions. The acquirers, in turn, will be urged to refinance their investment by generating money through corruption schemes established in the public sector. In this context, money is generated at different levels, starting from the grassroots, for example, in the education and health sectors. An arena particularly prone to such practices is traditionally the traffic police. At the same time, there is evidence that in some countries parts of the money flow upwards in the informal power structure. Our own field research in Kiev in February/March 2013 revealed, for example, that under Yanukovych many interview partners regarded corruption as a political phenomenon rooted at all levels. There was also a widespread view that corruption was controlled centrally or even part of a pyramid system, with the Yanukovych family and close allies as final beneficiaries on the top. Our interview partner confirmed this, but stressed that the true picture would probably be more complex.

At the same time, systemic corruption regularly turns out to be a political risk for international enterprises, as in such environments they are confronted with demands for entrance fees, kickbacks as well as payments necessary to ward off unjustified claims by tax authorities, the security service or the customs authority. In this context, corruption is always associated with a significant degree of arbitrariness and lack of predictability, even if certain "going rates" apply (Meissner 2017b, p. 21).

Another form of corruption characteristic of state capture is favoritism. The difference between this and the aforementioned form of corruption is that this version refers to cases in which members of the ruling elite strive for gains that are not immediately limited to monetary nature, although they will indirectly lead to material benefits as well. It means that private actors systematically use public office to foster the business interests of the ruler himself or the ruler's clientele, while impeding initiatives by actors not part of the ruler's network. In this context, rulers distribute licenses, contracts and public projects to their own business networks. As a result, political elites accumulate ever more wealth. Systematic favoritism is a major risk to international businesses as certain markets are freely accessible only at first sight. The businesses of the ruling elite are favored while international companies suffer from difficulties in gaining any access to such networks. Moreover, they are often disfavored, as ruling elites want to keep the market free from competitors[1] (Meissner 2017b).

A third factor that materializes as political risk under constellations of state capture is more a consequence of corruption interests and practices than a form of corruption itself. We call it "institutional ambiguity." It goes back to the fact that in such environments, laws, regulations and procedures are often unclear, contradictious or full of loopholes. Such deficiencies are often due to pure calculation by the ruling elites, as institutional ambiguity is a fundamental part of money-raising schemes. As people or companies cannot comply with the legal framework, legal violations become unavoidable. On this occasion, public servants can ask for illegal payments to unofficially solve the problem. However, in such environments, ambiguity furthermore appears between the formal institutions and the informal sphere. A classic example for this is bribes. According to informal rules, bribes might be rational in certain situations and constellations. However, corruption may also entail legal prosecution. In the worst case, it might even give ground for blackmailing[2] (Meissner 2017b).

When looking at the firm level since the Maidan Revolution, an interesting question has been whether managers of international companies that are active in this post-Soviet country are more concerned about the Crimea crisis and the armed conflict in Eastern Ukraine than they are about the consequences of state capture and its impact on their businesses. Considering that Crimea and the Donbass region have been totally isolated from Ukraine in economic terms, the assumption was that for international companies the loss of these markets might be a risk that is more serious than the old-fashioned problems of state capture. The results of a recent empirical investigation on that topic indicate a different picture however (Leitner and Meissner 2017).

The negative consequences of the conflict between Russia and Ukraine and the economic downturn that followed the conflict seemed to be a major issue for managers immediately after these negative events had commenced. Nevertheless, those political risk factors that are rooted in the country's internal political and institutional system again prevailed

---

[1] For empirical evidence on how international firms were confronted with (dis-)favoritism under Yanukovych, see Meissner (2017a).

[2] For empirical evidence on how international firms were confronted with institutional ambiguity under Yanukovych, see ibid.

over those external political risk factors such as the war. A Ukrainian manager of an international company that produces construction materials explained the most important three risks in the country:

1. The absence of rule of law,
2. The judicial system and the courts, and
3. Corruption. (Leitner and Meissner 2017, p. 87)

This assessment corresponds well with the results provided by the fifth round of the Business Environment and Enterprise Performance Survey (BEEPS), a project which has been initiated jointly by the European Bank for Reconstruction and Developments (EBRD) and the World Bank. The fifth round of BEEPS was carried out between 2011 and 2014, and covered roughly 16,000 enterprises in 30 countries, including 4,220 enterprises in 37 regions in Russia. The survey intends to portray the greatest challenges for firms – both local and international – operating in Ukraine relating to the business environment. Corruption is mentioned as the second most concerning obstacle in Ukraine, surpassed only by "political instability," and followed by "informal sector." Corruption contains informal payments to secure a government contract, but also informal payments to obtain electrical connections, construction-related permits, operating licenses, tax authorities, and to obtain an import license (BEEPS 2017).

In the context of political risk, a crucial question is whether the pervasiveness of corruption has changed since the inauguration of the new president Petro Poroshenko after the Maidan Revolution. According to the 2015 round of the "Corruption in Ukraine" survey of the Kiev International Institute of Sociology, two-thirds of the respondents still are of the opinion that the president is the person responsible for overcoming corruption in Ukraine. The president is followed by the Verkhovna Rada (the parliament) (41.7 percent), and the prime minister and Cabinet of Ministers (37.7 percent). The institutions that are least attributed the responsibility to fight corruption are non-governmental and civic organizations (1.4 percent) and volunteer movements (0.5 percent) (Kiev International Institute of Sociology 2016, p. 21).

With regards to improvements concerning corruption and state capture, the performance of Petro Poroshenko is quite unimpressive. In the perception of managers active in Ukraine, Petro Poroshenko is only a new face who nevertheless follows the old logic of state capture. Although there are anti-corruption reforms on the way, a major drawback is that the judiciary and the court system have not yet changed, which substantially impedes any further progress beyond measures taken in reforming the process for public tenders and the traffic police. A manager of an international insurance company put it this way:

> They [the real reformers] have put new people in power to fight corruption, but they often do not survive very long. This is due to the fact that the judiciary and the prosecution have not changed yet, though a new law has just been signed to do so until 2019. When corrupt people are dismissed, the appeal at court and the old lawyers bring them back to power. (Leitner and Meissner 2017, p. 91)

SPATIAL DIFFERENCES IN THE SCOPE OF CORRUPTION IN UKRAINE

Recent research on the spread of corruption in Ukraine has shown that there are regional differences. Although Ukraine remains a highly centralized state, the country's regions and their population have diverse historical traits and often different political preferences. The major dividing line in Ukraine is between Eastern Ukraine and Western Ukraine. Although these attributions of "East" and "West" are more ideological markers than geographical indicators, the Eastern Ukrainian region comprises the area of Donetsk, Luhansk, Kharkiv, Zaporizhzhia, and Dnipropetrovsk. Most of this territory is part of the combat zone of Ukraine's east, where Russian backed forces fight for the territory's independence from Ukraine. This part of Ukraine has the longest shared history with Russia. Also, the eastern parts of Ukraine host the country's steel, chemical and machinery industries, which were heavily linked to the Russian markets and accounted for more than half of the country's total exports. Western Ukraine refers to all the four remaining regions of Ukraine, with the North, which was under Russian control from the 1650s to the end of the 1700s, and the Southern region, which was under Ottoman rule until the late 1700s, when it was taken over by Russia. It also includes the center of Ukraine, which was under Poland's rule until the late 1700s before falling under Russian control, and the Western region, which were part of the Austro-Hungarian monarchy until 1918 (Barrington und Herron 2004).

A common feature across all Ukrainian regions seems to be a general distrust of the population against authorities and state institutions. However, while local authorities at the village and city government levels tend to be the most trusted in all regions, a closer look reveals that in the East, only 10.4 percent of the respondents trust those authorities, while in the West almost a quarter (24.7 percent) do so (Kiev International Institute of Sociology 2016). However, it is important to consider that in 2015, due to the military conflict in large parts of Eastern Ukraine, the survey could only be conducted in those parts of the east where Ukrainian forces were in control. Even more interesting are the results with regards to the president and his administration. In the east of Ukraine the survey shows for 2011 that 21.6 percent of the respondents trusted the president and his administration. In 2011, Viktor Yanukovych was president of Ukraine and had to flee the country after the Maidan Revolution. Viktor Yanukovych was rooted in the Eastern Ukrainian Donetsk region and also began his political career there. In the 2015 survey, trust in the president and his administration in Eastern Ukraine dropped to 5.4 percent. Petro Poroshenko, the current president, is often seen as a politician who is in favor of the West, of the European Union, and the United States rather than Russia and the East. In the west of Ukraine, the pattern is reversed but not as accentuated. Viktor Yanukovych was trusted by 7.9 percent in 2011, while Petro Poroshenko is trusted by 11.3 percent of the survey participants. (Kiev International Institute of Sociology 2016, p. 20).

At the firm level, discrepancies among different regions are evident. Firms from the Eastern region perceive higher levels of corruption in the fields of tax inspections and customs as well as with the police, whereas firms from Western Ukraine report lower levels of perceived corruption in the police (Denisova-Schmidt and Huber 2014).

The fight against corruption has been a major argument of politicians in power, regardless of party affiliation, region or governance level. This final section provides a brief overview of the state of reforms, highlighting the most prominent institutions that have been established to fight corruption in Ukraine.

While numerous initiatives and organizations have been founded, a recent issue of the Ukrainian daily the *Kyiv Post* headlined that "Yanukovych-era criminal cases stand unsolved." The reason why the investigations are on hold is that criminal cases from the Yanukovych-era were ordered to be transferred from the Office of the Prosecutor General to a newly founded State Investigation Bureau, which, however, never started its operations. A second reason is the shortage of staff for the National Anti-Corruption Bureau (NABU). The newspaper also comments that President Petro Poroshenko and his fellows do not seem to be concerned with the investigations on hold (Kyiv Post 2017).

NABU was established in 2015 through a decree signed by Ukraine's President Petro Poroshenko. Within the first year of its operations NABU filed more than 20 lawsuits in corruption cases. By September 2016, 206 investigations were started and 1.5 million euros of stolen money were refunded to state accounts. Certainly, the existence and the success of this truly independent anti-corruption institution in Ukraine are disputed. Consequently, barriers were erected to counteract the NABU. A first strike against NABU was observed in the media. As soon as NABU was close to finishing a corruption case, the media – most of which were under the control of different oligarchs – started to discredit the institution and its prosecutors. Another strategy seems to be to congest the NABU by forwarding a much higher number of cases than NABU can actually process. The attorney general frequently shifts cases to NABU even though they are not connected to corruption, thereby blocking the institution from effectively being able to focus on its core mission (Zacharow 2016).

REFERENCES

Anders, Aslund. 2015. *Ukraine. What Went Wrong and How to Fix It.* Washington, DC: Peterson Institute for International Economics.

Barrington, L. and Herron, E. 2004. One Ukraine or many? Regionalism in Ukraine and its political consequences. *Nationalities Papers* **32**(1), 53–86.

BEEPS. 2017. BEEPS V Ukraine. Hg. v. EBRD und World Bank. http://ebrd-beeps.com/reports/beeps-v/ukraine/ (accessed November 30, 2017).

Denisova-Schmidt, Elena and Huber, Martin. 2014. Regional differences in perceived corruption among Ukrainian firms. *Eurasian Geography and Economics* **55**(1), 10–36.

Denisova-Schmidt, Elena and Prytula, Yaroslav. 2017. Trust and perceived corruption among Ukrainian firms. *Eastern European Economics* **55**(4), 324–41.

Denisova-Schmidt, Elena, Huber, Martin, and Prytula, Yaroslav. 2017. Corruption among Ukrainian businesses. Do firm size, industry and region matter? In Johannes Leitner and Hannes Meissner (eds), *State Capture, Political Risks and International Business: Cases from Black Sea Region Countries*, pp. 108–19. London and New York: Routledge.

Dollbaum, Jan. 2017. Die Ukraine in Internationalen Korruptions- und Transparenz-Rankings. Hg. v. Heiko Pleines. Forschungsstelle Osteuropa der Universität Bremen Ukraine Analysen.

Fairbanks, C. 1996. Clientelism and the roots of post-Soviet disorder. In R. Suny (ed.), *Transcaucasia, Nationalism, and Social Change, Essays in the History of Armenia, Azerbaijan, and Georgia*, pp. 341–74. Ann Arbor, MI: University of Michigan Press.

Fisun, Oleksandr. 2003. Developing democracy or competitive neopatrimonialism? The political regime of Ukraine in comparative perspective. http://www.utoronto.ca/jacyk/Fisun-CREES-workshop.pdf (accessed July 24, 2015).

Fisun, Oleksandr. 2012. Electoral laws and patronage politics in Ukraine, http://www.ponarseurasia.org/sites/default/files/policy-memos-pdf/pepm_229_Fisun_Sept2012.pdf (accessed November 25, 2017).

Gligorov, Vladimir, Grieveson, Richard, Havlik, Peter, and Podkaminer, Leon. 2017. CESEE back on track to convergence. Economic analysis and outlook for Central, East and Southeast Europe. The Vienna Institute for International Economic Studies. Vienna (Forecast Report). https://wiiw.ac.at/wiiw-forecast-reports-ps-50.html (accessed November 30, 2017).

Hale, Henry. 2015. *Patronal Politics: Eurasian Regime Dynamics in Comparative Perspective.* Cambridge: Cambridge University Press.

Heinritz, K. 2008. *"Defekte Demokratisierung" – ein Weg zur Diktatur? Turkmenistan und die Republik Sacha (Jakutien) in der Russischen Föderation nach dem Ende der Sowjetunion.* Frankfurt am Main: Peter Lang.

Huang, Chiung-Ju. 2016. Is corruption good or bad for economic growth? Evidence from Asia-Pacific countries. *North American Journal of Economics and Finance* **35**, 247–56.

Kiev International Institute of Sociology. 2016. Corruption in Ukraine: Comparative analysis of national suveys: 2007, 2009, 2011, and 2015. Kiev International Institute of

Sociology. Kiev, Ukraine. http://www.kiis.com.ua/?lang=eng&cat=reports&id=595&page=11 (accessed October 17, 2017).

Kyiv Post. 2017. Special Investigations Office in danger of losing control over Yanukovych corruption cases. *Kyiv Post.* https://www.kyivpost.com/ukraine-politics/special-investigations-office-danger-losing-control-yanukovych-corruption-cases.html (accessed December 1, 2017).

Laruelle Marlene. 2012. Discussing neopatrimonialism and patronal presidentialism in the Central Asian context. *Demokratizatsiya* **20**(4), 301–24.

Leitner, Johannes. 2017. Political risk and international business: where they interfere, consequences, and options. In Johannes Leitner and Hannes Meissner (eds), *State Capture, Political Risks and International Business: Cases from Black Sea Region Countries*, pp. 11–25. London and New York: Routledge.

Leitner, Johannes and Meissner, Hannes. 2017. Politische Risiken in der Ukraine: Die größten Probleme sind hausgemacht. *Wirtschaft und Management* **25**, 77–94.

Marples, David. 2015. Introduction. In David Marples and Frederick Mills (eds), *Ukraine's Euromaidan: Analyses of a Civil Revolution*, pp. 9–6. Stuttgart: ibidem Press.

Mauro, Paolo. 1995. Corruption and growth. *Quarterly Journal of Economics* **110**(3), 681–712.

Meissner, Hannes. 2010. The resource curse and rentier states in the Caspian region: a need for context analysis. http://www.giga-ham-burg.de/dl/download.php?d=/content/publikationen/pdf/wp133_meissner.pdf (accessed November 10, 2017).

Meissner, Hannes. 2011. Informal politics in Azerbaijan: corruption and rent-seeking patterns. http://www.isn.ethz.ch/isn/Digital-Library/Publications/Detail/?id=94386 (accessed November 11, 2017).

Meissner, Hannes. 2017a. State capture, political risks and international business: evidence from Ukraine under Yanukovych. Conference Paper presented at the MPSA Annual Conference, Chicago. https://ccbsr.fh-vie.ac.at/state-capture-political-risks-and-international-business-evidence-from-ukraine-under-yanukovych-analysis-by-hannes-meissner/ (accessed November 11, 2017).

Meissner, Hannes. 2017b. Corruption, favouritism and institutional ambiguity as political risks: insights from the concept of neopatrimonialism. In Johannes Leitner and Hannes Meissner (eds), *State Capture, Political Risks and International Business: Cases from Black Sea Region Countries*, pp. 11–25. London and New York: Routledge.

Menon, Rajan and Rumer, Eugene (eds). 2015. *Conflict in Ukraine: The Unwinding of the Post-Cold War Order*. Cambridge, MA: MIT Press.

Mo, Pak Hung. 2001. Corruption and economic growth. *Journal of Comparative Economics* **29**, 66–79.

Pleines, Heiko. 2012. From competitive authoritarianism to defective democracy: political regimes in Ukraine before and after the Orange Revolution. In Susan Stewart, Margarete Klein, and Andrea Schmitz (eds), *Presidents, Oligarchs and Bureaucrats: Forms of Rule in the Post-Soviet Space*, pp. 125–38. Farnham: Ashgate.

Robinson, Neil. 2013. Economic and political hybridity: patrimonial capitalism in the post-Soviet sphere. *Journal of Eurasian Studies* **4**(2), 136–45.

Rose-Ackermann, Susan. 1999. Political corruption and democracy. *Connecticut Journal of International Law* **14**(2), pp. 363–78. http://digitalcommons.law.yale.edu/fss_papers (accessed November 11, 2017).

Sindzingre, Alice. 2010. The concept of neopatrimonialism: divergences and convergences with development economics. "Neopatrimonialism in Various World Regions", International Workshop, GIGA, Hamburg, 23 August.

Schneider-Deters, Winfried. 2013. Die Ukraine nach der Parlamentswahl 2012 – "Die Familie" übernimmt die Regierung, *Ukraine-Analysen* **115**, 2–7.

Zacharow, Jewgenij. 2016. Korruptionsbekämpfung auf ukrainisch: Neue Elemente. In Heiko Pleines (ed.), *Ukraine Analysen*. Forschungsstelle Osteuropa der Universität Bremen.

12. Corruption in Russia
Günther G. Schulze and Nikita Zakharov

> Taking bribes is indissolubly interlaced with the whole system and political life.
> (Berlin 1910, p.48)

Corruption in Russia is more than the abuse of public power for private gain, as the standard definition by the World Bank suggests (World Bank 1997, p.102), it is an integral part of the power configuration (Dawisha 2015; Pavroz 2017) and, as such, systemic (Charap and Harm 1999). Corruption serves two main functions: rent extraction and securing loyalty of subordinates in the administrative hierarchy.

The first function is straightforward and has been the center of interest in economic literature on the topic.[1] Illegal rents are extracted from the economy by public officials, who use their administrative and political power in the form of bribery, kickbacks, patronage or direct embezzlement. In Russia, the rationale behind this rent extracting is not only greed, but also need – illegal income from corruption compensates for low official wages. Zhuravleva (2013, 2016) finds that public employees in Russia enjoy the same level of wealth and expenditures as their counterparts in the private sector even though the salaries in the civil service are lower. Gorodnichenko and Sabiryanova Peter (2007) found a similar effect in Ukraine. Schulze et al. (2016) find that corruption declines with rising relative income of public officials. The inadequate pay makes corruption almost universal in public services.

The second function of corruption is to secure loyalty and to promote stability of the political regime through subordination of the administrative hierarchy (Darden 2008). As long as lower levels of the administrative system remain loyal, illegal self-enrichment through the abuse of public office is informally tolerated and rarely prosecuted. Ledeneva (2009, p.278) notes that loyalty to the political regime in Russia is "an essential operating principle in public administration with rewards distributed through the system of perks and informal payments." Corruption thus creates a powerful leverage against those who engage in it since it is officially illegal. The punishment for disloyalty is then masqueraded as a prosecution of

---

[1] See, for example, Olken and Pande (2012), Kis-Katos and Schulze (2013), and Dimant and Tosato (2018) for literature reviews.

corruption. Wages are kept artificially low to induce corruption, which allows disciplining disloyal public officials and rewarding loyal ones at will. This view is opposed to the popular perception of corruption as a sign of a malfunctioning state. In fact, it is creating stability in autocratic regimes.

This argument is supported by empirical evidence. Fjelde and Hegre (2014) found that corruption is positively associated with political stability in countries with autocratic and hybrid regimes, but not in democracies. At the country level, the disciplining function of corruption has been established for Indonesia (McLeod 2008, 2012), the Philippines (Quimpo 2009), and for African countries (Arriola 2009), among others. The most recent illustration of this mechanism in Russia is the arrest of Economic Development Minister Aleksey Ulyukaev in an intricate criminal case of extorting a bribe from the powerful CEO of the state oil company, Rosneft, and close friend of President Vladimir Putin, Igor Sechin. In the light of contradictory evidence, most experts on Russian politics suggest that it is *more likely a personal conflict of loyalty.*[2] Before the arrest, Ulyukaev was a trusted public official, who served in different offices for over 25 years and, despite his modest official salary, managed to accumulate considerable wealth in the form of real estate and business including offshore companies.[3] While it is not possible to say whether the minister profited from his office illegally before or whether the allegation was justified, the unprecedented arrest of an official as high-ranked as a minister sent a message to the bureaucracy that there is no political immunity when subordination is compromised.

These functions have made the practice of corruption extremely successful for the political elite in Russia at every level of the bureaucratic hierarchy. At the top, it has guaranteed an unchallenged monopoly on political power of President Putin and his inner circle for the last 17 years

---

[2]　See the interview of Julius von Freytag-Loringhoven, director of the Moscow offices of the Friedrich Naumann Foundation, with Deutsche Welle: http://www.dw.com/en/russian-ministers-arrest-a-fight-against-corruption-or-liberalism/a-36404692 (accessed October 1, 2017). The official allegation states that the minister threatened Rosneft and extorted a bribe in exchange for sanctioning a purchase from Bashneft, another oil company, a $5 billion deal that was already approved by Putin. The most likely cause of conflict was the initial objection of Ulyukaev and several other senior officials to the Bashneft deal as it would have resulted in Rosneft dominating the Russian oil market. For more information see http://foreignpolicy.com/2016/11/15/why-was-russias-economic-minister-ulyukaev-putin-rosneft-corruption-russia/ (accessed October 1, 2017).

[3]　For example, the official monthly salary of a minister in Russia is about $6,000, but in 2015 Ulyukaev declared an income of over $1 million (average estimates of officials' salaries are published by RBC at http://www.rbc.ru/politics/09/03/2015/54fdbb749a794705c04ffc03 (accessed October 1, 2017). The data from the official declaration can be accessed from a declaration database of public officials at http://declarator.org/person/4491/ (accessed October 1, 2017).

and has supported them financially. The lower levels of bureaucratic hierarchy, such as the regional elites, hold their offices on the implicit condition of providing favorable electoral support (often via voter mobilization or through electoral fraud) and to relinquish personal political autonomy. In return they enjoy administrative rents and strengthen their loyalty network by further informally sanctioning rent-seeking among subordinates (Reuter and Robertson 2012; Bliakher 2013).

The origins of the Russian corruption model have been broadly discussed in the political science literature. Dawisha (2015) argues that it has been designed and implemented by Vladimir Putin himself as it was central to his success in rising to power and, once he had become president, as his main goal was to build an authoritarian state in order to satisfy his plutocratic interests. Her argument is supported by a rich account of news reports, official documents, memoirs, WikiLeaks and witness testimonies collected by Russian and foreign journalists. In contrast, Zygar (2016), based on dozens of interviews with key Kremlin figures, portrays the political elite as much less structured around Putin as usually believed. He argues that the political life in Russia is not guided by an authoritarian masterplan, but by incentives that are chaotic and short term, and thus opportunistic in their nature.

In this chapter, we argue that the spontaneous order is derived as a product of institutional path dependence: the system of corruption may not have been consciously planned by specific political actors but, nevertheless, is a consequence of the historical evolution of the Russian state. The historical perspective demonstrates how both functions of corruption persisted across centuries and played very similar roles in state governance under all political systems (monarchy, socialism or formal democracy).

We also review the empirical evidence on corruption in modern Russia and show that corruption intensity is very unequally distributed across the Russian regions and that these differences are responsible for a heterogeneous development of the regions in terms of investment, foreign direct investment and even road safety. Finally, we look at the main determinants of corruption in Russia and inquire how they might suggest a solution to the corruption problem.

THE HISTORY OF RUSSIAN CORRUPTION

Corruption in Russia shares considerable similarity with systemic corruption in countries such as Mexico, the Philippines, and Indonesia (Johnston 2008; O'Hara 2014), which inherited their institutional backwardness from their colonial heritage. Angeles and Neanidis (2015) argue that countries in which European colonizers constituted only a small minority in the population suffer from higher levels of corruption. European settlers formed powerful local elites who engaged in corruption to procure benefits for themselves at the expense of the population at large without running the risk of being penalized by law. The patronage system once installed by Europeans to exploit their colony was later adopted by the following generation of elites, who reproduced this extractive system of rent-seeking.[4]

While Russia was not subject to European colonization, it has a similar chapter in its history. In 1240, long before Russia was a centralized state, the country was conquered by the Mongols and became a dominion of the Golden Horde (1240–1480).[5] The new rulers demanded tribute and installed a dual administration, where the Russian grand princes were entrusted with tax collection and in return were allowed to keep their share (Ostrowski 2002). The system was profitable for the grand princes since they could exploit their position to extract surpluses from the population for private use. The Mongols were indifferent to this rent-seeking as long as their share was paid and entrusted grand princes showed their loyalty. There was no guarantee that the right to collect taxes would remain in the hands of the same grand prince as the Horde rulers exercised the policy of divide and rule by preventing any princedom from becoming strong enough to contest their political power (Hartog 1996). This produced an environment where local elites were accountable only to the supreme power of the Horde and the subdued population was subject to oppressive taxation and exploitation by both levels of administration.

The end of Mongol rule in 1480 did not dismantle the extractive practices; instead, they were successfully adopted by the Russian grand princes to establish control over their extensive territories. The practice of taxation under dual administration then became an institution known as *kormlenie* ("feeding" or "nourishment" in Russian): a vicegerent received a province to supply ("feed") him and his servants as a reward for service and tax collection. However, it was not a typical feudal system in a sense of "a formalized, hierarchical set of relationships" (Hosking 2000, p. 302) as peasants were free to change their landlord and even viceregents could quit the service of one grand prince and move to another.[6] The right of

---

[4]   This is a variant of the more general argument that the form of colonization – extractive or inclusive – was largely determined by the relative number of European settlers in the colony and that the quality of institutions so determined has persisted (Acemoglu et al. 2001; Acemoglu and Robinson 2012).

[5]   The Golden Horde was originally the northwestern part of the Mongol empire but later became a separate entity.

[6]   The mobility was suppressed in the second half of the sixteenth century.

kormlenie was only temporary and could be revoked at any time, eroding the system of private and public property and propelling the abuse of authority. The rate of extracted resources and its limits were not determined officially, but emerged as a result of informal negotiations between an official and the population under his rule. Often *kormlenie* is seen as the earliest Russian practice associated with corruption.[7] Kovalevsky (1902, p. 83) calls it "a legalized bribery" that "preceded the evolution of Russian institutions."

Hedlund (2005) provides a convincing account of path dependence in institutional development from the early times of Kievan Rus through the collapse of the Soviet Union to the current electoral authoritarian system and finds that despite its official abolition in the sixteenth century, *kormlenie* has persisted over time in various forms. It persisted under different regimes because Russian rulers could not afford to pay adequate official salaries and because it was expedient for them to secure loyalty among subordinate officials as a reward for sanctioning income that was officially illegal.

The historical evidence provides curious accounts on the intensity of corruption. For example, the complaints of the local population in the Dvina region, one of the northwestern regions of Russia, at the end of the seventeenth century suggest that the governors who were appointed to the region used to extract about 2,050 rubles per year, which amounted to a third of the annual tax collection (Kopanev 1984, p. 201). The other source of data on corruption in imperial Russia is found in the record books from large agricultural provincial estates of nobility. Using this source, Korchmina and Fedyukin (2018) find that almost every interaction with a representative of the state was accompanied with some sort of "gift," creating significant extra incomes for public officials. Their estimates suggest that extralegal payments to officials amounted up to 70 percent of their legal income and that high-ranking officials received the largest shares of almost twice their salary. This is only a lower bound of corrupt income since the data from Korchmina and Fedyukin (2018) cover only big agricultural estates and do not include other sources of corruption such as trade.

The institution of *kormlenie* not only survived the 1917 revolution, it also played a prominent role in the demise of the Russian monarchy. Corruption was widespread during the revolution and civil war among both conflicting sides, as depicted in Brovkin (2003). However, there

---

[7]   There is an abundant historical literature on the topic of *kormlenie*, see, for example, Davies (1997).

was a crucial difference in the corrupt practices that eventually played an important role in the victory of the Bolsheviks. Brovkin (1994, p. 199) argues that "the Whites believed that the population had to be so grateful to them for the liberation from the Bolshevik rule that it should be willing to pay for the maintenance of the army and the new administration. This view generated immense corruption among the White civil service." At the same time, the Whites did not engage the local population in governance, as did the Reds: "The Bolsheviks divided the local community by promoting some over others, delegating administrative authority to them, knowing full well that they would use that authority to serve their own interests" (Brovkin 1994, p. 200). This reincarnation of the familiar *kormlenie* system was once again "a method of governance" (Brovkin 1994, p. 200).

Corruption as a method of governance continued and strengthened after the revolution. Anderson (2012, p. 72) writes, "The Communist (Bolshevik) takeover of the state in October 1917 ushered in a radical top-down transformation of social, political, and economic life that created ideal conditions for the proliferation of corruption. Both opportunity and incentive for corrupt behavior were embedded in the very political and economic structure of the Soviet Union." The Communist Party of the Soviet Union (CPSU) held the monopoly on political power and established a strong hierarchical system (*nomenklatura*), in which main official positions were appointed by the higher-level administration. Members of the *nomenklatura* enjoyed social status and privileges and could not be prosecuted on criminal charges without approval from the party. The absence of political competition and the full control of state institutions precluded a system of checks and balances as accountability mechanism (Harasymiw 1969) and the rule of the party was above the law (Simis 1982). This gave rise to extensive informal networks that systematically abused political power for personal gain and for even more political power. Vaksberg (1991) documents the activities of these networks of public officials and finds their stark resemblance to organized crime syndicates or the mafia. Just like the mafia, the Soviet *nomenklatura* valued loyalty and subordination of its members the most, as opposed to competence or honesty.

While opportunities for corruption soared with an increasing role of the government in economy, politics and even everyday life (Kramer 1977), the incentives for rent-seeking were still very similar to those of the past regime: public officials were inadequately paid, especially outside of the top ranks of the *nomenklatura* (Anderson 2012). The constant severe shortages in goods and services were only partially compensated by a complex system of privileges, and some privileges such as personal cars or *dachas*, if provided, were not granted permanently but could be withdrawn (Matthews 2011). That essentially reproduced the system of *kormlenie* once again.

The fight against corruption in the Soviet Union was ineffective and often politically motivated. Kramer (1977) observes that despite official concerns publicly expressed by higher authorities and anti-corruption propaganda, corruption remained generally unpunished and occasional convictions often resulted in mild punishments, such as job loss. Kramer (1977, p. 222) notices that "many dismissed officials manage to be reappointed to important positions where they again engaged in corruption." Kramer suggests that exposing corruption was undesirable in the Soviet system simply because of the risk of being criticized for allowing it to occur in the first place (Simis 1982). Nevertheless, some corruption cases received wide coverage in Soviet newspapers. Clark (1993) gathered data on all convictions of public officials reported by the press during 1965–90 and analysed it with respect to timing and geographical distribution. He found that a dramatic increase in convictions of local officials in the Soviet republics was often preceded by changes in important party positions. These findings suggest that convictions were politically motivated and anti-corruption measures were used as a weapon within the *nomenklatura*. In fact, this mechanism was a part of the bigger informal practice of *kompromat*, short for "compromising materials," under which information on illegal or otherwise compromising behavior was collected and preserved in secrecy as leverage (Ledeneva 2006). *Kompromat* was efficient in enforcing subordination because officials with "guilty secrets" could be easily controlled (Harrison 2011), and corruption was probably the most common of all "guilty secrets" used for that purpose.

Corruption eventually played an important role in the demise of the Soviet Union. In the face of a dysfunctional economy and heavy costs of the Cold War, increasing corruption in the Soviet Union became a "substitute for reform of the institutional structure" (Schwartz 1979). While formal reforms would have sooner or later challenged the exclusive political status of the elite, corruption offered the Soviet leadership a stable informal alternative to accommodate a diverse array of social interests but naturally at significant costs. Besides the obvious economic damage from corruption, a decrease in vertical mobility caused by corruption was one reason behind the growing dissatisfaction with the regime among the majority of the population that was excluded from privileged networks (Jowitt 1983). When the costs reached unbearable levels, the party had to initiate reforms – *perestroika* and *glasnost* – but they were not able to stop the disintegration of the Soviet Union.

When the formal Soviet institutions were demolished, informal practices took a leading role in all spheres of Russian life. Clientelistic networks formed out of existing branches of the *nomenklatura*, practices of *kormlenie*, patronage and *kompromat* were applied with even growing regularity (Kryshtanovskaya and White 1996). Brovkin (2003) provides an extensive account of corruption during the 1990s, particularly in the area of privatization, the military sector and banking. Nationwide, privatization became an instrument to transfer public property to the political elite, creating a class of rich oligarchs (Barnes 2006). Black et al. (2000) suggest that the corrupt nature of privatization laid the foundation for a kleptocratic regime as it provided lucrative windfall gains, which were reinvested into buying off politicians and media outlets to secure political power.

The corrupt practices that shaped Russian politics in the early 1990s made a U-turn to authoritarian rule seem logical. The reverse in democratization started with the presidential election of 1996, which was won by Boris Yeltsin as a result of a biased media campaign (Brovkin 1997), a politicized distribution of public money (Treisman 1998; Treisman and Gimpelson 2001) and electoral fraud (Myagkov and Ordeshook 2008). The development put Russian society in a familiar situation as corruption was used to supplement low official salaries and was tolerated as long as the subjects expressed loyalty and political support. However, this time the market economy was a more fertile soil for the practice of *kormlenie* than the rigid planning system of the past (Oleinik 2011). The persistence of the Soviet legacy on corruption was corroborated in a study by Libman and Obydenkova (2013), who find that regional corruption levels in 2010 strongly correlate with the number of Communist Party members who were registered in the region in 1976.

After the unsuccessful attempt at democratization (Evans 2011), it was only natural that the country came under the rule of a former KGB officer, Vladimir Putin. In the Soviet Union, the KGB was the main state security organ and was also responsible for the supervision of corruption and for reinforcing subordination within the Communist Party (Ledeneva 2006). In modern Russia, the KGB has evolved under a different name (Federal Security Bureau, FSB) into a powerful organization that established control over major spheres of the economic and political life via extensive corruption networks of its former and current members (Cheloukhine and King 2007). Putin and his inner circle profited enormously from the political monopoly – according to *Forbes* in 2017, the three billionaires who increased their wealth the most had the closest ties to Putin.[8] The personal wealth of Putin remains secret, but recent leaks

---

[8]   See a news article in *Forbes* about Russian billionaires at https://www.forbes.com/sites/danalexander/2017/03/29/putin-vladimir-donald-trump-russia-billionaires-oligarchs/#3ea2eb9 543f9 (accessed October 1, 2017).

of financial documents from Panama offshore companies suggest that it is by no mean modest.[9]

RUSSIAN CORRUPTION TODAY

Today, corruption is rampant in Russia. In the 2016 Corruption Perception Index (CPI) by Transparency International, Russia scored a low 29 points out of 100 points, placing it 131th out of 176 countries surveyed, comparable to other highly corrupt countries such as Iran, Nepal and Kazakhstan.[10] This ranking suggests that Russia is the most corrupt country among the members of the G-20.

Since empirical studies often find the CPI score useful for explaining differences in economic growth across countries (Mo 2001; Méon and Sekkat 2005), we use it for a back-of-the-envelope extrapolation of its effect on the Russian economy. Using a data set of 101 countries over the last 15 years and accounting for endogeneity and time-invariant heterogeneity, D'Agostino et al. (2016) estimated that a drop of CPI by one point leads to a loss of 0.056 percentage points in annual growth of the gross domestic product (GDP). If corruption in Russia were to improve to the level of Poland, a large post-communist neighbor, which scored 62 points in 2016, Russia would have had a positive GDP growth rate of 1.65 percent instead of its current negative rate of –1.0–0.2 percent.[11] In absolute numbers, the cost of corruption in one year is almost 24 billion dollars, or roughly the GDP of Estonia.

While being informative, cross-national comparisons disregard a distinct heterogeneity of corruption within large countries such as Russia. The available data at the regional level tell an interesting story of substantially divergent corruption levels across Russian regions. A good illustrative example is provided by the public opinion survey across Russian regions, carried out in 2011 by the Public Opinion Foundation (POF).[12] In this large survey, over 54,000 citizens in 74 regions were asked about their

---

[9]   The connections between Putin's friends and Panama offshore companies are elaborated in the investigation by Süddeutsche Zeitung. See http://panamapapers.sueddeutsche.de/articles/56fec05fa1bb8d3c3495adf8/ (accessed October 1, 2017).

[10]   Data are available at the Transparency International website, https://www.transparency.org/news/feature/corruption_perceptions_index_2016 (accessed October 1, 2017).

[11]   Data on GDP growth rate are from the World Bank, https://data.worldbank.org/indicator/NY.GDP.MKTP.CD?cid=GPD_29&locations=RU (accessed October 1, 2017). The results are relatively similar if we use data and the estimates from D'Agostino et al. (2016) for the World Bank index.

[12]   The full survey is available at the website of the Public Opinion Foundation, http://bd.fom.ru/pdf/d18pkr11.pdf (accessed October 1, 2017).

experience of corruption over the last two years. On average, 15 percent of respondents answered positively to the question: "during the last 1–2 years have you encountered a situation when a public official asked for or expected an unofficial payment or a gift for his/her services?" This share varies significantly across regions, ranging from 5 percent of positive answers in the Jewish Autonomous Oblast to a maximum of 35 percent in the Kabardino-Balkar Republic. The spatial distribution of corruption experiences is illustrated in Figure 12.1.[13]

Zakharov (2018) uses the differences in corruption across the regions to explain variations in local investment. He finds a negative correlation between corruption and investment by private companies and even more so for investment of companies with foreign ownership. As public opinion data can be biased, especially in the case of Russia (Sharafutdinova 2010), Zakharov employs data from police authorities on registered corruption cases as the preferred corruption measure (used previously by Schulze et al. 2016). The number of registered cases of bribe acceptance per 100,000 population is a good proxy for corruption as it reflects a number of corrupt incidents, is much less politicized than conviction rates, which are commonly used in the literature on corruption in the USA (Alt and Lassen 2012), and reflects the timing of corruption incidents more accurately. Because corruption incidence data are available for all regions for the period 2004–13, Zakharov (2018) is able to run fixed effects regressions that include region-specific and year fixed effects as well as a comprehensive set of controls and thus controls adequately for unobserved heterogeneity between regions. As causality between corruption and investment can run both ways, Zakharov establishes a causal effect of corruption on investment through an instrumental variables approach, in which the idiosyncratic development of regional freedom of press is used to instrument for (potentially endogenous) corruption levels. Again, he finds that corruption is a significant determinant of local under-investment for private, and in particular foreign, investment: periods when a region had a relatively free press and consequently experienced less corruption were periods with higher local investment. Private investment as one of the most important determinants for economic growth is thus extremely sensitive to corruption also in Russia.

Other studies analyse the effect of regionally different corruption levels on other economic outcomes. Kuzmina et al. (2014) used a survey of small and medium-sized firms in 40 Russian regions and found that the effect

---

[13]   This correlation has been previously found in Dininio and Orttung (2005) for earlier years.

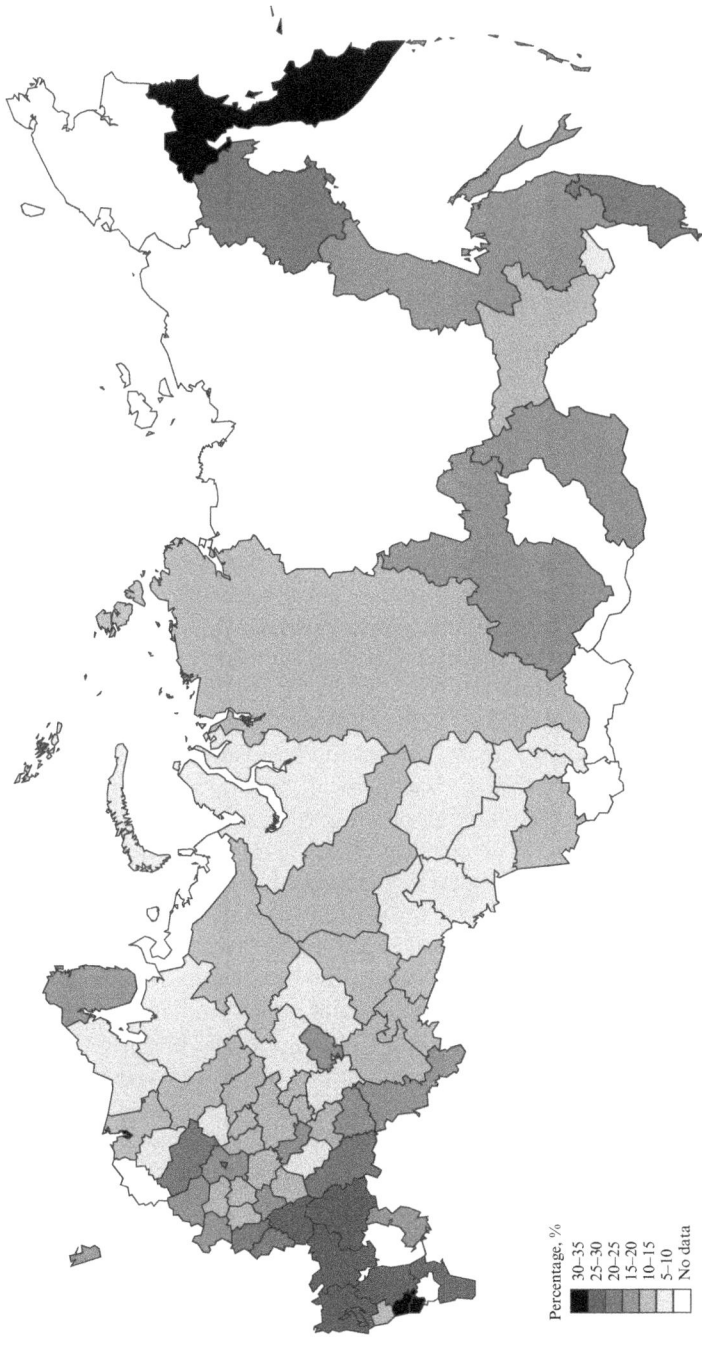

Legend:
Percentage, %
30–35
25–30
20–25
15–20
10–15
5–10
No data

Source: Public Opinion Foundation (http://bd.fom.ru/pdf/d18pkr11.pdf).

Figure 12.1 *Geographical distribution of corruption experiences of the Russian population in 2011 (percentage of respondents with corruption experience)*

of corruption on (non-offshore) foreign direct investment is significantly negative. They address a potential endogeneity of corruption by using historical determinants of regional corruption as instruments.

Corruption has also been found to be detrimental to bank lending. Using survey data on regional corruption perceptions and experiences from a survey by Tranparency International and the Information for Democracy Foundation and data on the volume of bank lending by 882 banks in Russia in 2002, Weill (2011) found that higher levels of corruption, especially self-reported amounts of bribes given, is strongly associated with fewer loans granted to households and firms, but not with loans to the government.

Even routine areas of everyday life, such as driving, are affected by corruption. Oleinik (2016) establishes a causal negative relationship between petty corruption by traffic police and road safety in Russian regions. He points out a vicious circle of increasing severity of traffic laws to enforce compliance and, as a result, even more widespread corruption to avoid increasing traffic fines.

The literature provides clear evidence that corruption in Russian regions remains politically motivated. At its core, corruption continues the tradition of *kormlenie*: appointment to the region as a governor comes with an opportunity to engage in illegal but financially rewarding activities, which are often tolerated by the federal center if coupled with loyalty and electoral support. Rochlitz (2014) finds that corruption among local public officials associated with criminal corporate raiding is positively correlated with regional results in the elections of the president and his ruling party. His findings suggest that federal authorities turn a blind eye to corrupt activities as long as the governor provides favorable electoral support at national elections. In line with this argument, Reuter and Robertson (2012) found that good electoral results for the national elections as a proxy for loyalty are more important for future reappointment of the governors than good governance as measured by economic development.

Since corruption remains a source of enrichment for governors (and their teams) only as long as they hold office, the end of their terms is accompanied by a sharp increase in rent-seeking. Sidorkin and Vorobyev (2017) identify these political cycles using surveys of business people and their timings with respect to end of the governors' terms. The cycle persists only when governors learn about the likelihood of their reappointment being low.

In the regions, rents from corruption are often invested in securing political power. Mironov and Zhuravskaya (2016) detect political cycles of corruption with respect to regional elections by using extensive financial data of Russian private firms in the period 1999–2004. They identify numerous "fly-by-night" firms, legal entities that exist for a very short duration, pay no taxes and employ no personnel, and show that these firms were often used to channel cash transfers associated with corruption. Fly-by-night firms received cash transfers from firms with government procurement contracts, and this money was used to fund electoral campaigns of governors. After the elections, even more procurement contracts were allocated in return for funding. The strength of these cycles is associated with the intensity of corruption in the region. These findings suggest that local corruption networks were able to undermine gubernatorial elections.[14]

CONTROL OF CORRUPTION

Understanding the determinants of corruption is crucial for formulating an effective anti-corruption strategy. We have argued that the historical institution of *kormlenie*, which gave rise to systemic corruption, has endured because of traditionally low official salaries in the public sector. Would a better remuneration solve the prevalence of corruption? This question is addressed by Schulze et al. (2016) in the most extensive empirical study of determinants of corruption in modern Russia. The authors use police data on bribe-acceptance incidents for the period 2004–13, which is prior to the annexation of Crimea and Western sanctions to avoid any political bias arising from these events. Using detailed regional salary data from the Russian Federal State Statistics Service (FSSS), Schulze et al. construct a measure of salaries of public officials in the region *relative* to the business counseling sector, which is the most comparable one to the civil service in terms of skill requirements and responsibilities. Employing both OLS (ordinary least squares) and fixed effects regressions, they find that corruption declines as relative salaries rise, but at diminishing rates. This implies that increases in salary promote more honest behavior especially among severely underpaid officials, and become less effective as salaries approach competitive levels. If salary levels reach about 176 percent of the average in business counseling, further increases in civil servants' pay will produce no anti-corruption effect, and could even be counterproductive. These findings are corroborated when cases of experienced corruption or corruption convictions are used instead of registered incidents of bribe acceptance, and they are robust to alternative

---

[14]   Gubernatorial elections existed in Russia until 2005, when they were replaced by a system in which the heads of the regional governments were appointment by the president. After 2012, the elections were reintroduced.

reference salaries of white-collar workers in manufacturing. The increase in remuneration of public officials is a costly anti-corruption measure and, as Schulze et al. (2016) show, effective only to a certain extent – it is by no means a silver bullet.

Stricter laws are often considered as an effective deterrent to corruption, and Schulze et al. (2016) tested if the introduction of new anti-corruption legislation at the regional level affected registered bribe acceptance. Regional laws were introduced shortly after the introduction of a federal anti-corruption law came into force in 2009. Their main purpose was to establish legislative instruments for fighting corruption, such as audits of the newest legislative acts against their potential vulnerability to corruption, creating intolerance for corruption among the population, and strengthening selection processes of new public officials. Schulze et al. (2016) show that there was a steady increase of over 10 percentage points in the number of registered bribe-acceptance incidents when regions installed new anti-corruption laws and attribute this effect to the increased opportunity to report corruption when the law is in place. Their findings indirectly speak in favor of controlling corruption by legislative initiatives.

Higher local unemployment rates reduce corruption levels as the opportunity costs of corruption increase – job loss and costs of finding an alternative job in case of detection. The same negative correlation with corruption is found for the share of economically active population with a university degree – a better education provides some safeguard against corrupt practices. While educational profile and unemployment rates are negatively correlated with corruption, it is hard to argue that the relationship is clearly unidirectional since corruption could make finding a new job more difficult and it could also discourage people from investing in education. Educational profile and unemployment are not immediately actionable parameters (and would be undesirable in the case of unemployment) and thus provide no entry point for fighting corruption.

(Relative) press freedom is an important determinant of corruption that does provide a second entry point for corruption control. Professional and independent journalism poses significant risks for corrupt officials of being exposed and prosecuted, as their actions can no longer be covered up. In many countries, this press freedom is curtailed by state capture of the media. In Russia, press freedom has been in decline since Putin came to power, however, some regions have preserved critical and independent media longer than the rest of the country, which has affected corruption levels. Schulze et al. (2016) show this using data from regional surveys of expert opinion from a Russian non-governmental organization, Glasnost Defense Foundation (GDF), which seeks to protect independent journalism and freedom of expression. The GDF surveys, which cover the period

2006–10 and 78 regions, rank local press freedom on a four point scale as "free," "relatively free," "relatively unfree" and "unfree." Because no single region is characterized as free, Schulze et al. (2016) analysed whether "relatively free" is good enough to curtail corruption. Interestingly, they find that when a region becomes "relatively free," there are 8 percent fewer cases of registered corruption incidents, and the converse effect materializes when a region loses its relative press freedom.

Zakharov (2018) looks into this relationship in more detail and notices that often the status of relative press freedom can coexist with violence or censorship against journalists, which reduces journalists' ability to effectively limit the bureaucracy; the effect of press freedom as found in Schulze et al. (2016) was reduced by two-thirds. The effect is most likely causal as the development of press freedom in Russia was idiosyncratic and exogenous to corruption, as discussed in Zakharov (2018).

The main policy implication of these findings is that corruption in Russia may be confined by sustaining freedom of press and allowing journalists to do their job independently without the risk of being censored, oppressed or physically harassed. A recent study by Enikolopov et al. (2018) shows that even one independent blogger investigating corruption can make a difference: they show that the anti-corruption blog posts by Aleksei Navalny, a popular Russian anti-corruption activist, affected market returns of state-controlled companies and their management turnover and lowered conflicts with minority shareholders, indicating an overall disciplining effect.

Evidence-based policy recommendations would thus include an encompassing civil service sector reform with higher salaries, strengthened accountability mechanisms, a non-corrupt judiciary and a clean central law enforcement agency. It would also strongly advocate an independent free press, free also from harassment by public or private agents. Yet such recommendations disregard the systemic nature and the incentives that have led to corruption. In a recent survey of the state of corruption research, Lambsdorff and Schulze (2015, p. 109) write, "Yet, as systemic corruption affects the entire political economic system, its understanding is crucial for designing successful anti-corruption policies. This includes a comprehension of factors that have led to corruption being systemic (and not only frequent) and of the way the incentive structures were designed to keep the system corrupt."

We have sketched the systemic nature of corruption in Russia and the incentives of the ruling elite to preserve corruption as a method of governance and enrichment. Any demand for truly enhanced accountability and effective, far-reaching anti-corruption reforms would challenge the political monopoly of the elites and thus would most likely evoke repressive policy responses rather than improved control of corruption. Yet, pervasive corruption has the potential to eventually suffocate the current regime economically and politically as it did with the Soviet Union, and may once again open an opportunity to build a more accountable and democratic state in Russia.

REFERENCES

Acemoglu, D. and J. Robinson. 2012. *Why Nations Fail: The Origins of Power, Prosperity, and Poverty.* New York: Crown Publishers.

Acemoglu, D., S. Johnson, and J. Robinson. 2001. The colonial origins of comparative development: an empirical investigation. *American Economic Review* **91**(5), 1369–401.

Alt, J. and D. Lassen. 2012. Enforcement and public corruption: evidence from the American states. *Journal of Law, Economics, and Organization* **30**(2), 306–38.

Anderson, L. 2012. Corruption in Russia: past, present, and future. In C. Funderburk (ed.), *Political Corruption in Comparative Perspective: Sources, Status and Prospects*, pp. 71–94. Burlington, VT: Ashgate.

Angeles, L. and K. Neanidis. 2015. The persistent effect of colonialism on corruption. *Economica* **82**(326), 319–49.

Arriola, L. 2009. Patronage and political stability in Africa. *Comparative Political Studies* **42**(10), 1339–62.

Barnes, A. 2006. *Owning Russia: The Struggle over Factories, Farms, and Power.* Ithaca, NY: Cornell University Press.

Berlin, P. 1910. Russian corruption as a social-historical phenomenon. *Contemporary World* **8**, 46–56 (in Russian).

Black, B., R. Kraakman, and A. Tarassova. 2000. Russian privatization and corporate governance: what went wrong? *Stanford Law Review* **52**(6), 1731–808.

Bliakher, L. 2013. The regional barons. *Russian Politics & Law* **51**(4), 30–9.

Brovkin, V. 1994. *Behind the Front lines of the Civil War: Political Parties and Social Movements in Russia, 1918–1922.* Princeton, NJ: Princeton University Press.

Brovkin, V. 1997. Time to pay the bills: presidential elections and political stabilization in Russia. *Problems of Post-Communism* **44**(6), 34–42.

Brovkin, V. 2003. Corruption in the 20th century Russia. *Crime, Law and Social Change* **40**(2), 195–230.

Charap, M. and C. Harm. 1999. Institutionalized corruption and the kleptocratic state. IMF Working Paper 99/91. International Monetary Fund, Washington, DC.

Cheloukhine, S. and J. King. 2007. Corruption networks as a sphere of investment activities in modern Russia. *Communist and Post-Communist Studies* **40**(1), 107–22.

Clark, W. 1993. Crime and punishment in Soviet officialdom, 1965–90. *Europe-Asia Studies* **45**(2), 259–79.

D'Agostino, G., J. Dunne, and L. Pieroni. 2016. Government spending, corruption and economic growth. *World Development* **84**, 190–205.

Darden, K. 2008. The integrity of corrupt states: graft as an informal state institution. *Politics & Society* **36**(1), 35–59.

Davies, B. 1997. The politics of give and take: kormlenie as service remuneration and generalized exchange, 1488–1726. In A. Kleimola and G. Lenhoff (eds), *Culture and Identity in Muscovy, 1359–1584*, pp. 39–67. Moscow: ITs-Garant.

Dawisha, K. 2015. *Putin's Kleptocracy: Who Owns Russia.* New York: Simon & Schuster.

Dimant, E. and G. Tosato. 2018. Causes and effects of corruption: what has past decade's empirical research taught us? A survey. *Journal of Economic Surveys* **32**(2), 335–56.

Dininio, P. and R. Orttung. 2005. Explaining patterns of corruption in the Russian regions. *World Politics* **57**(4), 500–29.

Enikolopov, R., M. Petrova, and K. Sonin. 2018. Social media and corruption. *American Economic Journal: Applied Economics* **10**(1), 150–74.

Evans, A. 2011. The failure of democratization in Russia: a comparative perspective. *Journal of Eurasian Studies* **2**(1), 40–51.

Fjelde, H. and H. Hegre. 2014. Political corruption and institutional stability. *Studies in Comparative International Development* **49**(3), 267–99.

Gorodnichenko, Y. and K. Sabiryanova Peter. 2007. Public sector pay and corruption: measuring bribery from micro data. *Journal of Public Economics* **91**(5), 963–91.

Harasymiw, B. 1969. Nomenklatura: the Soviet communist party's leadership recruitment system. *Canadian Journal of Political Science/Revue canadienne de science politique* **2**(4), 493–512.

Harrison, M. 2011. Forging success: Soviet managers and accounting fraud, 1943–1962. *Journal of Comparative Economics* **39**(1), 43–64.

Hartog, L. 1996. *Russia and the Mongol Yoke*. London: British Academic Press.

Hedlund, S. 2005. *Russian Path Dependence*. New York: Routledge.

Hosking, G. 2000. Patronage and the Russian state. *Slavonic and East European Review* **78**(2), 301–20.

Johnston, M. 2008. Japan, Korea, the Philippines, China: four syndromes of corruption. *Crime, Law and Social Change* **49**(3), 205–23.

Jowitt, K. 1983. Soviet neotraditionalism: the political corruption of a Leninist regime. *Soviet Studies* **35**(3), 275–97.

Kis-Katos, K. and G. Schulze. 2013. Corruption in Southeast Asia: a survey of recent research. *Asian-Pacific Economic Literature* **27**(1), 79–109.

Kopanev, A. 1984. *Peasantry of the Russian North in the 17th Century.* Leningrad: Nauka (in Russian).

Korchmina, E. and I. Fedyukin. 2018. Extralegal payments to state officials in Russia, 1750s–1830s: assessing the burden of corruption. *The Economic History Review*, forthcoming.

Kovalevsky, M. 1902. *Russian Political Institutions*. Chicago, IL: University of Chicago Press.

Kramer, J. 1977. Political corruption in the USSR. *Western Political Quarterly* **30**(2), 213–24.

Kryshtanovskaya, O. and S. White. 1996. From Soviet nomenklatura to Russian elite. *Europe-Asia Studies* **48**(5), 711–33.

Kuzmina, O., N. Volchkova, and T. Zueva. 2014. Foreign direct investment and governance quality in Russia. *Journal of Comparative Economics* **42**(4), 874–91.

Lambsdorff, J. and G. Schulze. 2015. What can we know about corruption? A very short history of corruption research and a list of what we should aim for. *Journal of Economics and Statistics (Jahrbücher für Nationalökonomie und Statistik)* **235**(2), 100–14.

Ledeneva, A. 2006. *How Russia Really Works: The Informal Practices that Shaped Post-Soviet Politics and Business.* Ithaca, NY: Cornell University Press.

Ledeneva, A. 2009. From Russia with "blat": can informal networks help modernize Russia? *Social Research* **76**(1), 257–88.

Libman, A. and A. Obydenkova. 2013. Communism or communists? Soviet legacies and corruption in transition economies. *Economics Letters*, **119**(1), 101–3.

Matthews, M. 2011. *Privilege in the Soviet Union: A Study of Elite Life-styles under Communism*. Abingdon, UK: Routledge.

McLeod, R. 2008. Inadequate budgets and salaries as instruments for institutionalizing public sector corruption in Indonesia. *South East Asia Research* **16**(2), 199–223.

McLeod R. 2012. Endemic corruption in Indonesia during the Soeharto era. In S. Khoman (ed.), *A Scholar for All: Essays in Honour of Medhi Krongaew*, pp. 165–79. Bangkok: Thamasat University Press.

Méon, P. and K. Sekkat. 2005. Does corruption grease or sand the wheels of growth? *Public Choice* **122**(1), 69–97.

Mironov, M, and E. Zhuravskaya. 2016. Corruption in procurement and the political cycle in tunneling: evidence from financial transactions data. *American Economic Journal: Economic Policy* **8**(2), 287–321.

Mo, P. 2001. Corruption and economic growth. *Journal of Comparative Economics* **29**(1), 66–79.

Myagkov, M. and P. Ordeshook. 2008. Russian elections: an oxymoron of democracy. VTP Working Paper No. 63. California Institute of Technology, Pasadena, CA, and Massachusetts Institute of Technology, Cambridge, MA.

O'Hara, P. 2014. Political economy of systemic and micro-corruption throughout the world. *Journal of Economic Issues* **48**(2), 279–308.

Oleinik, A. 2011. *Market as a Weapon: The Socio-economic Machinery of Dominance in Russia.* New Brunswick, NJ: Transaction Publishers.

Oleinik, A. 2016. Corruption on the road: a case study of Russian traffic police. *IATSS Research* **40**(1), 19–25.

Olken, B. and R. Pande. 2012. Corruption in developing countries. *Annual Review of Economics* **4**(1), 479–509.

Ostrowski, D. 2002. *Muscovy and the Mongols: Cross-cultural Influences on the Steppe Frontier.* Cambridge: Cambridge University Press.

Pavroz, A. 2017. Corruption-oriented model of governance in contemporary Russia. *Communist and Post-Communist Studies* **50**(2), 145–55.

Quimpo, N. 2009. The Philippines: predatory regime, growing authoritarian features. *Pacific Review* **22**(3), 335–53.

Reuter, O. and G. Robertson. 2012. Subnational appointments in authoritarian regimes: evidence from Russian gubernatorial appointments. *Journal of Politics* **74**(4), 1023–37.

Rochlitz, M. 2014. Corporate raiding and the role of the state in Russia. *Post-Soviet Affairs* **30**(2–3), 89–114.

Schulze, G., B. Sjahrir, and N. Zakharov. 2016. Corruption in Russia. *Journal of Law and Economics* **59**(1), 135–71.

Schwartz, C. 1979. Corruption and political development in the USSR. *Comparative Politics* **11**(4), 425–43.

Sharafutdinova, G. 2010. What explains corruption perceptions? The dark side of political competition in Russia's regions. *Comparative Politics* **42**(2), 147–66.

Sidorkin, O. and D. Vorobyev. 2017. Political cycles and corruption in Russian regions. *European Journal of Political Economy*, **52**, 55–74.

Simis, K. 1982. *USSR – the Corrupt Society: the Secret World of Soviet Capitalism.* New York: Simon & Schuster.

Treisman, D. 1998. Dollars and democratization: the role and power of money in Russia's transitional elections. *Comparative Politics* **31**(1), 1–21.

Treisman, D. and V. Gimpelson. 2001. Political business cycles and Russian elections, or the manipulations of "Chudar." *British Journal of Political Science* **31**(2), 225–46.

Vaksberg, A. 1991. *The Soviet Mafia.* New York: St. Martin's Press.

Weill, L. 2011. How corruption affects bank lending in Russia. *Economic Systems* **35**(2), 230–43.

World Bank. 1997. *World Development Report: The State in a Changing World.* New York: Oxford University Press.

Zakharov, N. 2018. Does corruption hinder investment? Evidence from Russian regions. *European Journal of Political Economy,* forthcoming.

Zhuravleva, T. 2013. Corruption measurement: the case of the Russian Federation. Working Paper, Gaidar Institute for Economic Policy, Moscow.

Zhuravleva, T. 2016. Social benefits, job security and corruption: what "fine" state employees. The Russian Presidential Academy of National Economy and Public Administration Working Paper No. 3051 (in Russian).

Zygar, M. 2016. *All the Kremlin's Men: Inside the Court of Vladimir Putin.* New York: PublicAffairs.

13. Turkey's fight against corruption: current state and the road ahead

Alfredo Jiménez, Secil Bayraktar and Mesut Eren

Corruption is a global phenomenon affecting both developed and developing economies. While its clandestine nature prevents the existence of accurate statistics, the World Bank estimates that the costs of corruption amount to at least $1 trillion annually. Corruption is considered critical since it reflects the economic, political, cultural, and legal situation in a country (Godinez and Liu 2015; Peng et al. 2008; Svensson 2005).

There are multiple definitions of corruption. For example, the World Bank (2000) defines corruption as the abuse of power to acquire private benefits. Accordingly, some forms of corruption include bribery, nepotism, embellishment, and inappropriate payments or use of influence in public contracting. Other authors define corruption as "monetary payments to agents (both public and private) to induce them to ignore the interests of their principals and to favor the private interests of the bribers instead" (Rose-Ackerman 2006, p. 14) or "the abuse of entrusted power for private gain" (Cuervo-Cazurra 2016, p. 36). Despite these multiple definitions, researchers have a common understanding that corruption denotes a serious indication of poor institutional quality and disrespect for rules and regulations (Kaufmann et al. 2003; Lee et al. 2010).

Previous research investigating the effect of corruption on investment success has found contradictory results (Cuervo-Cazurra and Genc 2008; Eren and Jiménez 2015). Some researchers have argued that corruption has a positive effect on efficiency, following the "grease the wheels" perspective. They claim that corruption increases efficiency by allowing the most interested firms to pay to speed up bureaucratic processes or by providing an opaque environment in which disruptive technology can be hidden from imitating competitors (Egger and Winner 2005; Leff 1964; Olson 1993). On the other hand, proponents of the "sand the wheels" perspective hold that corruption has a negative effect on investment success since corruption can increase costs and uncertainty (Brouthers et al. 2008; Mauro 1995; Wei 2000a, 2000b). Corrupt countries are usually perceived as less attractive to foreign investors because of their lack of protection for firm assets, including brands, patents, and technology (Brada et al.

2012). However, other studies have not found such an impact (Alesina and Weder 2002; Busse and Hefeker 2007; Wheeler and Mody 1992).

Although the corruption literature provides evidence for both positive and negative effects of corruption on investments, the most recent empirical studies have shown that the effect is mainly negative, supporting the "sand the wheels" approach (Godinez and Liu 2015; Lee and Weng 2013). Possible reasons for this negative effect are inequality of income distribution, enhanced poverty levels, distortion in public policies, and slower information dissemination caused by corruption (Ades and Di Tella 1997; Chen et al. 2010).

In line with the studies concluding that corruption has a negative impact on the foreign direct investments in a country, many studies focused on Turkey have also demonstrated the negative influence of corruption (e.g., Akan and Arslan 2007; Eren and Jiménez 2015; Tosun et al. 2014). As corruption reduces investment, which in turn inhibits economic growth (Mauro 1995), it is a necessity to treat corruption as a major issue. In order to analyse the impact of corruption in Turkey, we first review the socio-economic and cultural roots of Turkey and then the current state of corruption in Turkey. We then discuss what has been done, including the measures taken, against corruption what remains a concerns. Finally, before concluding, we elaborate on the road ahead and what needs to be done.

THE CONTEXT: SOCIAL, ECONOMIC, AND CULTURAL ROOTS

The phenomenon of corruption cannot be studied in isolation from the socio-cultural environment in which it emerges (Baran 2000; Pena Lopez and Sanchez Santos 2014; Warf 2016). In what follows we aim to provide a description of the social, economic, and cultural features that define the Turkish context.

Turkey in a Snapshot

Turkey is located in both southeastern Europe and western Asia, with cultural roots going back to the ancient Greek, Roman, Byzantine, Persian, and Ottoman Empires. Turkey offers an unprecedented example of an imperial past and successive modern nation state, both experienced in a single century. The governmental system in Turkey is a republic, and its constitution declares it to be a secular, democratic, and parliamentary system (Kabasakal et al. 2012). Roughly 99 percent of the population adheres to the Islamic religion and the official language is Turkish, spoken by 90 percent of the people (Kabasakal and Bodur 2007).

According to data from the World Bank, Turkey is currently the 18th largest economy in the world, with an expected growth rate of 4.7 percent in 2018, becoming the fastest growing among Organisation for Economic Co-operation and Development (OECD) countries. Turkey is negotiating its admittance to the European Union. Its population exceeds 80 million (with more than 50 percent under 31 years old) and foreign trade peaked at $341 billion at the end of 2016 (Turkish Statistical Institute 2017).

Economic Development and Corruption

Bedirhanoglu (2007) describes Turkey as a neo-liberalizing state, where corruption accompanies social and economic change. According to Cuervo-Cazurra and Genc (2008), low salaries of public employees provide a suitable environment for corrupt practices (e.g., bribes) in developing countries. Bribery is a prevalent issue in Turkey. Turkey was ranked 19th out of 28 countries on the 2011 Bribe Payers Index (Transparency International 2011); Turkish companies' propensity to engage in bribery while doing business had a score of 7.5 on a scale of 10. According to the *Global Enabling Trade Report* (GETR 2014), corruption at the borders (i.e., customs administration) was common especially due to lack of efficiency and time-consuming bureaucratic restrictions limiting exports and imports, which open the way for public officials to demand bribes.

Central Role of the State

The state is the central institution shaping business practices in Turkey (Kabasakal and Bodur 2007). The bureaucratic centralized state of the Ottoman Empire continued in the Republic (Soyaltin 2017a). The centrality of power in the state may facilitate corruption. For example, patrimonial characteristics of the state facilitate the development of patron-client relationships between public officials and key players in the private domain, which increase illegitimate actions such as favoritism (Ozen and Akkemik 2012). Ozen (2014) further posits that a strong state leads to weak institutions, and consequently gives opportunities for elite groups to engage in informal politics. Bodur and Madsen (1993) found that governmental regulations and personal relations with the powerful governmental officials were crucial for Danish investors to do business in Turkey. Furthermore, the unpredictable and frequent policy changes introduced by the government may increase the uncertainty of the business environment, and thus pose challenges to private companies, which are predominantly dependent on the state for financial incentives (Bugra 1990). Although the extensive liberalization process diminished the role of the state in recent years, newly established institutions could not effectively implement the policies; neopatrimonial practices gave way to new forms of informal networks rather than being transformed; and institutional legacies inherited from the old regime continues to have an impact (Soyaltin 2017b).

Geography and Culture

Are the level and the extent of corruption different from one country to another? Are countries the prisoners of their geographic location? Does location have a bearing on the corruption level of a country? These are questions that are frequently asked. According to Dastmalchian and Kabasakal (2001), geography paves the way for commonality in some variables like language, ethnicity, climate, and religion. Turkey falls into the Middle Eastern/North African (MENA) cluster according to the GLOBE study (House et al. 2004). However, it should also be noted that even though MENA region countries have commonalities in their societal norms, they also have many dissimilar demographic, socio-economic, and ethnic dynamics (Kabasakal et al. 2012). Therefore, it would be fair to analyse each country both in relation to regional commonalities and its unique characteristics.

The GLOBE study (House et al. 2004) defines culture as shared motives, values, beliefs, identities, and meanings of significant events that result from common experiences of members of collectives and are transmitted across generations. Many studies have shown that values and underlying belief systems lead to different business practices (Hayes and Prakasam 1989; Newman and Nollen 1996). Lee and Guven (2013) assert that it is surprising that cultural dimensions have been largely neglected in the corruption literature, given the fact that culture has a significant impact on business practices. They further claim that in order to comprehend how corruption becomes the norm, one needs to understand the processes by which people feel more or less comfortable about engaging in corruption, and thus there is the need to analyse corruption through the lens of culture as well. Along the same lines, Bardhan (1997) posits that something perceived as corrupt in one culture can be acceptable as a routine transaction in another one, and vice versa. For example, gift exchanges or small tips (*baksheesh*) may be a common practice for doing business in Middle Eastern countries, including Turkey. Similarly, Warf (2016) states that geographic variations of corruption can be attributed in part to differences in national cultures; however, he also advises against deeming a country

Table 13.1   *Turkey's scores in power distance and in-group collectivism cultural practices in comparison to the MENA region and the world*

	Turkey	MENA region average	World average
Power distance	5.57	5.19	5.15
In-group collectivism	5.88	5.52	5.12

Source: GLOBE project data (House et al. 2004); MENA region and world average scores calculated by the authors.

corrupt by claiming that the culture is corrupt. Acar (2015) points to the need for studies in Turkey that analyse the culture-corruption linkage.

In the following we give a brief introduction to the cultural context in Turkey and explain how certain cultural dimensions can facilitate normalization or legitimization of some corrupt practices.

Power distance and in-group collectivism
According to GLOBE (House et al. 2004), power distance is the degree to which members of a society expect and accept unequal distribution of power and in-group collectivism is the degree to which individuals express pride, loyalty, and cohesiveness in their organizations or families. Turkish society is characterized by high power distance and high collectivism (Hofstede 1980; House et al. 2004; Kabasakal and Bodur 2007). According to the results of the GLOBE study, Turkey's scores in power distance (5.57) and in-group collectivism (5.88) are higher than the world average (5.15 and 5.12, respectively) and the MENA region (5.19 and 5.52, respectively), as displayed in Table 13.1 (Kabasakal et al. 2012). Hierarchical practices and a strong group orientation prevail in the society. High societal power distance in Turkey results in inequality and stratification of members according to their prestige, status, authority, and wealth (Pellegrini and Scandura 2006).

Pena Lopez and Sanchez Santos (2014) found that power distance and collectivism are two major cultural dimensions that have a positive linkage to corruption practices. They assert that especially in cultures characterized by dependency relations and closed communities (e.g., high power distance and high collectivism), corruption practices are legitimized more easily, thus setting the conditions for development of rent-seeking behavior. Collectivist cultures place more emphasis on maintaining relationships and the identity of members comes from the groups to which they belong

(Pellegrini and Scandura 2006; Sullivan et al. 2003). In such cultures, in-group members (e.g., close family members or friends) are often given privileges over out-groups in business life, which may lead to nepotism.

Trust

Research has found that societies with high levels of trust rank better in corruption scores (Uslaner 2008a). Turkey has a very low level of inter-personal trust according to international surveys (e.g., the World Values Survey), where it ranks towards the bottom of the list (Kayaoglu 2017); low trust levels set a fertile ground for corruption. Kayaoglu (2017) asserts that high levels of distrust in Turkey are a hindrance to socio-economic development, and suggests that determinants of trust must first be understood to formulate proper social and economic policies. She suggests that improving labor market conditions, investing in education, and enhancing society's level of civic engagement are crucial to increase trust levels in Turkey.

Paternalism

Turkey scores high on paternalistic values (Aycan et al. 2000). In paternal-istic cultures, there is a reciprocated form of relationship where political authorities take on a parental role and provides care and protection to those under their power, and subordinates show compliance and loyalty in return (Pellegrini and Scandura 2006). Paternalistic cultures are linked to corrupt practices such that authority figures tend to build patronage relationships with their subordinates, which may promote employment practices based on patronage rather than professional links (Kabasakal and Bodur 2007).

Culture and legitimization of corruption

Acar (2015) investigated corruption "excusing" or "legitimizing" behav-iors in Turkey in a qualitative study composed of 55 face-to-face inter-views. He refers to the most common practices of corruption in Turkey as nepotism/favoritism in recruitment and promotion in the public sector, misuse of public resources, rent seeking in city planning, gift giving and bribery in the public sector, public procurement, and using "middle-men" or "strong men." He further links the enduring corruption to the country's socio-cultural roots, drawing attention to how traditional idioms and sayings reflect the normalizing view of corruption. Acar (2015) further states that many common corruption practices originated in the Ottoman Era, such as gift giving (a form of bribery), and are thus deeply ingrained in the culture. He holds that people tended to normalize or justify cor-ruption with reasons such as "everybody does it," "they steal but at least they work hard," "corruption rotates between people over the eras, they did it before, now it is our turn," and "it is a period of transition and transformation coming along with being a developing country." He argues that the normalizing-legitimizing reasons reflect highly culture- and country-specific tones and that the resilience of corruption in the country can be attributed to the perceived role of social and cultural norms as well as commonality of these normalization efforts.

STATE OF CORRUPTION IN TURKEY

The World Economic Forum's *Global Competitiveness Report* (2014) found that corruption was the ninth most problematic factor for doing business in Turkey. According to other research (Transparency International Global Corruption Barometer, GCB), citizens shared the perception that corruption is a serious concern in the country and 57 percent of the respondents believed that it had increased in the last three years (between 2007 and 2010) (GCB 2010). To assess the business world's perception towards corruption, a study conducted by TUSIAD in 2014 showed that 801 business men and women, who participated in the study, viewed corruption as one of the foremost problems impacting their business conduct; 46 percent of the respondents believed that corruption would follow an increasing trend in Turkey. Caha (2009) refers to finance, health, defense, and energy as the most corrupt sectors in Turkey as well as bureaucratic and political reasons and economic policies applied by governments as constituting the main reasons behind corruption.

Table 13.2 illustrates the evolution of Turkey's scores and country ranking over the last 20 years (1997–2016). The scores are taken from the Corruption Perception Index, the most widely used measure of corruption in the world, compiled by Transparency International (http://www.transparency.org), a non-governmental organization that monitors corruption in the private and public sector. This measure is a composite index combining surveys and assessments of corruption based on how corrupt a country's public sector is perceived to be.

Figure 13.1 illustrates the evolution of Turkey's perceived corruption scores. In 2016, Turkey had a score of 41 and was listed as the 75th least corrupt among 176 countries. After an increasing trend between 2011 and 2013, Turkey's scores started to decline after 2013, decreasing from 50 in 2013 to 45, 42, and 41, respectively, in the following years. In line with this trend, in the ranking of countries from the least to the most corrupt, Turkey's ranking fell from 53 in 2013 to 75 in 2016. An economic analysis by ISMMMO (Istanbul Muhasebeciler ve Mali Musavirler Odası) in 2007 of corruption scores estimated that a one point loss in country score resulted in a loss of 10.2 million Turkish liras and a 11 percent decrease in

Table 13.2   Turkey's perceived corruption scores, 1997–2016

	Score	Ranking	Number of countries in the ranking
2016	41	75	176
2015	42	66	167
2014	45	64	175
2013	50	53	177
2012	49	54	176
2011	42	61	183
2010	44	56	178
2009	44	61	180
2008	46	58	180
2007	41	64	179
2006	38	60	163
2005	35	65	158
2004	32	77	145
2003	31	77	133
2002	32	64	102
2001	36	54	91
2000	38	50	90
1999	36	54	99
1998	34	54	85
1997	32	38	52

Note: The scores of the Corruption Perception Index range from 0 representing a completely corrupt state to 10 for an entirely corruption-free state.

Source: Transparency International.

foreign direct investment (FDI). Furthermore, using a gravity estimation approach, De Groot et al. (2004) argue that by improving the institutions and transparencies through a regulatory framework and thus reducing corruption, trade would increase by 17–27 percent.

Using the data from Corruption Perceptions Index (CPI) of Transparency International, Warf (2016) categorized countries according to their level of corruption. Turkey falls into the moderately corrupt category (third most corrupt group with scores ranging from 40 to 59), which consists of states with a growing middle class, increasing life standards, and high levels of inequality (Warf 2016). Uslaner (2008b) points to a vicious circle that comes with inequality such that higher levels of inequality lead to lower levels of trust in the society, which in turn leads to more corruption and bigger gap in income levels, and thus resulting in even more inequality.

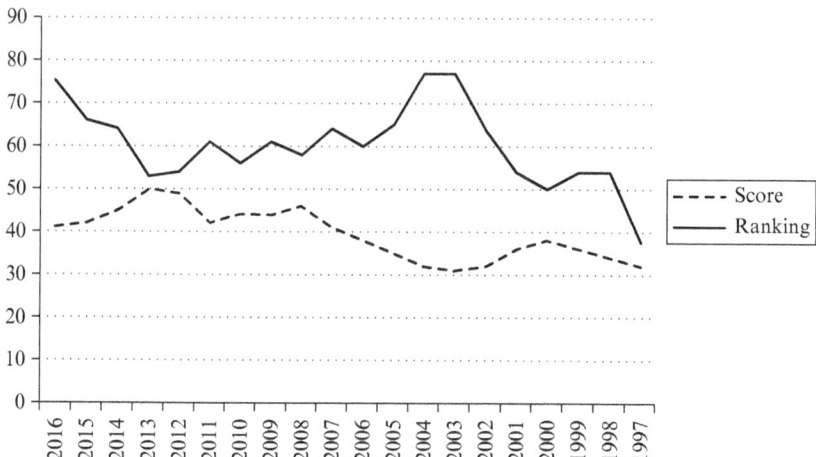

Source: Authors based on data from Transparency International's Corruption Perception Index.

Figure 13.1 Turkey's perceived corruption scores chart, 1997–2016

WHAT HAS BEEN DONE? MEASURES TAKEN AGAINST CORRUPTION

After Turkey applied for admission to the European Union (EU) in 1999, the EU requirements challenged it to take serious steps to fight corruption (Ulusoy 2014). The financial crisis in 2001, together with the demands of the International Monetary Fund, triggered the planning and implementation of more severe, organized, and comprehensive reforms against corruption, including many international conventions (Duzgit and Carkoglu 2009; Soyaltin 2017a). An action plan was designed in 2002 to enhance transparency, improve governance, modernize audit systems, and increase sanctions, which was followed by an agreement between the EU and Turkey in 2003 aimed at raising public awareness (Chene 2012). Turkey also became a member of the Group of States against Corruption (GRECO) in 2004. Especially after 2010, major efforts were undertaken to reduce corruption with the realization that a more collaborative and coordinated effort involving other institutions was needed. In line with that intent, the Financial Crimes Investigation Board, Coordination Board for Combating Financial Crimes, Council of Ethics for the Public Service, and Turkish Court of Accounts took on more active roles. Other efforts included simplifying company establishment procedures, reducing permit requirements, facilitating company registration through a single form, and establishing the Investment Support and Promotion Agency to facilitate the use of public services (Business Anti-corruption Portal 2011). Among the specific measures taken, the introduction of e-government online applications was a significant step forward, as it made it possible to monitor public expenditures via the Internet (Bozkurt 2010). Furthermore, the new Turkish Commerce Law that effectuated as of July 1, 2012 entailed the company annual reports, which is an important indicator of transparency and accountability. In 2016, the Turkish Prime Ministry announced an action plan on "Increasing Transparency and Strengthening the Fight against Corruption" (Global Legal Insights 2017). The action plan focused on: (1) prevention; (2) enforcement of sanctions; and (3) enhancement of social awareness. While the fight against corruption continues to be declared a major priority of the government, the changes taking place in the last decade unfortunately have not been sufficient to make significant progress.

WHAT REMAINS A CONCERN?

Despite the efforts undertaken to fight corruption since the early 2000s, Turkey still endures this problem. The consecutive fall of CPI scores between 2013 and 2016 have also led to criticisms from international organizations, mainly regarding issues such as lack of an independent anti-corruption body, excessive political influence on the judiciary and law system, and ineffective investigation, prosecution, and conviction procedures (Kalayci et al. 2017). Okuyucu-Ergun (2007) suggests that stand-alone reforms remain insufficient for effective outcomes. The European Union reported in 2007 reported insufficient progress in Turkey and demanded improvement in regulations and a central governance body (European Commission 2007). The European Commission progress report for 2010–14 mentioned concerns about lack of legislation, inadequate judicial statistics, and the proper implementation of anti-corruption strategies (Tosun and Yurdakul 2016).

A "decoupling" issue emerges as a major problem in the ineffective results of anti-corruption reforms such that there is a gap between the formal rules and regulations and how they are adopted in real life practices. Soyaltin (2017b) addressed the issue of decoupling by using both empirical data from policy documents and reports and 45 semi-structured interviews with policy makers and experts. She argued that the practice-regulation gap in terms of enforcing laws can only be decreased if favorable domestic conditions are met. In other words, anti-corruption laws can be successfully implemented only with strong administrative coordination and low bureaucratic costs, which constitute the favorable domestic conditions. She further stated that decoupling varies across three sectors, namely, civil administration, public finance management, and public procurement because of differential conditions regarding the bureaucratic costs and administrative coordination in each sector. Accordingly, the civil administration in Turkey has seen the most effective institutional change against corruption among the three because it could enforce both legal and administrative change with having both favorable conditions present. However, public finance management lacked strong coordination, in which case legal change was barely followed by administrative changes. On the other hand, in public procurement, high bureaucratic costs associated with the required legal reforms made legal change more difficult and caused selective institutional change. Soyaltin (2017b) draws attention to the fact that the decoupling problem needs to be handled by considering domestic conditions for reforms to be implemented effectively across different sectors.

THE ROAD AHEAD

Despite the major efforts undertaken to reduce corruption and improve the business environment, there is still a dire need for further progress. Chene (2012) mentions some points to take into consideration, such as the need for a central organization that is in charge of developing and evaluating policies against corruption, inadequate coordination of institutions involved, and an independent body to monitor anti-corrupt policy implementation. According to Bozkurt (2010), possible solutions include reducing the resources used by the public sector, increasing the administrative capacity of the Council of Ethic for Public Awareness, and increasing awareness in both the public and private sectors.

Pustu (2017) and Berkman (2010) assert that in the fight against corruption, government has not sought the support of other relevant institutions, including the non-governmental organizations and the media. Pustu further suggests that the media should be used as an effective tool to fight corruption since the media can help in ways such as starting a public investigation, forcing corrupt politicians to resign, encouraging the publication of public audit reports, and improving policies that set the ground for corruption.

Finally, a culture of ethics and ethical leadership in organizations may be a priority. Ethical leadership is defined as "the demonstration of normatively appropriate conduct through personal actions and interpersonal relationships, and the promotion of such conduct to followers through two-way communication, reinforcement, and decision-making" (Brown et al. 2005, p. 120). Implementing a culture of ethics among top managers can be influential (Bozkurt 2010). Setting the tone at the top with a commitment to honesty, integrity, and ethical behavior has significant impacts on the culture of an organization, just as political leaders' actions have significant influence on people's values and behaviors in the society (Kalaycı et al. 2017).

Consequently, when the legal regulations are insufficient, a culture of ethics that has been implemented from the top can act as a control mechanism itself.

REFERENCES

Acar, M. 2015. Report legitimization and normalization of corruption in the public administration in Turkey: cultural and institutional perspectives. *Anti-Corruption Policies Revisited, Global Trends and European Responses to the Challenge of Corruption*. http://anticorrp.eu/publications/legitimation-and-normalization-of-corruption-in-the-public-administration-in-turkey-cultural-and-institutional-perspectives (accessed September 1, 2007).
Ades, A. and Di Tella, R. 1997. National champions and corruption: some unpleasant interventionist arithmetic. *Economic Journal* **107**, 1023–43.
Akan, Y. and Arslan, I. 2007. The effect of corruption on foreign direct investment, the case of Turkey. *Lex ET Scientia International Journal* **XIV**, 200–6.
Alesina, A. and Weder, B. 2002. Do corrupt governments receive less foreign aid? *American Economic Review* **92**(4), 1126–37.
Aycan, Z., Kanungo, R., Mendonca, M., Yu, K., Deller, J., Stahl, G., and Kurshid, A. 2000. Impact of culture on human resource management practices: a 10-country comparison. *Applied Psychology: An International Review* **49**(1), 192–221.
Baran, Z. 2000. Corruption: the Turkish challenge. *Journal of International Affairs* **1**(54), 127–46.
Bardhan, P. 1997. Corruption and development: a review of issues. *Journal of Economic Literature* **35**(4), 1320–46.
Bedirhanoglu, P. 2007. The neoliberal discourse on corruption as a means of consent building: reflections from post-crisis Turkey. *Third World Quarterly* **28**(7), 1239–54.
Berkman, U. 2010. Yolsuzlukla mücadelede yeni strateji arayışı: Devlet merkezli yaklaşımdan toplum ve paydaş merkezli stratejiye yöneliş. *İş Ahlakı Dergisi* **3**(6), 81–93.
Bodur, M. and Madsen, T. 1993. Danish foreign direct investments in Turkey. *European Business Review* **93**(5), 28–44.
Bozkurt, C. 2010. Yolsuzlukla mücadele boyutuyla etik ve etik kültürü. *Denetişim* **5**, 44–53.
Brada, J.C., Drabek, Z., and Perez, M. 2012. The effect of home-country and host-country corruption on foreign direct investment. *Review of Development Economics* **16**(4), 640–63.
Brouthers, L., Gao, Y., and McNicol, J. 2008. Corruption and market attractiveness influences on different types of FDI. *Strategic Management Journal* **29**(6), 673–80.
Brown, M., Treviño, L., and Harrison, D. 2005. Ethical leadership: a social learning perspective for construct development and testing. *Organizational Behavior and Human Decision Processes* **97**(2), 117–34.
Bugra, A. 1990. The Turkish holding company as a social institution. *Journal of Economics and Administrative Studies* **4**(1), 35–51.

Business Anti-corruption Portal. 2011. Turkey country profile. http://www.business-anti-corruption.com/countryprofiles/europe-central-asia/turkey (accessed August 28, 2007).

Busse, M. and Hefeker, C. 2007. Political risk, institutions and foreign direct investment. *European Journal of Political Economy* **23**(2), 397–415.

Caha, H. 2009. Türkiye'de yolsuzluk: Yapısal boyutlar ve uygulama. Amme İdaresi. *Dergisi* **42**(1), 105–37.

Chen, C., Ding, Y., and Kim, C. 2010. High-level politically connected firms, corruption and analyst forecast accuracy around the world. *Journal of International Business Studies* **41**(9), 1502–24.

Chene, M. 2012. *Overview of Corruption and Anti-corruption in Turkey*. U4 Expert Answer Report no. 313, Transparency International.

Cuervo-Cazurra, A. 2016. Corruption in international business. *Journal of World Business* **51**(1), 35–49.

Cuervo-Cazurra, A. and Genc, M. 2008. Transforming disadvantages into advantages: developing-country MNEs in the least developed countries. *Journal of International Business Studies* **39**, 957–79.

Dastmalchian, A. and Kabasakal, H. 2001. Leadership and culture in the Middle East: norms, practices and effective leadership attributes in Iran, Kuwait, Turkey and Qatar. *Applied Psychology: An International Review* **50**(4), 479–595.

De Groot, H., Linders, G., Rietveld, P., and Subramanian, U. 2004. The institutional determinants of bilateral trade patterns. *Kyklos* **57**(1), 103–23.

Duzgit, A.S. and Carkoglu, A. 2009. Reforms for a consolidated democracy: Turkey. In L. Morlino and A. Magen (eds), *International Actors, Democratization Rule of Law: Anchoring Democracy?* pp. 120–56. UACES Contemporary European Studies. New York: Routledge.

Egger, P. and Winner, H. 2005. Evidence on corruption as an incentive for foreign direct investment. *European Journal of Political Economy* **21**(4), 932–52.

Eren, T.M. and Jiménez, A. 2015. Institutional quality similarity, corruption distance and inward FDI in Turkey. *Journal for East European Management Studies* **20**(1), 88–101.

European Commission, 2007. *Turkey 2007 Progress Report*, SWD (2007) 1436. Brussels: European Commission.

GCB. 2010. Transparency International, 2010 Global Corruption Barometer. http://www.transparency.org/policy_research/surveys_indices/gcb/2010 (accessed September 12, 2007).

GETR. 2014. *The Global Enabling Trade Report 2014*. M. Hanouz, T. Geiger, and S. Doherty. http://www3.weforum.org/docs/WEF_GlobalEnablingTrade_Report_2014. pdf (accessed September 6, 2007).

Global Legal Insights. 2017. Turkey: bribery and corruption 2017. https://www.globallega-linsights.com/practice-areas/bribery-and-corruption/global-legal-insights---bribery-and-corruption/turkey (accessed August 31, 2007).

Godinez, J. and Liu, L. 2015. Corruption distance and FDI flows into Latin America. *International Business Review* **24**(1), 33–42.

Hayes, J. and Prakasam, R. 1989. Culture: the efficacy of different modes of consultation. *Leadership and Organization Development Journal* **10**, 24–32.

Hofstede, G. 1980. *Culture's Consequences: International Differences in Work-related Values.* London: Sage.

House, R., Hanges, P., Javidan, M., Dorfman, P., Gupta, V., and Globe Associates. 2004. *Leadership, Culture and Organizations: The Globe Study of 62 Societies*. Thousand Oaks, CA: Sage.

Kabasakal, H. and Bodur, M. 2007. Leadership and culture in Turkey: a multifaceted phenomenon. In J. Chhokar, F. Brodbeck, and R. House (eds), *Culture and Leadership Across the World: The GLOBE Book of In-depth Studies of 25 Societies*, 2nd edn, pp. 835–74. Mahwah, NJ: Lawrence Erlbaum.

Kabasakal, H., Dastmalchian, A., Karacay, G., and Bayraktar, S. 2012. Leadership and culture in the MENA region: an analysis of the GLOBE project. *Journal of World Business* **47**(4), 519–29.

Kalaycı, B., Esin, F.T., and Bayman, O. 2017. Analysis of the corruption perception in Turkey. *Compliance and Ethics Professional*, August, 47–52.

Kaufmann, D., Kraay, A., and Mastruzzi, M. 2003. Governance matters III: governance indicators 1996–2002. World Bank Working Paper 3106. Washington, DC.

Kayaoglu, A. 2017. Determinants of trust in Turkey. *European Societies* **19**(4), 492–516.

Lee, W.S. and Guven, C. 2013. Engaging in corruption: the influence of cultural values and contagion effects at the microlevel. *Journal of Economic Psychology* **39**, 287–300.

Lee, S.H. and Weng, D.H. 2013. Does bribery in the home country promote or dampen firm exports? *Strategic Management Journal* **34**(12), 1472–87.

Lee, S.H., Oh, K., and Eden, L. 2010. Why do firms bribe? *Management International Review* **50**(6), 775–96.

Leff, N. 1964. Economic development through bureaucratic corruption. *American Behavioral Scientist* **8**, 8–14.

Mauro, P. 1995. Corruption and growth. *Quarterly Journal of Economics* **110**(3), 681–712.

Newman, K. and Nollen, S. 1996. Culture and congruence: the fit between management practices and national culture. *Journal of International Business Studies* **27**(4), 753–79.

Okuyucu-Ergun, G. 2007. Anti-corruption legislation in Turkish law. *German Law Journal* **8**(9), 903–14.

Olson, M. 1993. Dictatorship, democracy, and development. *American Political Science Review* **87**, 567–76.

Ozen, H. 2014. Informal politics in Turkey during the Özal era (1983–1989). *Alternatives: Turkish Journal of International Relations* **4**(12), 77–91.

Ozen, S. and Akkemik, A. 2012. Does illegitimate corporate behaviour follow the forms of polity? The Turkish experience. *Journal of Management Studies* **3**(49), 516–37.

Pellegrini, E. and Scandura, T. 2006. Leader–member exchange (LMX), paternalism, and delegation in the Turkish business culture: an empirical investigation. *Journal of International Business Studies* **37**, 264–79.

Pena Lopez, J. and Sanchez Santos, J. 2014. Does corruption have social roots? The role of culture and social capital. *Journal of Business Ethics* **122**, 697–708.

Peng, M., Wang, D., and Jiang, Y. 2008. An institution-based view of international business strategy: a focus on emerging economies. *Journal of International Business Studies* **39**, 920–36.

Pustu, Y. 2017. Yolsuzlukla mücadele ve medya. *Gazi Üniversitesi Sosyal Bilimler Dergisi* **9**, 3–23.

Rose-Ackerman, S. 2006. *International Handbook on the Economics of Corruption*. Cheltenham, UK and Northampton, MA, USA: Edward Elgar.

Soyaltin, D. 2017a. Transformation of corporate governance in Turkey: eliminating or accommodating political risks for doing business? In J. Leitner and H. Meissner (eds), *State Capture, Political Risks and International Business: Cases from Black Sea Region Countries*, pp. 137–54. New York: Routledge.

Soyaltin, D. 2017b. Public sector reforms to fight corruption in Turkey: a case of failed Europeanization? *Turkish Studies* **18**(3), 1–20.

Sullivan, D., Mitchell, M., and Uhl-Bien, M. 2003. The new conduct of business: how LMX can help capitalize on cultural diversity. In G. Graen (ed.), *Dealing with Diversity*, pp. 183–218. Greenwich, CT: Information Age Publishing.

Svensson, J. 2005. Eight questions about corruption. *Journal of Economic Perspectives* **19**(3), 19–42.

Tosun, U. and Yurdakul, M. 2016. An inquiry on the likely effects of corruption on the foreign direct investments in Turkey. *Hacettepe University Journal of Economics and Administrative Sciences* **34**(4), 71–96.

Tosun, M.U., Yurdakul, M.O., and Iyidoğan, P.V. 2014. The relationship between corruption and foreign direct investment inflows in Turkey: an empirical examination. *Transylvanian Review of Administrative Sciences* **42E**, 247–57.

Transparency International. 2011. Bribe Payers Index. http://bpi.transparency.org/results/

Turkish Statistical Institute. 2017. http://www.turkstat.gov.tr (accessed September 12, 2007).

Ulusoy, K. 2014. Turkey's fight against corruption: a critical assessment. Global Europe in Turkey. *Commentary* **19**, 1–5.

Uslaner, E. 2008a. *Corruption, Inequality, and the Rule of Law*. New York: Cambridge University Press.

Uslaner, E. 2008b. The foundations of trust: macro and micro. *Cambridge Journal of Economics* **32**, 289–94.

Warf, B. 2016. Global geographies of corruption. *GeoJournal* **81**(5), 657–669.

Wei, S. 2000a. How taxing is corruption on international investors? *Review of Economics and Statistics* **82**(1), 1–11.

Wei, S. 2000b. Local corruption and global capital flows. *Brookings Papers on Economic Activity* **2**, 303–54.

Wheeler, D. and Mody, A. 1992. International investment location decisions: the case of US firms. *Journal of International Economics* **33**(1–2), 57–76.

World Bank. 2000. *Helping Countries Combat Corruption: Progress at the World Bank since 1997*. Washington, DC: World Bank.

World Economic Forum. Geneva. 2014. *Global Competitiveness Report*. http://www3.weforum.org/docs/WEF_GlobalCompetitivenessReport_2014-15.pdf (accessed September 18, 2017).

14. *Wasta* in the Arab world: an overview
Marcus Marktanner and Maureen Wilson

The Arab uprisings that began in Tunisia in 2010 and quickly spread across the region drew global attention to the increasing dissatisfaction among citizens with their governments in the Arab world. Major issues cited as triggers of the wave of protests, and later revolutions, include youth unemployment, economic decline, human rights violations, and corruption (Lynch 2014; Peleg and Mendilow 2014). Despite the revolutions that have occurred across the region since 2010, corruption remains a pervasive problem in the Arab world. Five of the top ten most corrupt states in the world are in the Middle East and North Africa region according to a 2016 Transparency International survey (Heinrich 2017). Syria, Sudan, Yemen, Libya, and Iraq are all currently facing severe political and economic crises, highlighting the need to address in greater depth the issue of good governance in general, and of corruption in particular.

Corruption is not limited to the power players of the world. In the Arab world petty corruption is the most common form of corruption in the region (Alissa 2008). Distinct from grand or systemic corruption, petty corruption takes place on a smaller scale and often involves citizens seeking public goods and services. For example, recent studies have concluded that 30 percent of people in the Middle East and North Africa who accessed public services over a 12-month period paid a bribe and more than 60 percent of those surveyed believe that corruption has increased in recent years (Bearak 2016; Pring 2016).

Bribery and petty corruption take on a culturally distinct flavor in the Arab world in the form of *wasta*. *Wasta* refers to "the phenomenon of using 'connections' to find jobs and obtain government services, licenses or degrees that would otherwise be out of reach or would take a long time or effort to obtain" (Ramady 2016, p.vii). In many parts of the Arab world, it is believed that *wasta* is necessary to have a successful life (Ma'ayeh 2011). People without *wasta* will do whatever it takes to "cultivate *wasta*," often in an effort to access opportunities for themselves or family members (Hessler 2012).

In this chapter, we will discuss the concept of *wasta* from various perspectives. The chapter begins with a discussion of the many dimensions of the term *wasta*. This is followed by a game-theoretical analysis of the phenomenon of *wasta* as an evolutionary stable strategy. We then continue with an attempt to conceptualize and measure *wasta* empirically and conclude with a brief summary of our main arguments and discussion of the future of *wasta* in the Arab world.

DEFINING *WASTA*

While *wasta* does not refer to corruption in the traditional sense, often the practice of *wasta* can be perceived as a form of petty corruption. This is especially true when people receive benefits or opportunities without merit. The term *wasta* in Arabic loosely translates to nepotism. It is a term that describes personal connections people have that can ultimately help them "get things done." *Wasta* stems from the root word for "middle" in Arabic, implying the role of an intermediary or intercession on someone's behalf (Cunningham and Sarayrah 1993). While the Arabic language contains several other words that express corruption (*fassad*) and bribe (*rashwah*) in their truer meanings, *wasta* is best interpreted as the use of personal connections for personal gains. As opposed to the term corruption, which has an exclusively negative connotation, *wasta* has both culturally acceptable and unacceptable connotations (Lackner 2016). In other words, there is good and bad *wasta*. An example of good *wasta* would be the consultation of a tribal leader in the settlement of a dispute. An example of bad *wasta* would be the promotion of a less qualified person over a more qualified one only because the less qualified person had a closer connection to the boss (Adwan n.d.; Ramady 2016).

Despite various attempts to narrow down the term *wasta*, there is no generally accepted definition and, at least to our knowledge, no systematic study that attempts to identify how *wasta* is similar to or different from social networks in other cultures. This is surprising because social networks play an important role in other cultures as well and the general concept of cronysism often has regionally distinct variations. For example, in Cuba, *socialismo* allows people to use their social networks to get around bureaucratic roadblocks and gain access to scarce goods and services (Díaz-Briquets and Pérez-López 2006). In Russia, the term *blat* refers to exchanging favors. During the era of the Soviet Republics, *blat* was a means to acquire access to scarce state-controlled resources. Even after the collapse of the Soviet Union, *blat* remains a way to gain employment and access to goods and services that one might normally have a difficult time acquiring (Ledeneva 1998). Finally, *guanxi* is a concept that permeates Chinese society (Li et al. 2016). Personal connections are leveraged to gain economic favors in both social and business contexts; this is considered a generally ethical practice. In sum, the use of social networks to gain access to scarce resources is not unique to the Arab world. While there is variation among these social network systems, they all involve calling on personal connections in times of need. It is questionable that any taxonomy attempt of the similarities and differences of various cultural social networks could identify clear demarcation lines between *wasta* and other regional and cultural uses of social networks.

It is important to note that while the use of personal networks for personal gain may be viewed as a form of corruption in many cultural settings, the Arabic language distinguishes between the use of personal and non-personal networks to gain access to public goods and benefits. Thus, what may look like two different cases of gaining advantaged access to scarce public resources in the Arab world may not be linguistically and conceptually distinguishable by a non-Arabic speaker.

What is commonly accepted, however, is the fact that one is very likely to be confronted with discussions about *wasta* in the Arab world, possibly more so than with other forms of social networks in other cultural settings. The omnipresence of *wasta* in the Arab world seems to suggest that social, political, and economic structures require regular citizens to rely heavily on personal networks. This may be because the Arab world still wrestles with a legacy of tribalism and an unfinished colonial dialectic (Henry and Springborg 2001).

While the existence of "bad *wasta*" is generally acknowledged in public life, it is in contradiction to the scripture of the Quran. Like every other holy scripture, the Quran mandates virtuous behavior and condemns favoritism, nepotism, bribery, and other forms of corruption. For example, the Quran says, "And do not devour your wealth among yourselves through falsehood, and offer it not as bribe to the authorities that you may knowingly devour a part of wealth of other people with injustice" (Quran 2, verse 189). Similarly, the Bible commands that "You shall not take a bribe, for a bribe blinds the clear-sighted and subverts the cause of the just" (Exodus 23:8).

Just as the Roman Empire in the early days of Christianization struggled with aligning the political reality with the godly ideal, so did political Islam. In medieval scholastics, this dilemma was most famously described by Saint Augustine (354–430 AD) who lamented that "two cities have been formed by two loves: the earthly by the love of self, even to the contempt of God; the heavenly by the love of God, even to the contempt of self" (Augustine 2009, p. 430). Likewise, the famous medieval Islamic scholar Ibn Khaldun (1332–1406) noted towards the end of the Golden Age of Islam in the fourteenth century that regimes fall because of the corruption of their leaders. Specifically, he describes the rise and fall of political regimes in three stages with corruption being the final nail in good governance's coffin. He notes that the first political generation "retains the desert qualities, desert toughness, and desert savagery" before the second generation moves "from privation to luxury and plenty" and a third generation of ruler becomes corrupted. This third generation "takes many clients and followers. They help the dynasty to some degree, until God permits it to be destroyed, and it goes with everything it stands for" (Ibn Khaldun 1967, chapter III, section 12).

Yet, while the use of personal networks for personal gains is generally equated with corruption and its negative societal impacts, many people in the Arab world agree that there are instances when *wasta* may serve a legitimate reason. An example would be someone that uses their *wasta* to help their grandmother see a doctor. Yet, such justifications for the use of a social network for personal gains raises many other questions – the most important being why would anyone need *wasta* to see a doctor in the first place?

From the previous discussion follows that *wasta* – as the use of a personal network to obtain a personal benefit – may or may not be culturally accepted and may or may not involve a payment. As such, within these three dimensions, *wasta* can be visualized as the shaded space of Figure 14.1. The unshaded area represents corruption. As illustrated in Figure 14.1, there are four possible distinct *wasta* classifications, all of which involve the use of a personal network. Alternatively, there are four distinct corruption types that do not use a personal network.

The four types of *wasta* and four corruption types illustrated in Figure 14.1 are summarized in Table 14.1. The column headings in Table 14.1 describe different *wasta* types that correspond to the shaded area of Figure 14.1. The first dimension addresses whether a certain corruption or *wasta* incident is culturally accepted. The second dimension captures the nature of payment, which can be either money or an in-kind transfer. Finally, and most importantly, the use of a personal network in the context of Arab culture determines whether the interaction is classified as *wasta* or corruption.

To further clarify the various *wasta* and corruption types, consider again the example of someone attempting to get a doctor's appointment for their grandmother. In the first scenario this person knows one of the nurses at the hospital and pays them 50 Dinars to schedule an appointment with the doctor for their grandmother. Assuming that this occurs in a society where there is widespread understanding of such behavior, this could be classified as culturally accepted monetized *wasta* because this individual used her personal network and paid a bribe for a legitimate and well-intentioned purpose.

Now consider the same person pays their friend who works as an admissions officer at a local university to get her child admitted to school despite the child having flunked an official admissions test. Assuming this

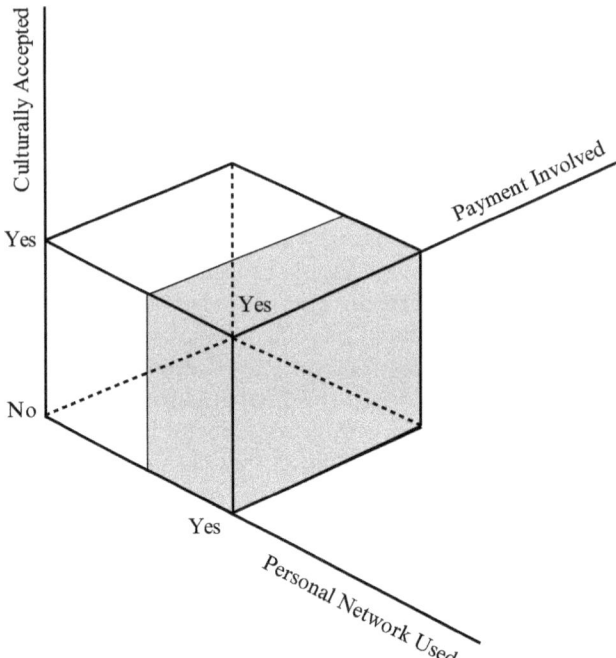

Source: Authors.

Figure 14.1   Dimensions of wasta

scenario occurs in a society that generally frowns upon such behavior, but no institutions are present to prevent such practices, then *wasta* is still playing an important role and a bribe is being paid. However, the behavior in this scenario is considered culturally unacceptable as university admission is supposed to be a competitive, merit-based process.

A different variation of the same scenarios would replace the payment of bribes with the promise of future non-monetized favors (in-kind gifts). In other words, if the person trying to get medical help for her grandmother makes an in-kind gift to the nurse in the form of, for example, jewelry, then this transaction could be considered culturally accepted barter *wasta*. Likewise, if the person bribes the admission officer with the same gift to have her child admitted to university, then this could be called culturally unacceptable barter *wasta*.

Next, reconsider the same scenarios except there is no personal network used. Now we are faced with examples of corruption that are either considered to be culturally acceptable (getting a doctor's appointment for

Table 14.1 Classification of wasta *and corruption*

Culturally accepted	Payment involved	Personal network used	Classification
•	•	•	culturally accepted monetized *wasta*
•	×	•	culturally accepted barter *wasta*
×	•	•	culturally non-accepted monetized *wasta*
×	×	•	culturally non-accepted barter *wasta*
•	•	×	culturally accepted monetized corruption
•	×	×	culturally accepted barter corruption
×	•	×	culturally non-accepted monetized corruption
×	×	×	culturally non-accepted barter corruption

grandmother) or culturally unacceptable (circumventing a competitive, merit-based application process). Again, any of these scenarios can be facilitated by a monetary or barter transaction.

Finally, *wasta* is not necessarily a quid pro quo transaction. For example, a person trying to get a doctor's appointment for their grandmother or admission to university for her child could beg and plead with the nurse or admission officer without any transfer of money or in-kind gifts. Such a case, if successful and even if no quid pro quo transaction takes place, would also be classified as *wasta* as long as a personal network is involved.

WASTA: A GAME-THEORETICAL PERSPECTIVE

What we seem to observe in the Arab world is that while people sometimes reject *wasta* and sometimes support it, it is a persistent characteristic of Arab society. Game theory may provide an interesting explanation for this. Specifically, game theory would suggest that the pursuit of *wasta* is an evolutionary stable strategy in a mixed strategy game. Since this may not be immediately obvious, we illustrate this idea in the following in more detail.

Consider the following stag-hunt game in Table 14.2 (see, e.g., Wydick 2007). A stag-hunt game is a game used to illustrate the evolution of a

Table 14.2   Stag-hunt game: no mixed strategy

		Player B	
		Support *wasta*	Reject *wasta*
Player A	Support *wasta*	<u>3</u>, <u>3</u>	5, 0
	Reject *wasta*	0, 5	<u>6</u>, <u>6</u>

social contract in which social cooperation creates a social optimum. In this game there are two players, A and B. Each player can either support or reject the use of *wasta* as a strategy. In this game no player has a dominant strategy and the game results in two Nash equilibria (cells with both payoffs underscored). A Nash equilibrium is defined as a situation in which no player has an incentive to change their strategy unless the other player possibly changes their strategy as well. In the example below, the first Nash equilibrium is given when both players support *wasta* and the second when both players reject *wasta*. If for example, the Nash equilibrium is both players supporting *wasta*, both players could be better off by agreeing to reject *wasta*. Both players rejecting *wasta* is said to be Pareto-dominant to both players supporting *wasta*. In the equilibrium in which both players reject *wasta*, no player has an incentive to leave the equilibrium unilaterally and neither would it make sense for both players to change their strategies jointly.

This game is played as follows. Imagine *wasta* is used to gain access to medical services. If Player A thinks Player B supports *wasta*, Player A will also choose to support *wasta* (underscored payoffs always represent one player's best response strategy to whatever strategy the other player chooses). This is because without *wasta* nothing can get accomplished, unless both players reject *wasta*. Likewise, if Player A thinks Player B rejects *wasta*, Player A's best strategy is also to reject *wasta*. If both players reject *wasta*, both players can see a doctor without having to pay a bribe which translates into a higher payoff for a doctor's visit. The best strategy for both players is to make the same decision in which they both eventually benefit, while the risk of choosing different strategies could result in one player receiving nothing at all.

Another way to consider the same scenario is as follows. If everyone pays a doctor a bribe, eventually no one gets preferential treatment and everyone has to pay more for a doctor's visit. Therefore, the social optimum, which is the highest aggregate payoff, would be for both players to reject *wasta*. Similarly, if one player pays a doctor the bribe while the other does not, the bribe payer may receive better treatment than would

Table 14.3   *Stag-hunt game: mixed strategy with a* wasta *support*
                probability of 50 percent

		Player B	
		Support *wasta*	Reject *wasta*
Player A	Support *wasta*	4, 4	4, 3
	Reject *wasta*	3, 4	3, 3

be the case if both players paid the bribe (this may be because the doctor
dedicates more time to the bribe-paying patient). In a simple stag-hunt
game, the strategy to reject *wasta* is Pareto-dominant and predicted to
emerge as the only equilibrium over time.

But what if people play different strategies in different circumstances?
For example, assume that each player supports *wasta* 50 percent of the
time and rejects it 50 percent of the time. This may be because using *wasta*
in order to get a doctor's appointment for one's grandmother is socially
accepted, while the same person may not be willing to use their *wasta* to
get his child admitted to university. If both players support *wasta* 50 percent of the time and reject *wasta* 50 percent of the time, the payoff matrix
of Table 14.2 becomes the new payoff matrix that is illustrated in Table
14.3. The results of Table 14.3 show that in a mixed strategy game with a
wasta support probability of 50 percent, supporting *wasta* is the only Nash
equilibrium. Supporting *wasta* is also a dominant strategy that is played
irrespective of the other player's anticipated choice, and both players supporting *wasta* is also a social optimum.

When does reject *wasta* become a dominant strategy? Applied to the
payoff matrix of Table 14.2, the answer is that *wasta* becomes rejected
when the following condition is met:

$$p(3) + (1-p)5 < p(0) + (1-p)(6) \tag{14.1}$$

where
p = probability of supporting *wasta* and
$(1 - p)$ = probability of rejecting *wasta.*

Solving for p then yields $p < 1/4$. This means that rejecting *wasta* only
evolves as an evolutionary stable strategy when both players play "support
wasta" less than 25 percent of the time. Applying, for example, a probability of $p = 0.2$ to "support *wasta*," the payoff matrix of Table 14.2 becomes
the payoff matrix of Table 14.4. In Table 14.4, both players rejecting *wasta*
is now the only Nash equilibrium. Rejecting *wasta* is now also a dominant
strategy and both players rejecting *wasta* the social optimum.

Table 14.4   Stag-hunt game: mixed strategy with a wasta *support probability of 20 percent*

		Player B	
		Support *wasta*	Reject *wasta*
Player A	Support *wasta*	4.6, 4.6	4.6, 4.8
	Reject *wasta*	4.8, 4.6	4.8, 4.8

This game can therefore explain why people in the Arab world both take advantage of *wasta* and condemn it, at least occasionally. Despite the social optimum being for all to reject *wasta* at all times, such a situation will not evolve as long as institutions prevail that at least partially encourage the practice of *wasta*. Ultimately, it is bad or corrupt institutions that invite culturally acceptable and unacceptable forms of *wasta*.

This result begs the question of what explains the pursuit of a mixed strategy game of *wasta* in the Arab world. One possible answer might be that the process of national political consolidation has not yet been fully completed. Henry and Springborg (2001), for example, argue that the Arab world consists largely of states that are stuck in the second stage of the colonial dialectic. The colonial dialectic's first stage is called the assimilation phase and describes a period in the early stages of colonial rule during which a social elite evolves and aligns itself with the colonial rulers. This social elite often consists of urban commercial players. At the same time, during this first stage of the colonial dialectic, other segments of society become marginalized. These marginalized groups are often rural communities supported by intellectual elites. The second stage of the colonial dialectic is called the counter-revolutionary movement. In this stage, the marginal segments of society manage to overthrow colonial rule and lead the country into independence. This independence, however, is characterized by strong divisions between the old and new elites. The third stage of the colonial dialectic is called the reconciliation stage; to complete this stage, the conflict that divides the old and new elites must be resolved.

The colonial dialectic suggests that Arab nation-building has been characterized by various patron-client relationships. The first was between the colonial ruler and commercial elites; the second between the leader of independence and the predominant supporters among rural and military communities. Within these various patron-client relationships (as opposed to equal opportunity), personal networks become much more of a lifeline than in societies that have successfully completed the

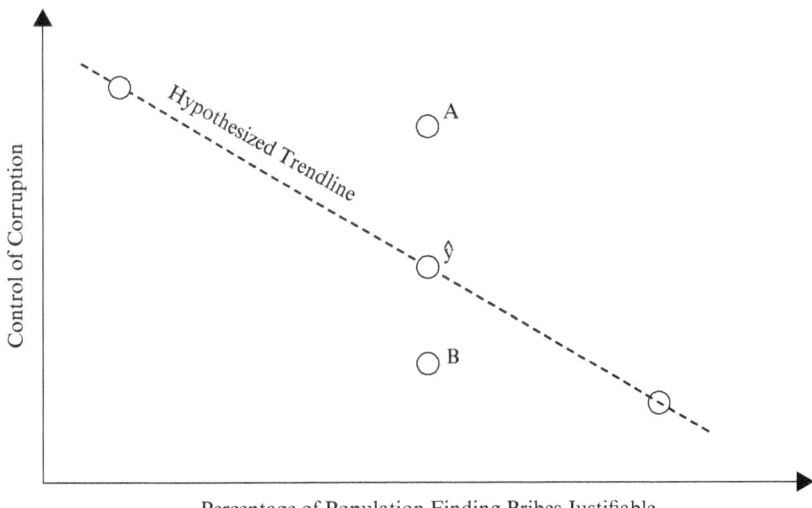

Figure 14.2   Control of corruption versus percentage of population finding bribes justifiable

third stage of the colonial dialectic. It remains to be seen whether the Arab uprisings that began in 2010 will succeed in accomplishing national consolidation by completing the third stage of the colonial dialectic. Ideally, this would render obsolete the reliance on personal networks by creating nation-states that are characterized by transparency and equal opportunity.

CONCEPTUALIZING *WASTA*

We interpret the cultural phenomenon of *wasta* as a weight that the societal attitude towards bribes exercises on the actual control of corruption (e.g., Marktanner and Wilson 2016). This is shown in Figure 14.2.

The graph shows a hypothesized inverse relationship between a country's attitude towards bribes being justifiable on the x-axis and a control of corruption score on the y-axis. One would expect that the more a society is penetrated by a positive attitude towards bribes, the more this public sentiment is reflected in a lower control of corruption.

In order to approach the weight that a societal attitude towards bribes has on the control of corruption, it is important to see that two countries with the same "Bribe-Justifiable" attitude may have very different control of corruption levels. Compare, for example, countries A and B in Figure

14.2. Both countries have the same "Bribe-Justifiable" attitude, but country A has much better control of corruption than country B. In fact, country A's control of corruption is much better than the predicted value on the line of best fit, while country B performs much worse. This suggests that the same attitude towards bribes carries a greater political infiltration weight in country B than in country A.

Alternatively, when thinking of the public attitude towards bribery as a cultural phenomenon, then the ratio of the predicted control of corruption score (y-hat) to the actual control of corruption can be interpreted as the weight that a given cultural acceptance of bribes executes on the political system. We therefore offer the following interpretation of the culturally determined weight of bribe acceptance on the political system:

$$Cultural\ Bribe\ Weight\ =\ \frac{control\ \widehat{of}\ corruption}{Actual\ control\ of\ corruption} \tag{14.2}$$

According to this formula, scores greater than one indicate an above average cultural bribe weight and scores less than one indicate a below average cultural bribe weight. In light of this conceptualization, we would argue that *wasta* is indeed a distinct Arabic phenomenon, if the cultural bribe weight scores of Arab countries are statistically significantly higher than in other regions of the world.

WASTA IN THE ARAB WORLD: MYTH OR REALITY?

In this section, we explore the conceptual framework of the cultural bribe weight as a proxy for *wasta* from an empirical perspective. For this purpose, we work with two variables. The first is "Justifiable: Someone accepting a bribe in the course of their duties," taken from the World Value Survey's Longitudinal Data Series (World Values Survey 1981–2014). In this survey, the respondent provides a rating on a scale between one (never justifiable) and ten (always justifiable). Our variable is the average of all available scores for each country between 1981 and 2014. We abbreviate this variable as "BribeJust."

The second variable, "Control of Corruption" is taken from the World Bank's Worldwide Governance dataset. This variable is the average of all observations for each country between 1996 and 2014. The country averages were then indexed on a scale between one and 101. This transformation was necessary because the original control of corruption score includes negative values and in order for the cultural bribe weight formula to generate plausible results, all actual and predicted control of corruption scores must be greater than zero. We abbreviate this variable as "ConCorrX."

While the two variables' observations do not exactly overlap time-wise, we argue that the "BribeJust" variable is rather time invariant. As such, we conclude that the absence of an exact time match of the observations does not jeopardize the legitimacy of the empirical exercise.

Our final dataset consists of 95 countries. Table 14.5 lists these countries by region.

In terms of the average attitude towards bribes, the Arab region's score does not stand out as particularly bribe-favoring. There is also no meaningful difference between Arab monarchies and Arab republics (Figure 14.3).

However, when looking at the actual Control of Corruption score, the Arab world in general – and the Arab republics in particular – exhibit comparatively low control of corruption scores (Figure 14.4).

The low control of corruption scores in the Arab world are also illustrated in Figure 14.5, which displays the "BribeJust" score on the x-axis and the "ConCorrX" score on the y-axis for all countries in our dataset. It shows that the Arab world's "ConCorrX" score as predicted by the "BribeJust" value is characterized by negative residuals with the exception of the three monarchies Bahrain, Kuwait, and Qatar (Figure 14.5).

Figure 14.6 shows the regional average of the "Cultural Bribe Weight" scores as the ratio of predicted to actual Control of Corruption values. It illustrates that within the Arab world, public attitudes towards finding bribes justifiable is characterized by the strongest political infiltration among all regions. Figure 14.6 also shows, however, that this regional phenomenon is heavily driven by the Arab republics, while the Arab monarchies generally have institutions in place that are much more resistant to the infiltration of public attitude towards bribes.

Finally, in order to test the statistical significance of the regional fixed effects, we ran three simple Ordinary Least Square Regressions of all 95 "BribeWeight" observations against an Arab Republic, Arab Dummy, and Arab Monarchy, respectively. This is

$$Bribe\,Weight_i = \gamma_0 + \gamma_1 Arab\ Republic_i + u_i$$

$$Bribe\,Weight_i = \beta_0 + \beta_1 Arab\ Dummy_i + u_i$$

$$Bribe\,Weight_i = \phi_0 + \phi_1 Arab\ Monarchy_i + u_i \qquad (14.3)$$

Table 14.5   Regional classification of countries in the dataset

East Asia and the Pacific (n = 12)	Eastern Europe and Central Europe (n = 25)	Sub-Saharan Africa (n = 11)	Middle East and North Africa (n = 16)
Australia	Albania	Burkina Faso	Algeria (A, R)
China	Armenia	Ethiopia	Bahrain (A, M)
Hong Kong	Azerbaijan	Ghana	Egypt (A, R)
Indonesia	Belarus	Mali	Iran (NA, R)
Japan	Bulgaria	Nigeria	Iraq (A, R)
South Korea	Croatia	Rwanda	Israel (NA, R)
Malaysia	Czech Republic	South Africa	Jordan (A, M)
New Zealand	Estonia	Tanzania	Kuwait (A, M)
Philippines	Georgia	Uganda	Lebanon (A, M)
Singapore	Hungary	Zambia	Libya (A, R)
Thailand	Kazakhstan	Zimbabwe	Morocco (A, M)
Vietnam	Kyrgyz Republic		Qatar (A, M)
Western Europe (n = 12)	Latvia	**Latin America and the Caribbean (n = 14)**	Saudi Arabia (A, M)
	Lithuania		Tunisia (A, R)
Andorra	Macedonia	Argentina	West Bank and Gaza (A, R)
Cyprus	Moldova	Brazil	Yemen (A, R)
Finland	Montenegro	Chile	
France	Poland	Colombia	**South Asia (n = 3)**
Germany	Romania	Dominican Republic	Bangladesh
Italy	Russian Federation	Ecuador	India
Netherlands	Serbia	El Salvador	Pakistan
Norway	Slovak Republic	Guatemala	**North America (n = 2)**
Spain	Slovenia	Mexico	Canada
Sweden	Ukraine	Peru	USA
Switzerland	Uzbekistan	Trinidad and Tobago	
Turkey		Uruguay	
		Venezuela	
		Puerto Rico	

Note: A = Arab League Member, NA = Non-Arab League Member, R = Republic, M = Monarchy.

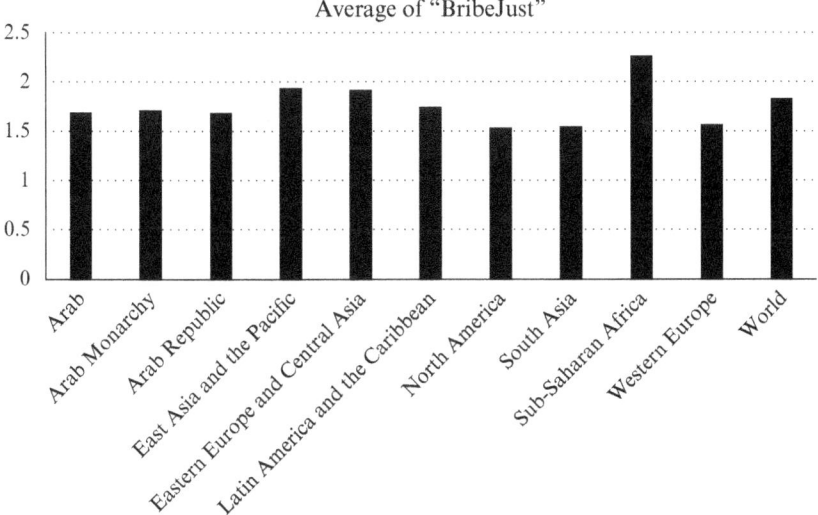

Figure 14.3   Regional attitudes towards finding bribes justifiable

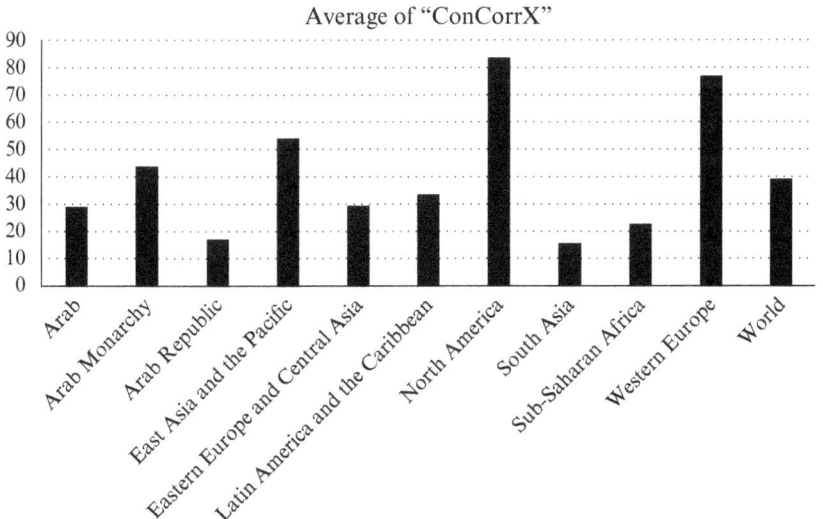

Figure 14.4   Control of corruption by region

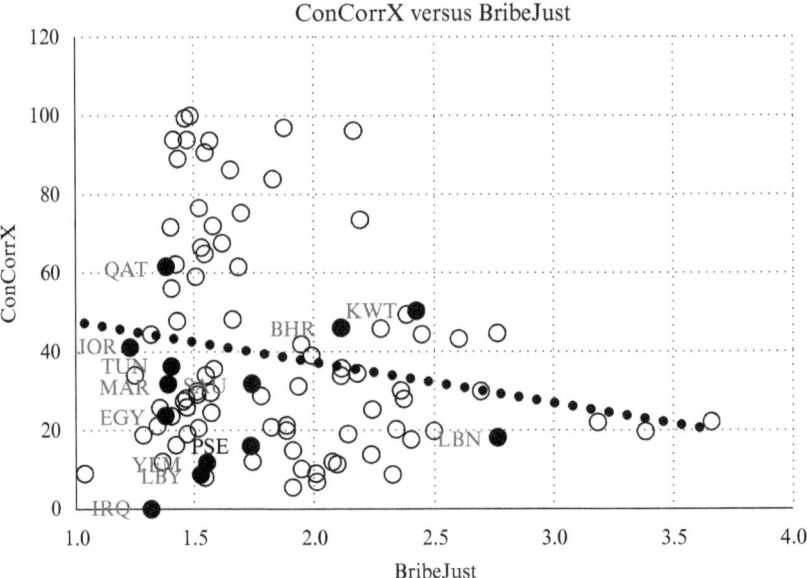

Note: BHR = Bahrain, EGY = Egypt, IRQ = Iraq, JOR = Jordan, KWT = Kuwait, LBN
= Lebanon, LBY = Libya, MAR = Morocco, PSE = West Bank and Gaza, SAU = Saudi
Arabia, TUN = Tunisia, QAT = Qatar, YEM = Yemen.

Figure 14.5   *Control of corruption versus attitude towards finding bribes*
           justifiable by region

Looking at Figure 14.6, we expect the coefficients γ_1 and γ_1 to be positive
and statistically significant, but not γ_3. Table 14.6 summarizes the regression results (omitting the constants).

The results suggest that one has to reject the null hypothesis that the
Arab world in general, and the Arab republics in particular, are not
any different when it comes to the political infiltration of attitudes on
justifiable bribery. Therefore, the phenomenon of *wasta* in the Arab world
seems to be more real than myth.

What explains the difference between republics and monarchies? The
concept of the colonial dialectic may again provide a useful explanation.
Europe's interest in the interaction with what are today the Gulf monarchies was largely limited to securing trade routes between Europe and
India. Before the discovery of oil, the Arab peninsula and Persian Gulf
states, as opposed to North Africa and the Levant, were of comparatively
little political, cultural, and economic interest to Europe. Today's Arab
republics, on the other hand, exhibited a much greater draw for Europe,

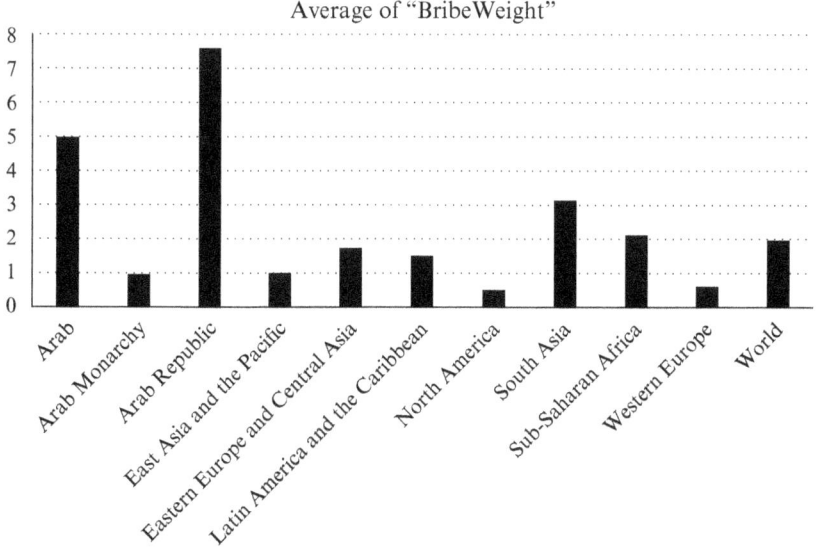

Average of "BribeWeight"

Figure 14.6   Cultural bribe weight by region

Table 14.6   Regression results (dependent variable = "Bribe Weight," n = 95)

Slope coefficient	Coefficient	Std. error	t-ratio
Arab Dummy (γ_1)	3.269**	1.279	2.557
Arab Republic (β_1)	6.143***	1.564	3.929
Arab Monarchy (ϕ_3)	–1.063	1.924	–0.553

Note: *** = significant at 1%, ** significant at 5%.

resulting in a much more disruptive colonial presence in the republics than in the monarchies.

CONCLUSIONS

In this chapter, we present an overview of *wasta* in the Arab world. While it is important to note that there is no commonly accepted definition, nor have there been previous attempts to classify and categorize *wasta*, we hope that this chapter can serve as a catalyst for future inquiry.

The chapter attempts to find the least common denominator of what constitutes *wasta*, which is the focus on the use of personal networks for personal gains and opportunities. For the non-Arab mind, every use of *wasta* is likely equated with corruption, but the same is not true for Arabs, for whom not every *wasta* is corruption. *Wasta* can be corruption, but it does not have to be. *Wasta* includes culturally accepted and unaccepted uses of social networks for personal gains, even though every use of *wasta* has some tinge of illegality.

As a second step, we asked what factors could explain the evolution of *wasta*. The fact that *wasta* is often both supported and rejected makes this concept naturally suited for a game-theoretical framework. We argue that games of mixed strategies provide an appropriate framework. This game-theoretical perspective, however, begged for a plausible explanation of why people in the Arab world resort to *wasta* in some instances, but not all the time.

We argue that the answer for this can be found in the colonial dialectic that has not yet been fully completed in many Arab nation-states. This colonial dialectic has always been characterized by strong patron-client relationships, first between colonial rulers and commercial elites, then between the governments leading Arab states into independence and their traditionally more rural supporters. In this context, people always trusted personal networks more than the rule of law. The fact that rule of law never won the upper hand over *wasta* practices may have played an important role in fueling the Arab uprisings as well.

The last section of this chapter is dedicated to assessing *wasta* in the Arab world from a quantitative perspective. Since *wasta* is presumably a strong cultural phenomenon in the Arab world, we asked whether a given attitude towards finding bribes justifiable has a greater political infiltration weight in the Arab world than elsewhere. When looking at the data, we find that while the attitude towards bribes in general does not stand out from a global comparative perspective, the Arab world in general, and the Arab republics in particular, lag behind in control of corruption. When dividing the corruption control scores predicted by countries' attitudes towards bribes by the actual control of corruption values, which can be interpreted as the political infiltration weight of attitudes towards bribes, the Arab world does stand out. Again, however, there is a strong difference between Arab republics and monarchies. The high political penetration weight of attitudes towards bribery in the Arab world is largely driven by the republics. This again is likely the result of a much less disruptive colonial legacy in today's Gulf monarchies than was the case in North Africa and the Levant. Today's Gulf monarchies are located largely outside the Fertile Crescent, where most of the economic opportunities were at home. The Levant and North Africa, on the other hand, experienced much greater colonial disruption as Europe tried to safeguard commercial activity on the Mediterranean Sea, which was regularly threatened by state-sponsored piracy, and to enforce a retreat of the Ottoman Empire and its influence on religion, culture, and commerce.

In summary, though *wasta* is not a new phenomenon in the Arab world, it has been differently shaped by history in the republics and monarchies. The colonial influence has put the Arab republics on a trajectory where *wasta* is still an important survival tool that has not yet been eliminated by rule of law institutions. Unless the societal divisions within the Arab republics become reconciled, *wasta* will likely continue to play a greater role in the lives of Arab citizens in the republics than in the monarchies.

REFERENCES

Websites last accessed October 27, 2017.

Adwan, C. n.d. Corruption terminology in the Arabic language. http://siteresources.world-bank.org/INTMNAREGTOPGOVERNANCE/Resources/CharlespieceonTerminology.pdf.

Alissa, S. 2008. Arab states: corruption and reform. Carnegie Endowment for International Peace. http://carnegieendowment.org/sada/21410.

Augustine of Hippo, Saint. 2009. Trans. Marcus Dodd. *The City of God*. Peabody: Hendrickson.

Bearak, M. 2016. 1 in 3 people in the Arab world had to pay a bribe for basic needs, survey reveals. *Washington Post*, May 4. https://www.washingtonpost.com/news/worldviews/wp/2016/05/04/as-if-the-arab-spring-never-happened-corruption-rampant-in-arab-nations-says-report/.

Cunningham, R. and Y. Sarayrah. 1993. *Wasta: The Hidden Force in Middle Eastern Society*. Santa Barbara, CA: Praeger.

Díaz-Briquets, S. and J. Pérez-López. 2006. *Corruption in Cuba: Castro and Beyond*. Austin, TX: University of Texas Press.

Heinrich, F. 2017. Corruption and inequality: how populists mislead people. Transparency International. https://www.transparency.org/news/feature/corruption_and_inequality_how_populists_mislead_people.

Henry, C. and R. Springborg. 2001. *Globalization of Development in the Middle East*. Cambridge: Cambridge University Press.

Hessler, P. 2012. Wasta. *New Yorker*, July 9. https://www.newyorker.com/magazine/2012/07/09/wasta.

Ibn Khaldun. 1967. *The Muqaddimah – an Introduction to History*, trans. F. Rosenthal. Princeton, NJ: Princeton University Press.

Lackner, Helen. 2016. Wasta: is it such a bad thing? An anthropological perspective. In M. Ramady (ed.), *The Political Economy of Wasta: The Use and Abuse of Social Capital Networking*, pp. 33–46. Berlin: Springer.

Ledeneva, A. 1998. *Russia's Economy of Favours: Blat, Networking and Informal Exchange*. Cambridge: Cambridge University Press.

Li, Y., J. Du, and S. Van de Bunt. 2016. Social capital networking in China and the traditional values of guanxi. In M. Ramady (ed.), *The Political Economy of Wasta: Use and Abuse of Social Capital Networking*, pp. 173–83. Cham, Switzerland: Springer International.

Lynch, M. 2014. Explaining the Arab uprisings. *Washington Post*, August 19. https://www.washingtonpost.com/news/monkey-cage/wp/2014/08/19/explaining-the-arab-uprisings/.

Ma'ayeh, S. 2011. Jordanians press for greater effort to wipe out wasta and defeat corruption. *The National*, June 22. https://www.thenational.ae/world/mena/jordanians-press-for-greater-effort-to-wipe-out-wasta-and-defeat-corruption-1.414717.

Marktanner, M. and M. Wilson. 2016. The economic cost of *wasta* in the Arab world: an empirical approach. In M. Ramady (ed.), *The Political Economy of Wasta: Use and Abuse of Social Capital Networking*, pp. 79–94. Cham, Switzerland: Springer International.

Peleg, I. and J. Mendilow. 2014. Corruption and the Arab Spring: comparing the pre- and post-Spring situation. In J. Mendilow and I. Peleg (eds), *Corruption in the Contemporary World: Theory, Practice, and Hotspots*, pp. 99–115. Lanham, MD: Lexington Books.

Pring, C. 2016. People and corruption: Middle East & North Africa survey 2016. *Issuu*. http://issuu.com/transparencyinternational/docs/2016_gcb_mena_en?e=2496456/35314511.

Ramady, M. 2016. *The Political Economy of Wasta: Use and Abuse of Social Capital Networking*. Berlin: Springer.

World Values Survey 1981–2014. Longitudinal Aggregate v.20150418. World Values Survey Association. http://www.worldvaluessurvey.org.

World Bank. Worldwide Governance Indicators. http://info.worldbank.org/governance/wgi/index.aspx#home.

Wydick, B. (2007). *Games in Economic Development*. Cambridge: Cambridge University Press.

15. Corruption and state capture in South Africa: will the institutions hold?
Karl Z. Meyer and John M. Luiz

Corruption is a phenomenon that has plagued various countries and societies since the beginning of organised human life. In seventh century BCE Greece, Hesiod condemned "bribe-swallowing" kings (Behrwald 2013), and in the Roman Empire, bribery and abuses had a long history and had become an institutionalised norm by the fourth century AD, where it was no longer a system of abuses, but an alternative system itself (Andrews 2014). Ultimately, corruption became a contributing factor to the empire's demise (MacMullen 1988), despite sophisticated legal institutions being present. Even though corruption is a phenomenon that appears to have been present since the beginning of organised human life, our grasp and understanding of the phenomenon and its foundations remains limited. Corruption research has increased substantially in recent years, and thus our knowledge and understanding of the phenomenon is growing, yet a review of the literature leaves one with more questions than answers, questioning not only the nature of corruption in various regions, but even the basis of the definition of the concept and what behaviours are or are not considered corrupt in the first place.

Most definitions centre on the abuse of public power for private gain (see Brown and Cloke 2004; Doh et al. 2003; Rose-Ackerman and Palifka 2016), but operationalising this in research proves problematic, as in many cases what constitutes "abuse" and "benefit" is not defined. Additionally, and adding further complexity, corruption has been shown to be culturally specific and context-dependent (Meyer 2015). In essence, what is considered corrupt in one country may be considered gift-giving or business-as-usual practices in another (Rodriguez et al. 2005). Thus, it is critical that both cross-sectional and time-series research on corruption acknowledge the nuances of both the context and the phenomenon itself. Corruption may also impact various economic regions in different ways, depending on the types and frequency of corrupt behaviours that are pervasive as well as the underlying levels of development. Corruption is frequently endemic in developing countries, which are least able to absorb the additional costs and burdens associated with it. For countries already suffering the effects of low levels of economic development, the costs of corruption can be overwhelming.

In this chapter, we examine the nature of corruption and its effects in South Africa, a developing and middle-income country, and detail the attempts to manage and curb corruption. South Africa is an interesting case because of its particular history and political economy, however its experience resonates with the wider African continent, and with middle-income countries more generally.

SOUTH AFRICA: CONTEXT AND CORRUPTION

The Union of South Africa was established in 1910, and although the formal system of apartheid was institutionalised with the National Party coming into power in 1948, the patterns of racial exclusion pre-date this time, and their origins are instead found in the colonial experience. Apartheid affected where people lived, who they married, the schooling available, what economic opportunities and job prospects were available to them, and every other facet of society. A complex arrangement of state institutions and legislation was put in place to ensure the functioning of the system. Within a decade of coming to power, the National Party was firmly entrenched as a dominant party within the already limited political landscape and the history of one-party domination commenced in South Africa. With the transition to democracy in 1994, the country has once again been dominated by one political party, the African National Congress (ANC). The lack of competitive pressures associated with the political system has played a role in the nature of corruption in the country both before and after the advent of democracy. Van Kersbergen and Van Waarden (2004) argue that concentration of power carries the danger of arbitrariness, of abuse, corruption and advancing self-interest by those in control, which emphasises the need to hold those in power accountable.

South Africa provides a rather unique set of socio-economic circumstances within which to research corruption, as it has a complex history (colonisation and apartheid), multi-cultural population (11 official languages and multiple ethnicities and religious affiliations), high unemployment (27.7 percent in 2017), and a rather unstable political climate, yet (counter-intuitively) also having a relatively well-developed institutional framework. In addition, South Africa's high level of income inequality further exacerbates the impact of corruption on the poor. South Africa has among the highest levels of inequality as measured by the Gini coefficient, and this provides an environment where the poor are further marginalised and are left with a lower probability of economic inclusion due to the lower quality and quantity provisions of healthcare, infrastructure, education and other public goods needed for the improvement of economic circumstances and quality of life.

Undoing the legacy of apartheid required major investments in both the social and physical infrastructure of the previously disadvantaged black population. Not only did this put enormous pressure on the state, but also on the institutions of governance which had previously not been geared towards this agenda. This makes it a challenge to discuss corruption in South Africa without recognising the burden that was placed on state institutions in its efforts to radically change South Africa's political economy.

Luiz (2014, p. 241) argues that the South African post-apartheid dispensation was a flawed negotiated settlement between elite groups that left the majority of the population without a voice:

> Those previously disadvantaged were given the opportunity to vote but in the presence of effectively a dominant single party they largely have no alternatives to express their dissatisfaction with the status quo. They were pacified through the introduction of an elaborate welfare system which does not seek to address their dispossession but rather seeks their silence. The black establishment was bought through the promise of black economic empowerment (BEE), which gave a small black elite the chance for rapid economic mobility, and big white capital was co-opted in return for the drop of nationalization and radical redistribution as a policy option.

The negotiated settlement thus failed to address the fundamental structural fault lines that were inherited and instead engaged in a process of accommodation and rent transfers.

Bhorat et al. (2017) identify key changes that have happened in South Africa since the transition. They refer to the first phase as that of "developmental welfarism" (which started during the Mandela presidency, 1994–99), though its specific institutional form took shape during the Mbeki era. It was organised around "deracialising" control of the economy through affirmative action policies, and transformation of white ownership of the economy through BEE. Mbeki moreover shifted political control away from the ANC itself to the presidency, which he built into a powerful apparatus of control and coordination at the centre of the state. This, in turn, created the conditions for the rapid expansion of corruption under the Zuma administration (post-2007) which took advantage of the developmental state discourse, and BEE, the emerging significance of the state-owned enterprises (SOEs) and state-investment institutions, and allowed a powerful elite centred on the president to capture state institutions as mechanisms for rent-seeking. Bhorat et al. (2017, p. 12) state that "To facilitate this, they needed brokers to help bypass regulatory controls and shift money around (through local and international

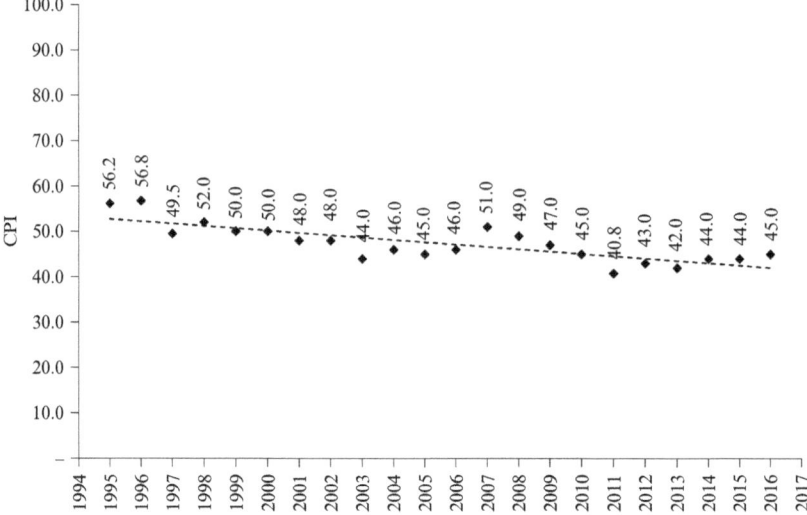

Source: Transparency International (2017).

Figure 15.1   South Africa normalised CPI score, 1995–2016

financial institutions) to finance deals as well as the transformation of the ANC into a compliant legitimating political machine." Thus developed the pervasiveness of the system in which state institutions and organisations are at least partially repurposed for the aim of rent-seeking.

By its nature, data on corruption in South Africa (and elsewhere) is concealed and as such is hard to measure, although there is growing evidence of its rifeness (Bhorat et al. 2017). However, Transparency International (2017) calculates the Corruption Perception Index (CPI), which provides an indication of the perceived pervasiveness of public corruption in a country. The CPI has been measured in South Africa since 1995, and thus only provides an indication of corruption in the post-apartheid years. Figure 15.1 provides the CPI measures for South Africa from 1995 to 2016; the method of calculation was modified in 2011 and changed from a 10 point scale to a 100 point scale, however we normalised these onto one scale. The data (note that lower values indicate greater levels of perceived corruption pervasiveness) shows that the level of perceived public corruption has increased over this time period. South Africa is ranked 64th out of 176 countries in 2016 in which the CPI is measured. (It should be noted that the CPI is calculated as a poll-of-polls, meaning that multiple surveys are aggregated to form a single index measure. There are methodological issues with this as the same surveys are not available for each region every

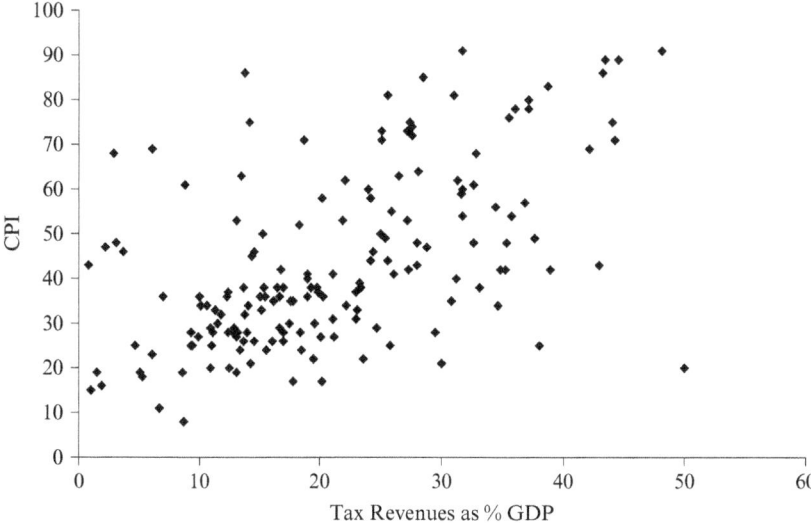

Source: Luiz (2016, p.212).

Figure 15.2   Scatterplot of Corruption Perceptions Index (CPI) and tax revenues as percentage of GDP

year and the measurement methods for each of the underlying surveys can change between years. Consequently, this measure does not allow for an entirely accurate comparison of corruption levels over time, or even between regions as the same survey sets are not used. However, as actual data on corrupt transactions in a region is so challenging to obtain, this measure is often operationalised in research.)

An interesting challenge that arises from corruption is the lack of trust that develops around the state and therefore its ability to extract revenue. This is critical in the South African context, as the state is attempting to undo the legacy of apartheid and redistribute economic resources. Rothstein et al. (2009) stress the importance of trustworthy, reliable, impartial and reasonably uncorrupted government institutions as a precondition for citizens' willingness to support policies of redistribution. They find that the quality of government positively affects the size and generosity of the welfare state. Figure 15.2 illustrates the relationship between perceived corruption and tax revenue as a percentage of gross domestic product (GDP). It demonstrates a strong correlation with countries with less perceived corruption having higher tax burdens as a percentage of GDP. Countries with high levels of corruption find it difficult to convince the citizenry to pay higher taxes because the efficacy of that revenue usage is questioned. Luiz (2016, p. 212) argues that the "argument about a developmental, welfare state for South Africa is moot in the context where people are increasingly questioning its capacity to deliver clean government."

There is a common misperception that most of the public corruption in South Africa began as the ANC government came to power in 1994. However, the assumption that a government transition involves the abrupt halting of the policies, inner workings and methods of operating of the previous government, and the instantaneous commencement of the new government's policies, inner workings and methods cannot be left unchallenged, as the time needed to sign new policies and laws into being, let alone operationalise them, is considerable. On this note, 40,000 documents that were recently made available under the Promotion of Access to Information Act indicate that the apartheid government not only abused its power, but also partook in organised crime, money laundering and corruption, and that many of the allegedly corrupt individuals of the old regime formed relationships with members of the new regime (Van Vuuren 2017). Therefore, we need to take heed of the intertwined nature of the old and the new governments and cannot definitively separate pre-1994 apartheid South Africa from the democratic post-1994 South Africa, at least not in terms of the presence of corruption as is commonly done in public discourse on the phenomenon.

While we acknowledge and in no way undermine that corruption and abuses of power took place during the apartheid era, and do acknowledge the intertwined nature of the two governments involved in the transition from apartheid to a more democratic South Africa, in this chapter, we focus more on the contemporary public and private corruption issues facing the country rather than the phenomenon's history in the country.

PUBLIC CORRUPTION IN SOUTH AFRICA

Following the first democratic election in 1994, corruption appeared relatively minor under the Mandela administration but escalated during the Mbeki administration, during which the multi-billion dollar arms deal (formally known as the Strategic Defence Package) was concluded. According to Corruption Watch (2017), a non-profit organisation focused on fighting corruption in South Africa, if the allegations of bribery and corruption currently under investigation are proven, it will be the largest corruption scandal in the country's history, although a currently proposed nuclear deal may overshadow this in terms of scale. Following the arms deal, it is alleged that South Africa's institutions were being purposely weakened to protect the companies and politicians that were involved. While some politicians have escaped relatively unscathed by the allegations surrounding their involvement, former President Jacob Zuma, although managing to have the charges dismissed initially, still faces multiple charges of corruption, fraud and racketeering, although he has yet to face his day in court. Political analysts argue that the appointment of his ex-wife, Nkosazana Dlamini-Zuma, as a Minister of Parliament, and potential successor as president of the ANC, prior to his removal from office (after the appointment of Cyril Ramaphosa as President) was a strategic move grounded in the need to shield himself from prosecution.

Former President Zuma has also been at the centre of multiple other corruption scandals during his tenure, not least of which were utilising USD18.4 million in public funds to upgrade his private home with "security upgrades" that included a visitor's centre, swimming pool, amphitheatre and a chicken run (Public Protector 2014), and being a key individual involved in the alleged state capture by the Gupta family. This family has amassed vast amounts of wealth from government contracts with the state and SOEs, many of which have been flagged as being improperly awarded, with procurement processes being skirted, checks and balances disregarded and state employees being bullied into awarding contracts to the Gupta-owned companies (Public Protector 2016). SOEs have also had their share of corruption-scandal limelight. As examples, Eskom (the state-owned power utility) and Transnet (state-owned rail, port and pipeline company) were allegedly involved in many corrupt transactions, some of which involve improperly awarding contracts to the Gupta family.

While South Africa had a relatively strong judiciary throughout this period, it was being undermined by purposive stifling of investigative institutions. Thus, it is alleged that the aforementioned cases have remained mere allegations due to the ineffective investigations carried out by both the South Africa Police Service and The Hawks (South Africa's Directorate for Priority Crime Investigation, which focuses on organised crime, corruption and other serious crimes). Due to delays and alleged maladministration, few of these serious cases of corruption made it past the investigation stage to reach the point where the judiciary would be able to take action regarding its mandated functions. Consequently, while South Africa's judiciary may be strong and uncorrupted, its effectiveness has been limited by the capability and efficiency of its investigative counterparts on which it relies.

Bhorat et al. (2017) chronicle a detailed and systematic network of corruption in South Africa, particularly in the last decade. They argue that the state capture is a betrayal of the promise that premised the transition to democracy in 1994. They outline how a power elite built around the former President has centralised control within seven broad areas:

1. Securing control overs SOEs by chronically weakening their governance and operational structures.
2. Securing control over the public service.
3. Securing access to rent-seeking opportunities by shaking down regulations.
4. Securing control over the country's fiscal sovereignty by capturing key institutions associated with the National Treasury.
5. Securing control over strategic procurement opportunities by intentionally weakening key technical institutions and formal executive processes.
6. Securing the loyalty of the security and intelligence services by appointing loyalists.
7. Securing parallel government and decision-making structures that undermine the executive – a government within the government.

The chronicle of corruption outlined in their report runs into potentially tens of billions of Rands, the scale of which would be unprecedented in South African history. A remarkable feature in the South African case is the fact that a robust civil society and parts of the fourth estate have documented these atrocities and continue to put pressure on the perpetrators to be exposed and prosecuted.

Since former President Jacob Zuma's days in office, there are signs that the tide is turning, both in terms of the presidential stance on corruption and rent seeking, as well as the extent to which institutions related to corruption are capacitated to effectively carry out their mandates. Since taking office in February, 2018, President Ramaphosa has restructured the boards of a number of South Africa's SOEs with improving governance structures being a main rationale. Additionally, the president has taken on the task of addressing leadership issues in the National Prosecuting Authority and the country's investigative institutions, and has begun to take action against many of those involved in the corruption and state capture cases that were previously ignored by investigative and prosecuting institutions. Thus, while a significant number of polities' rhetorical stance on corruption initially appears to have strong foundations, in many instances it remains mere rhetoric, President Cyril Ramaphosa appears to be following through on his anti-corruption stance. Although, we do note that it is still early days in his presidency.

PRIVATE CORRUPTION IN SOUTH AFRICA

While the impact of public corruption in South Africa is considerable, we cannot solely lay our focus there. While some authors (e.g., Rose-Ackerman 1999) have tended to focus primarily on public corruption, we cannot exclude private corruption from research and analysis as this would assume that corruption takes place exclusively in the public sector, while the private sector is driven by the ethics and ideals of competitive and efficient market theories. It stands to reason that in many cases of public corruption, a government cannot be corrupt on its own. A corrupt transaction involving bribery, for example, requires both a bribe-payer and a bribe-taker, and in the case of public corruption, the bribe-payer is in many instances a private organisation or individual. Supporting this, many researchers have made calls for the inclusion of private corruption in corruption research. For example, Sartor (2017) advocates the need for private corruption pervasiveness to be included in international business research, as it was found that the pervasiveness of private corruption influences the extent to which private organisations can be trusted as operational partners (a strategy employed by some multinationals to overcome the uncertainty involved in operating in pervasively corrupt foreign markets), and thus influences the entry mode strategies of multinationals (Sartor 2016). Additionally, one only needs to review the case studies of Enron, Worldcom or the Madoff investment scandal to see the types of private corruption that occur in developed economies. In South Africa (a developing economy), private organisations have had their fair share of corruption scandals (we provide examples and support for this below). Likewise, Luiz and Stewart (2014), who researched the strategic responses that multinational enterprises undertake when encountering corruption in Africa, found that avoiding corruption (through the use of both market and non-market strategies) was the most common strategic response, however instigating corruption was also identified as a strategic response employed by organisations. Interestingly, Luiz and Stewart identified that multinational enterprises (MNEs) tended to focus on themselves as victims of corruption – as institution takers – and neglected their impact on the environment of corruption and their influence – as institution makers.

In Meyer's (2015) survey of South African construction industry professionals (a large portion of whom service the government and SOE's procurement of infrastructure and capital equipment), 1 per cent of respondents stated that corruption was instigated by private sector individuals or organisations, 36.5 percent stated that corruption was instigated by public sector officials or organisations, and 62.5 percent of respondents stated that corruption was instigated by both public and private sector actors. Additionally, when asked how their organisation would respond if they encountered a contract that required a bribe, 22.9 per cent of respondents said that they would pay the bribe, and a further 11.5 percent said they would prefer not to say how they would respond. This potentially implies that 34.4 percent of respondents may actively partake in corruption, should the opportunity present itself. Taking a more ethical standpoint, 65.6 percent of respondents said they would: (1) lay a charge against the client; (2) inform the client's management; (3) withdraw immediately; or (4) report the incident to local and competition authorities. Hence, while private corruption is somewhat pervasive in South Africa, we cannot assume it is systemic, where all actors are corrupt.

A case which further evidences that private organisations partake in corrupt activities in South Africa is that of the FIFA 2010 Soccer World Cup. In the lead up to the competition, demand in the construction sector increased due to construction that was required on sports stadiums and transport infrastructure around the country. During this time, multiple construction organisations leveraged the opportunity to extract greater rents. Of these, 21 construction firms were identified to have allegedly collusively tendered for work, and, after being identified by the Competition Commission, and being given the opportunity to confess in exchange for lower fines and penalties than would have been imposed if the cases were prosecuted and tried in court, 15 companies admitted guilt (eNCA 2013).

THE RESPONSE TO CORRUPTION

Luiz and Stewart (2014) identify six main strategic responses that MNEs employ when encountering corruption in Africa. These are namely to (1) implement a corporate anti-corruption policy; (2) delay market entry or exit the market; (3) engage in host country partnerships; (4) try to enforce legal and regulatory compliance; (5) try to influence the institutional environment; and (6) participate in business groups. A seventh strategy which emerged in the study was to instigate corruption. While implementing a corporate anti-corruption policy was the most common strategic response employed, it should be noted that if corruption is curbed within the organisation, the difficulties of operating in pervasively corrupt environments do not disappear. In some markets, this may lead to the organisation fighting for survival. Local organisations (where the home country is the only country of operation) face a similar challenge, only for some of these organisations, market exit is not an option.

As the world's economies become more intertwined due to globalisation, so too does the reach of certain anti-corruption policies, tools and programmes. As an example, the United States implemented the Foreign Corrupt Practices Act, which criminalises corrupt acts carried out by US firms in other regions. In some cases, such interventions can compensate for legislative and judicial shortcomings in host countries. Additionally, member nations of the United Nations (UN), and its sub-programmes (e.g., the UN Global Compact), and the Organisation for Economic Co-operation and Development (OECD) have requirements placed on them to monitor and control corruption within the region. However, where corruption is pervasive, companies may ceremoniously comply with the requirements and/or guidelines of such initiatives, but decouple operations (Meyer and Rowan 1977) – in essence, masking their non-compliance with a façade of compliance.

In addition to initiatives such as those mentioned, the following are responses against corruption: (1) legal and administrative reforms in the state sector, which are only effective if the judiciary in the country is not weak or corruptible (Szeftel 2000); (2) corporate governance require-ments (e.g., those published by the OECD and King Committee); (3) attempts to encourage and aid the development of non-state institutions (e.g., independent media); (4) anti-corruption commissions (ACCs); (5) the OECD's International Convention on Combating the Bribery of Foreign Public Officials in International Business, which aims to bring all industrial economies in line with the United States; (6) establishing and strengthening institutions with the responsibility of financial oversight; and (7) corruption hotlines and whistle-blower protection initiatives.

However, in pervasively corrupt environments, the success of such responses have been called into question. Notably, ACCs have had limited success due to (1) procedural impediments; (2) fear of victimisation; (3) personal attitudes (Bowen et al. 2012); (4) that some ACCs face the same challenges faced by the country's institutions (Doig et al. 2007); (5) that the skills and capability levels of bureaucrats (and politicians) influencing policy decisions may not be adequate (Hambok 2001); (6) that one-size-fits-all approaches are commonly used to establish ACCs (Bale and Dale 1998; Doig et al. 2007; Laking 1999; Shah and Schacter 2004); (7) that local governments under-resource, add cumbersome processes and obstruct the implementation of solutions actioned by the ACC to undermine their effectiveness (Doig et al. 2007); (8) that corruption is treated as something that requires only technical solutions (e.g., economic policies and institutional reforms) to address, while excluding factors such as culture and context; and (9) a lack of research and analysis of the origins, dynamics and impacts of corruption in divergent settings (Brown and Cloke 2004). Evidence of the ineffectiveness of many ACCs can be seen throughout Africa. Tanzania's Prevention of Corruption Bureau produced only six convictions per year, most of which are low level officials involved in "petty" corruption (Shah and Schacter 2004) and in Kenya, their ACC resulted in only one successful conviction between 2003 and 2012 (Wanjala 2012).

The effectiveness of both mandatory and voluntary corporate governance requirements and guidelines (e.g., the King reports on corporate governance which is a South African code and the UN Global Compact) have also been questioned. In Meyer's (2015) survey of construction industry professionals, respondents were asked about the effectiveness of the King Code and Report. Only 26.0 per cent viewed King as being an effective corporate governance framework, while 5.2 per cent were unaware of the contents of King, and 68.8 per cent viewed it as ineffective. Respondents were also asked the extent to which they believed corporate governance was an effective way to direct and control organisations – 8.3 per cent of respondents stated that it was ineffective and wastes time and resources, 49.0 per cent stated that it was ineffective and needed improvement, and 42.7 per cent viewed it as effective. For example, the UN Global Compact has 114 organisations signed up from South Africa, including Eskom and Transnet and yet they are two of the parastatals that featured prominently in various reports of state capture and corruption in South Africa. This indicates some of the challenges that are faced in this regard.

Focusing on public corruption, we should highlight the extensive institutional protections enshrined in the Constitution of South Africa of 1996, in particular the so-called Chapter Nine institutions. These institutions include the Public Protector, the South African Human Rights Commission, the Auditor-General and the Independent Electoral Commission, among others. These institutions provide additional checks and balances to different levels of government and their independence, impartiality and powers are enshrined in the constitution. They were designed with strong constitutional checks to "constrain [the] arbitrary removal from office" of their heads and to "withstand the political aftershocks of principled action" (Calland and Pienaar 2016, p. 66). Calland and Pienaar (2016) discuss the evolution of these institutions and refer to the formative phase of 1995 to 1999 as the golden period of reform where Parliament was packed with talent and leadership and was overhauling the apartheid-era statute book. This was followed by a period of decline (1999–2009) that saw a weakening in the authority of these institutions and a diminution of Parliament's oversight strength. This went hand-in-hand with an increase in corruption in the public sector, especially around tender processes, and the scandal of the arms deal and the corruption associated with that. Attempts by these Chapter Nine institutions to rein in corruption saw their wings being clipped through the squeezing of their budgets and the questioning of their authority. Post-2009 represents a particular watershed where the independence of these institutions was severely tested. The Public Protector, in particular, under Madonsela, took on the growing tentacles of corruption all the way to the former President himself, insisting that cases of corruption which had been dropped be reinvestigated and charges laid. This led to confrontation between the Public Protector and both the legislature and the executive, with the Constitutional Court being called in repeatedly to provide guidance.

There are two ways of looking at this issue – the proverbial glass half full or half empty – and either one can be pessimistic about the extraordinary pervasiveness of corruption reaching into the very highest echelons of government, or one can celebrate the extraordinary institutions enshrined in the constitution and which have repeatedly exposed and doggedly pursued these cases of corruption. The strength of civil society, the judicial system and these Chapter Nine institutions may well end up being one of the most defining contributions of Mandela to South Africa.

CONCLUSION

Addressing corruption in South Africa will require a deeper connection with the country's political economy. It cannot *only* rely on building institutions of governance and policing corruption more effectively both within the public and private spheres, but also has to address the lack of a meaningful social contract. Bhorat et al. (2017, p. 62) point out there has "never really been a broadly shared and fully supported economic policy framework" and that such an economic consensus could unite different factions and get broad stakeholder support in the business community, labour sector and civil society, and that this will be the most effective bulwark against corruption in the long run. They conclude that

> What is clear is that state capture by shadowy elites has profound implications for state institutions. It destroys public trust in the state and its organs; it weakens key economic agencies that are tasked with delivering development outcomes; and it erodes confidence in the economy. When there is no trust in public institutions, there is little goodwill to express solidarity through tax, large companies are predisposed to sit on cash rather than reinvest profits towards productive use, criminality proliferates exploiting weaknesses in intelligence and crime enforcement authorities, and both capital and skills flee the country. The majority of South Africans will bear the brunt of these corrosive developments. Worryingly, large-scale corruption enables much wider corrupt

activities to go undetected at the lower tiers of government. Under such conditions, it is impossible to achieve transformative objectives that could improve livelihood of the majority of South Africans. (Bhorat et al. 2017, p. 63)

The South African case may resonate with the experience of both other African countries that gained independence from similar colonial experiences and have struggled with claims to broaden the inclusivity of their economic models, and that of other emerging middle-income countries. Many emerging market regions have experienced similar transitions to South Africa, both in terms of political and economic liberalisation, in recent decades: consider the transition in Eastern Europe after 1989, the transition to democracy in Latin America in the mid-1980s, and similarly in South Korea, Turkey and then in Southeast Asia. Many of these countries also faced the prospects of single party domination for decades post the transition to democracy due to weaknesses within the political system. Malaysia springs to mind and its recent struggles with massive corruption within a system of a single dominant party, and deliberate efforts at affirmative action in labour and other markets, bear a striking resemblance to South Africa.

REFERENCES

Andrews, E. 2014. 8 reasons why Rome fell. Retrieved 19 August 2017 from http://www.history.com/news/history-lists/8-reasons-why-rome-fell

Bale, M. and Dale, T. 1998. Public sector reform in New Zealand and its relevance to developing countries. *World Bank Research Observer* **13**(1), 103–21.

Behrwald, R. 2013. Corruption. In R.S. Bagnall, K. Brodersen, C.B. Champion, A. Erskine, and S.R. Huebner (eds), *The Encyclopedia of Ancient History*, pp. 1803–5. Malden, MA: Wiley-Blackwell.

Bhorat, H., Buthelezi, M., Chipkin, I., Duma, S., Mondi, L., Peter, C., and Swilling, M. 2017. *Betrayal of the Promise: How South Africa is Being Stolen*. Johannesburg: State Capacity Research Project.

Bowen, P., Edwards, P., and Cattel, K. 2012. Corruption in the South African construction industry: a thematic analysis of verbatim comments from survey participants. *Construction Management and Economics* **30**, 885–901.

Brown, E. and Cloke, J. 2004. Neoliberal reform, governance and corruption in the south: assessing the international anti-corruption crusade. *Antipode* **36**(2), 272–94. Blackwell Publishing Ltd.

Calland, R. and Pienaar, G. 2016. Guarding the guardians: South Africa's Chapter Nine Institutions. In D. Plaatjies, C. Hongoro, M. Chitiga-Mabugu, T. Meyiwa, and M. Nkondo (eds), *State of the Nation*, pp. 65–91. Cape Town: HSRC Press.

Corruption Watch. 2017. The arms deal: what you need to know. Retrieved 8 August 2017 from http://www.corruptionwatch.org.za/the-arms-deal-what-you-need-to-know-2/

Doh, J., Rodriguez, P., Uhlenbruck, K., Collins, J., and Eden, L. 2003. Coping with corruption in foreign markets. *Academy of Management Executive* **17**(3), 114–27.

Doig, A., Watt, D., and Williams, R. 2007. Why do developing country anticorruption commissions fail to deal with corruption? Understanding the three dilemmas of organisational development, performance expectation, and donor aid government cycles. *Public Administration and Development* **27**, 251–9.

eNCA. 2013. Construction companies fined R1.5bn. Retrieved 19 August 2017 from http://www.enca.com/money/construction-companies-fined-over-r1bn

Hambok, A. 2001. Governance and policy in Africa: recent experiences. Discussion Paper No. 2001/126. World Institute for Development Economics Research, United Nations University.

Laking, R. 1999. Don't try this at home? A New Zealand approach to public management reform in Mongolia. *International Public Management Journal* **2**(2), 217–35.

Luiz, J. 2014. Social compacts for long-term inclusive economic growth in developing countries. *Development in Practice* **24**(2), 234–44.

Luiz, J. 2016. The missing social contract for economic development in South Africa. In D. Plaatjies, C. Hongoro, M. Chitiga-Mabugu, T. Meyiwa, and M. Nkondo (eds), *State of the Nation*, pp. 205–19. Cape Town: HSRC Press.

Luiz, J. and Stewart, C. 2014. Corruption, South African multinational enterprises and institutions in Africa. *Journal of Business Ethics* **124**, 383–98.

MacMullen, R. 1988. *Corruption and the Decline of Rome*. New Haven, CT: Yale University Press.

Meyer, J. and Rowan, B. 1977. Institutionalized organizations: formal structure as myth and ceremony. *American Journal of Sociology*, **83**(2), 340–63.

Meyer, K. 2015. Constructing strategic responses to corruption: how ethical companies can strategically navigate corruption in the South African construction industry. Dissertation. Henley Business School, University of Reading.

Public Protector. 2014. *Secure in Comfort: Report on an Investigation into Allegations of Impropriety and Unethical Conduct Relating to the Installation and Implementation of Security Measures by the Department of Public Works at and in Respect of the Private Residence of President Jacob Zuma at Nkandla in the Kwazulu-Natal Province*. Report No. 25 of 2013/1114.

Public Protector. 2016. *State of Capture: Report on an Investigation into Alleged Improper and Unethical Conduct by the President and Other State Functionaries Relating to Alleged Improper Relationships and Involvement of the Gupta Family in the Removal and Appointment of Ministers and Directors of State-owned Enterprises Resulting in Improper and Possibly Corrupt Awards of State Contracts and Benefits to the Gupta Family's Businesses*. Report No. 6 of 2016/17.

Rodriguez, P., Uhlenbruck, K., and Eden, L. 2005. Government corruption and the entry strategy of multinationals. *Academy of Management Review* **30**(2), 383–96.

Rose-Ackerman, S. 1999. *Corruption and Government*. Cambridge: Cambridge University Press.

Rose-Ackerman, S. and Palifka, B. 2016. *Corruption and Government: Causes, Consequences and Reform*. Cambridge: Cambridge University Press.

Rothstein, B., Samanni, M., and Teorell, J. 2009. Social risks, institutional trust and the welfare state contract: quality of government versus the power resource theory. Paper presented at the workshop "Equality and Personal Responsibility in the New Social Contract". Oxford University.

Sartor, M. 2016. Host market corruption and foreign subsidiary investments in emerging markets. Paper Presented at the Academy of International Business Conference, New Orleans, LA.

Sartor, M. 2017. Host market corruption, strategy and exit. Paper presented at the Academy of Management Conference, Atlanta, GA.

Shah, A. and Schacter, M. 2004. *Combatting Corruption: Look Before You Leap. Finance and Development*. Washington, DC: International Monetary Fund.

Szeftel, M. 2000. Between governance and under-development: accumulation of Africa's catastrophic corruption. *Review of African Political Economy* **84**, 287–306.

Transparency International. 2017. Corruptions Perception Index. Retrieved 8 August 2017 from https://www.transparency.org/

Van Kersbergen, K. and Van Waarden, F. 2004. "Governance" as a bridge between disciplines: cross disciplinary inspiration regarding shifts in governance and problems of governability, accountability and legitimacy. *European Journal of Political Research* **43**, 143–71.

Van Vuuren, H. 2017. *Apartheid, Guns and Money: A Tale of Profit.* Johannesburg: Jacana Media.

Wanjala, S.C. 2012. *Fighting Corruption in Africa: Mission Impossible?* International Anticorruption Summer Academy. International Anticorruption Academy. Laxenburg, Austria.

16. Drugs and corruption in former Soviet Central Asia

Filippo De Danieli

The Pamir highway is among the highest driveways in the world. Built by the Soviets in the 1930s, it connects the capital of Tajikistan, Dushanbe, to the mountainous eastern part of the country, the Gorno-Badakshan region, and then continues up until Osh, in Kyrgyzstan, through the Pamir Plateau, at altitudes ranging from 3,500 to 4,500 meters. The first stretch, from Dushanbe to Khorog, runs for most of it length along the Pianji River, which marks the border between Tajikistan and Afghanistan. The contrast between the two sides of the river is stunning. On the Tajik side, a large road, 5 to 10 meters wide, connects villages of houses with concrete walls and asbestos roofs. On the Afghan side, the road is a narrow path carved into the cliff side, and the village houses are made of mud bricks and stones. These differences are the result of human actions. About 100 years ago, the Soviets decided to invest resources in developing this remote, but strategically important periphery. They brought electricity, roads, education, and health care. On the Afghan side of the frontier, on the contrary, things have remained quite the same as they were in the early twentieth century, and as they had been for centuries.

In parallel to such modernization and development efforts, the Soviets militarized the border. Border troop detachments were established along the River Pianji. Mobile patrols constantly monitored the stretches between the border posts. As a result, the movement of people and goods across the border gradually decreased, until coming to a complete halt at some point after World War II. Locals say that they could be shot at just for pointing their fingers at the Afghan villages across the Pianji. This started changing with the Soviet invasion of Afghanistan. Later, in the early 1990s with the collapse of the Soviet Union, the border opened up again. And the flow of narcotics, specifically opium and heroin, started.

At that point in time, the five Central Asian republics, Turkmenistan, Uzbekistan, Kazakhstan, Kyrgyzstan and Tajikistan, found themselves located between the world's largest opium producing country, Afghanistan, and the most dynamic opiates consumer market, Russia. For the past 20 years or more, about 80 or 90 percent of the global production of opium has come from Afghanistan. One-quarter of it, roughly 90 or

100 tons, is shipped out of the country through former Soviet Central Asia along the so-called Northern Route. If we consider that the Central Asian crime groups can buy a kilo of heroin from their Afghan counterparts for 2,000–5,000 dollars (depending on quality), and that the wholesale price of it in Russia is at least of 25,000–30,000 dollars, we realize how lucrative the drug trade can be. The United Nations estimated that the yearly profits of Central Asian crime groups deriving from drug trafficking are between 1.5 and 2 billion dollars (Fenopetov 2006). This is an enormous amount of money for a region like Central Asia; for comparison, the gross domestic product (GDP) of Tajikistan in 2016 was 6.95 billion dollars.

The chapter starts with an analysis of the factors that contributed to the establishment of drug routes in Central Asia. It then looks at the dynamics and actors of the Central Asian drug trade. Finally, it places this phenomenon in the broader context of corruption and state weakness.

SOVIET COLLAPSE AND THE DEVELOPMENT OF THE CENTRAL ASIAN DRUG ECONOMY

The Central Asian republics share a 2,300 kilometer-long border with Afghanistan. Under the Soviet Union, these borders were sealed, with the northern side heavily militarized. Cross-border movement was strictly regulated, and only very few selected individuals, mainly representatives of the Communist Party or law enforcement officials, could enter Afghanistan. In the early 1990s, with the collapse of the Soviet Union, the newly established states could not maintain such control over their frontiers. In order to secure its southern borders, the Soviet Union invested an enormous amount of money for infrastructural works and to keep thousands of border guards permanently stationed on the frontier. The five Central Asian republics simply had no resources to do something like that. Consequently, border security relaxed, and communities that had been divided for a long time by an externally imposed frontier could re-establish ties. With the flows of people and goods, a valuable commodity such as opiates started being traded across the border. In this initial phase the drug trade was poorly organized, and almost everybody who lived close to the border could cross to Afghanistan and exchange their most valuable belongings, such as clocks, carpets or golden jewelry, for opium.

Opiates production grew exponentially in Afghanistan during the 1990s. Political instability, widespread lawlessness, the end of economic support from the Soviet Union, and the advent of the Taliban were among the key factors that contributed to boost the drug trade. Opium became the most valuable crop for farmers, while large-scale trafficking turned into the main source of revenues for the Taliban, as well as for the other warring factions. Until the 1990s Afghan opiates were being smuggled out of the country either via Iran or Pakistan. But with the collapse of the Soviet Union, drug trade networks operating in Afghanistan had the opportunity to open up a new exporting channel.

Instability and conflict in Tajikistan, the southeasternmost Central Asian republic, and the one that shares the longest stretch of the border with Afghanistan, about 1,350 kilometers, provided fertile ground for the establishment of drug trading routes across the former Soviet territory during this initial stage. A bloody civil war erupted in Tajikistan soon after the country's independence in 1991. Two regionally based coalitions, with opposite political and religious views, contended for control over the capital of the country during 1992. Eventually, the secular-conservative bloc took power at the central level, while the Islamo-democratic opposition retired to its strongholds in the south and east of the country, from which it conducted an armed resistance which lasted until 1997. Both warring factions had close ties with Afghan warlords and relied on drug trafficking as a mean of financing the war effort (De Danieli 2010).

Thus, drug trafficking routes were established in former Soviet Central Asia thanks to a combination of conducive factors: geographical factors, such as proximity to Afghanistan or rugged borders, as well as political factors, in particular weakness of the state institutions and the Tajik civil conflict. Since then, this lucrative business has been consolidating and the region has been turned into a major hub along transnational drug trafficking routes.

THE GEOGRAPHIES OF DRUG TRAFFICKING

Afghanistan produces more than two-thirds of the world's opiates. Production increased five-fold between 1990 and 2013. There have been fluctuations over time, but the general trend remained constant. The extension of the area under poppy cultivation jumped from about 32,000 hectares in the second half of the 1980s to over 130,000 hectares in the beginning of the 2010s (UNODC 2005). Opium poppy fields are spread all across the country, with the major concentration in the southwestern provinces of Kandahar and Helmand. From the late 1990s, large-scale processing of opium into heroin started in Afghanistan. This implied a major added value for Afghan trafficking networks. The volume of Afghanistan's opium economy, including all the different stages of the value chain, from production to processing and export, was calculated for the year 2016 at over three billion dollars (UNODC 2017). Around 20–30 percent of the Afghan heroin is shipped out through the former Soviet Central Asian republics along the so-called Northern Route. According to estimates by the United Nations Office on Drugs and Crime (UNODC), of a total 360 tons of heroin manufactured in Afghanistan in 2010, 90 tons were trafficked on the Northern Route. Of these 90 tons, 80 supposedly went through Tajikistan, and the remainder through Uzbekistan and Afghanistan (UNODC 2012).

Three cities of the region, Dushanbe, Khorog and Osh, are the main trading hubs along the Northern Route. Dushanbe, the capital of Tajikistan, is the receiving point of drug shipments coming mainly from the northern Afghan districts of Balkh, Kunduz and Fariab. Here a variety of actors, including corrupt government officials, senior law enforcement personnel and wealthy businessmen are involved in the trade. Close ties between crime and state actors are in place since the time of the Tajik civil war. After the 1997 peace agreement, many of the former warlords obtained an informal "green light" on trafficking narcotics on the condition that they laid down arms (De Danieli 2010).

Khorog is the main center of the Gorno-Badakshan autonomous region of Tajikistan. Situated in the heart of the Pamir Mountains, this city of 30,000 inhabitants enjoys a strategic location along drug routes originating from northeast Afghanistan. An extremely rugged and remote terrain, together with a history of resistance to external domination and control, allowed for the creation of well-established, locally based, drug trafficking networks in this city. Commanders of military groups during the Tajik civil war became the main players in the narcotics business after the peace agreement. Interestingly, in the post-conflict scenario, the drug barons of Khorog have also been fulfilling certain social and economic functions. This became evident when, in the summer of 2012, an army of over 2,000 soldiers sent from Dushanbe to Khorog in order to arrest some of the Pamiri drug lords encountered fierce resistance by the local population. More than 30 Pamiris, and a number of soldiers, according to many sources at least 100, died during 48 hours of armed clashes (Peyrouse 2012).

Osh is the regional capital of southern Kyrgyzstan. Located in the eastern Fergana valley, this city lays at the crossroads of trade routes spanning Central Asia from south to north and from east to west. Indeed, this city has been one of the main centers along the historical Silk Road. Osh regained such a role in the 1990s when Central Asia, after the Soviet collapse, finally opened up to the global capitalist economy. Narcotics are shipped into Osh along three routes: from Dushanbe across the Garm valley; from Darvaz across the mountain passes of the western Pamirs;

and, finally, from Khorog through the Pamir highway, the 730 kilometer-long road built by the Soviets in the 1930s, which runs for most of its way at heights of more than 3,000 meters. Osh has the characteristic of a transnational drug trafficking hub with representatives of Tajik, Kazakh and Russian criminal groups, alongside those of Kyrgyz ones, present and active in the city (Townsend 2006).

At the consumer end of the commodity chain is Russia. According to the UNODC, out of the 90 tons shipped through CARs in 2010, 77 tons were bound for the Russian Federation. In Russia, starting in the 1990s, opiate consumption has been growing exponentially. Figures clearly show how this phenomenon has in many respects slipped out of control: from 1997 to 2009 the number of registered drug users in the Russian Federation increased five-fold, from 70,000 to more than 350,000. The UNODC estimates the number of heroin users in Russia at 1.5 million, equal to the aggregate of all other European countries, or 1.6 million (UNODC 2010, cited in Kramer 2011).

THE DRUG ECONOMY: ACTORS AND DYNAMICS

The groups active in the Central Asian drug trade are usually nationally or regionally based. Operations inside a country are carried out by indigenous criminal groups. Exchanges take place in border areas, where narcotic shipments are handed over to groups originating from the next country along the trafficking route. High-level criminals from the Central Asian republics, Afghanistan and Russia have meetings in towns like Osh or Khorog, which, as mentioned, are regional trafficking hubs. During such meetings, deals and business-related disputes are discussed and agreed upon. Therefore, connections exist among criminal actors from different countries, but still the majority of drug-related networks are of a national, rather than transnational, nature.

Crime groups operating in the narcotic business of Central Asia are primarily of small to medium size. According to the UNODC, in Kazakhstan the medium size of a group involved in drug trafficking is five members, in Kyrgyzstan, it is five to 15 members, while in Tajikistan the size tends to be larger as many groups were formed on the basis of military units that took part in the civil conflict (UNODC 2007). Often the groups have a strong leadership, which can be either an influential businessman or a local strongman. In none of the five Central Asian countries can an organization structured in a hierarchal way be observed, operating in a situation of monopoly, with clear internal rules and a clear division of labor between the various actors involved in the business. Rather, the Central Asian crime groups have the characteristics of decentralized criminal networks (De Danieli 2014).

Nevertheless, over the past 25 years the narcotics business has been consolidating, and the criminal groups involved in it have become more professional. In the early 1990s, drug trafficking started as a small, locally based trade, with village-level businessmen or law enforcement officers taking advantage of this new, and extremely valuable, illegal economic activity. But soon, under the push of increasing production in Afghanistan and booming consumption in Russia, Central Asian drug-related networks scaled up their operations. They improved relations with the Afghan suppliers, made international connections with groups of neighboring countries and Russia, and also invested in technology such as communication devices or concealment equipment.

Townsend (2006) identifies some of the key elements of this process of professionalization of narcotics trafficking in Central Asia. He argues that networks operating in Tajikistan, Kyrgyzstan and Kazakhstan had become increasingly streamlined, with the different nodes of the network more tightly knit. For example, he noted how drivers and couriers had started to be contracted on a regular, repetitive, basis, while mid-level players had made criminality their core business, that is to say, their main "job." Increased professionalism also meant that criminal groups established more systematic connections with state actors; indeed, the majority of law enforcement officials Townsend interviewed could record at least one episode in which they had been under pressure for halting an investigation or not arresting a drug lord. Golunov (2007) highlighted another important element of this consolidation process: the recruitment by the drug-related networks of professionals who are not part of the criminal world, but whose status or professional skills can be an asset for the organization. For example, in order to facilitate cross-border operations, crime groups hire railway workers, drivers of passenger buses, employees of logistic companies and express couriers, or simple villagers living in areas close to the borders.

Interestingly, the level of drug-related violence in Central Asia has remained quite low since the 1990s. No "turf war" involving large-scale violence among the various criminal groups, or against the state, has occurred so far in this part of the world. This might indicate, on the one hand, that the narcotics business has been florid and constantly expanding, thus competition is relatively low. On the other, this might represent a clear symptom of strong connections between state and criminal actors. A few examples from Kirghizstan help make this point clear. This country, unlike its neighbors, experienced significant political instability in the course of the past 15 years, in particular between the 2005 "tulip revolution" and the ethnic clashes in Osh in June, 2010. In those years, some of the most important criminal figures of the country were killed. However, as noted by various analysts, and also by the UNODC, this violence was not the result of a typical turf war among rival criminal groups; on the contrary, it seems that it was part of a broader strategy by high ranking officials who aimed to impose their control over the narcotics business (Marat 2012; UNODC 2012). The case of Aziz Batukaev is another good example. Batukaev, an ethnic Chechen, is a crime boss who played an important role in setting up the opiates trafficking business in Kyrgyzstan during the early 1990s. He was jailed in 2006, with a 16-year sentence for racketeering and murder of state officials. However, he mysteriously "walked out" of prison in 2013, and was escorted to a private plane that flew him to Chechnya, where, allegedly, he is still currently living (Vice News 2013).

STATE, CRIME AND CORRUPTION

All countries of the region are plagued by endemic corruption. According to the 2016 Corruption Perception Index of Transparency International, of 176 countries, Kazakhstan occupies the 131st position, Kyrgyzstan the 136th, Tajikistan the 151st, Turkmenistan the 154th and Uzbekistan the 156th. Corruption exists at all levels and permeates both the public and the private sectors of the economy. In order to understand the specific patterns of corruption in this part of the world it is necessary to look back at how the five republics were formed as part of the USSR and how the Soviet legacy has shaped the state-building process once they became sovereign states.

The five Central Asian republics obtained their independence in September 1991. None had previous experiences of statehood, in the modern sense. Before Russian rule (first tsarist and then Soviet), the territories of Central Asia were organized into emirates and khanates, which were, in many respects, feudal political systems. In the late 1920s, when the Soviet Union organized the administration of its territory, it decided to create five "national" republics in Central Asia. Each had its own dominant ethnic group, language and national symbols and, at the same time, its own government institutions and party structures. National institutions of the 15 republics that formed the USSR were only nominally independent, though de facto they were under strict control of Moscow.

At the moment of independence, the new Central Asia states had to confront the legacies of such limited sovereignty. They possessed a national territory, a national language, a national culture, national elites and national institutions; now they had to develop a "national interest," a concept that was totally new to them (Brill Olcott 1996). The nation-building process was made even more complicated by the fact that under Soviet rule, local-level identities based upon the territorial and administrative structures introduced by the Soviet authorities, like regions or districts, had become quite strong. For example, the region, according to Pauline Luong (2002), came to replace tribe as the pre-eminent political category for indigenous elites of Central Asia and regionalism was a key element in the internal political dynamics of the republics during Soviet rule. Strong solidarity ties developed also at lower administrative levels. According to the French author Olivier Roy, in the rural areas the *kolkhoz* (the collective farm) became the center around which traditional solidarity networks regrouped and reorganized. In the countryside, where the presence of state institutions was weak, all social and economic activities were conducted in the *kolkhoz*. The *kolkhoz* took care of such aspects as administrative issues, employment, social welfare and public works. But the *kolkhoz* also worked as a protective net for those members who had migrated to the cities for work or study: the head of the *kolkhoz* would use all its contacts in the Communist Party or in other state institutions in order to support them (Roy 2000).

Locally based identity networks that had developed in Central Asia during the Soviet Union played an important role in the post-1991 transitions. According to the above authors, regionalism and other types of local-level factionalism, in most cases, positively contributed to political order and stability, with the notable exception of Tajikistan. This apparent contradiction is well explained by Collins (2002), who argues that transitions in Central Asia reinvigorated existing informal systems of governance, according to what she calls the "the logic of clan politics." Clan politics is in her view "the politics of informal completion and deal making between clans in pursuit of clan interest." One of her main arguments is that weak states, especially those that experienced the sudden collapse of a central authoritarian system, can rely on deal making among clans as an alternative way to organize governance over key resources. This way is alternative in the sense that deal making and clan pacts belong to the sphere of informal politics and occur *outside* the official institutional framework. In the case of post-Soviet Central Asia, inter-clan deals have contributed to political stability in all countries across the region. Even in Tajikistan where in the first years of independence regional factionalism was a trigger of conflict, later it turned into a stabilizing factor, as informal bargains among regional clans over control of resources were fundamental for reaching the 1997 peace agreement (De Danielli 2010).

The analysis developed above is necessary in order to understand how drug-related networks found in post-Soviet Central Asia discovered a fertile ground for their establishment. When Central Asian republics obtained independence they lacked the historical and institutional foundations of sovereignty. Moreover, they lacked the economic means to put in place an effective system of governance. Thus, their political leaderships had to develop informal mechanisms to ensure control over resources and to maintain law and order. In this scenario, the narcotic trade suddenly emerged as a vehicle of capital accumulation and a new class of "violent entrepreneurs" formed around this business.[1] State actors could not afford the burden of confronting these the violent entrepreneurs directly, and so were left with the only options of either tolerating or co-opting them. The type of criminal networks that developed around the narcotics business resembles the characteristics of typical mafia formations. Mafias differ from other typologies of criminal groups in that they systematically seek to establish relations of collusion and connivance with state actors. Mafias are not anti-state entities. Their activities are often complementary, rather than antagonistic, to those of state institutions. The mafia-state relationship is usually based on an "exchange of favors": mafias commit themselves to the preservation of the status quo and allow state actors to participate in illegal economic activities in exchange for immunity from prosecution (Santino 1995; Armao 2000). In such a scenario, mafias can actively contribute to political order. If we apply this notion to the case of Central Asia, we realize how this region has enjoyed relative political stability for the past 25 years despite the fact that a massive narcotics industry has developed exactly in parallel to the state-building and nation-building processes.

CONCLUSION

Two main conclusions may be drawn in respect of the Central Asian drug trade. First, the narcotic business is extremely flexible and can be set up in any environment where conducive factors are in place. In a very short period of time, Central Asian drug-related networks built from scratch a complex system for moving narcotics shipments across borders, laundering their profits and protecting their activities from prosecution. Central Asian criminal groups were able to exploit their strategic geographical position,

---

[1] The term "violent entrepeneurs" was used by the Russian academic Vadim Volkov (2016) with reference to the Russian Federation in the 1990s.

especially the fact they shared "common languages" with both Afghan sellers and Russian buyers. They also exploited the phase of political transition after the Soviet collapse in which state actors were weak and had to build a political-economic coalition in support of the state-building process.

The second conclusion is that the drug trade is not in itself a factor of instability. If we take into consideration the volume of narcotics trafficked through Central Asia, that is, 20 percent of the global production of heroin, we might expect that the region would be torn by violence. But this is not the case. Overall, the region has remained stable for the past 25 years. The same ruling elites who emerged at the time of independence are still governing in Kazakhstan, Uzbekistan and Turkmenistan; while there has been a reshuffling of power in Tajikistan and Kyrgyzstan, caused by the civil war in the first case and by regime change in the second, this has not led to a real opening up of the system. As to the internal dynamics of the criminal networks involved in the narcotics trade, there has been no bloody turf war so far. Thus, the case of Central Asia shows that under certain conditions the development of a large narcotics industry may not necessarily lead to violence and instability. These conditions include an increase in both demand and supply over time; a relatively high number of players in the illegal trade market, with no cartel trying to monopolize it; state weakness and willingness of state actors to make deals with organized crime.

REFERENCES

Armao, F. 2000. *Il sistema mafia: Dall'Economia-Mondo al Dominio Locale*. Torino: Bollati e Boringhieri.

Brill Olcott, M. 1996. *Central Asia's New States: Independence, Foreign Policy, and Regional Security*. Washington, DC: United States Institute for Peace Press.

Collins, K. 2002. Clans, pacts, and politics in Central Asia. *Journal of Democracy* **13**(3), 137–52.

De Danieli, F. 2010. Silk Road mafias: the political economy of drugs and state-building in post-Soviet Tajikistan. PhD dissertation. School of Oriental and Asian Studies, London.

De Danieli, F. 2014. Beyond the drug-terror nexus: drug trafficking and state-crime relations in Central Asia. *International Journal of Drug Policy* **25**(6), 1235–40.

Fenopetov, V. 2006. The drug crime threat to countries located on the "Silk Road." *CEF Quarterly* **4**(1), 5–14. https://www.files.ethz.ch/isn/31925/Vol%204%20No%201_Narcotics_Full_Text.pdf

Golunov, S. 2007. Drug-trafficking through Russia-Kazakhstan border: challenge and responses. In T. Uyama (ed.), *Empire, Islam, and Politics in Central Eurasia*. Hokkaido: Slavic Research Center.

Kramer, J. (2011). Drug abuse in Russia: an emerging threat. *Problems of Post-Communism* **58**(1), 31–43.

Luong Jones, P. 2002. *Institutional Change and Political Continuity in Post-Soviet Central Asia: Power, Perceptions, and Pacts*. Cambridge: Cambridge University Press.

Marat, E. 2012. Bakiyev, the security structures, and the April 7 violence in Kyrgyzstan. *Central Asia-Caucuses Analyst.* https://www.cacianalyst.org/resources/pdf/issues/20100428Analyst.pdf

Peyrouse, S. 2012. *Battle on Top of the World: Rising Tensions in Tajikistan's Pamir Region.* Washington, DC: German Marshall Fund of the United States.

Roy, O. 2000. *The New Central Asia, the Creation of Nations.* London: I.B. Tauris.

Santino, U. 1995. *La Mafia Interpretata. Dilemmi, Stereotipi e paradigmi.* Palermo and Soveria Manelli: Rubettino.

Townsend, J. 2006. The logistics of opiate trafficking in Tajikistan, Kyrgyzstan and Kazakhstan. *CEF Quarterly* **4**(1).

UNODC. 2005. *Statistical Annex to the Note "The Opium Situation in Afghanistan."* Vienna: United Nations Office on Drugs and Crime.

UNODC. 2007. *An Assessment of Organized Crime in Central Asia.* Vienna and New York: United Nations Office on Drugs and Crime.

UNODC. 2010. *World Drug Report.* Vienna: United Nations Office on Drugs and Crime.

UNODC. 2012. *Opiate Flows through Northern Afghanistan and Central Asia. A Threat Assessment.* Vienna and Kabul: United Nations Office on Drugs and Crime.

UNODC. 2017. *Sustainable Development in an Opium Production Environment. Afghanistan Opium Survey Report 2016.* Vienna and Kabul: United Nations Office on Drugs and Crime.

Vice News. 2013. Kyrgyzstan is the latest victim of the global heroin trade. *Vice News,* October 28.

Volkov, V. 2016. *Violent Entrepreneurs: The Use of Force in the Making of Russian Capitalism.* Ithaca, NY: Cornell University Press.

17. Pakistan: a study in corruption
Feisal Khan

Pakistan came into existence on August 14, 1947 when the British Indian colony was divided into two new nations, India and Pakistan. Pakistan was originally split into two halves, East and West Pakistan, separated by almost a thousand miles of India, and in 1971, after a brutal civil war that culminated in an Indian military intervention, East Pakistan went its own way as Bangladesh. The creation of the two new countries in 1947 involved a mutual transfer of some 14–18 million people, with almost equal numbers of Muslims leaving India for Pakistan and Hindus and Sikhs leaving Pakistan for India. This was not a peaceful or orderly transfer of populations but an extremely violent, mutual ethnic cleansing in two of India's largest provinces – Punjab and Bengal, each with a population greater than Spain's – which were partitioned along the same Hindu-Muslim lines as British India. The Muslim-majority parts of each went to Pakistan and Hindu-majority parts to India. No accurate figures are available as to the extent of the slaughter that took place but conservative estimates put the dead at anywhere between 500,000 to one million, while other estimates go as high as two million.

While India inherited the entire bureaucratic apparatus of the British Indian Empire – the "administrative state" par excellence – virtually untouched except in the two partitioned provinces of Bengal and Punjab that bordered its much smaller new neighbor, Pakistan saw those same two provinces, by far its two largest, containing almost all of the new country's population, bloodily partitioned. Pakistan had to create an economy and an administrative bureaucracy essentially over-night since almost its entire commercial, professional and bureaucratic classes – overwhelmingly Hindu and Sikh – left en masse for India, and the skilled Indian Muslims it received in turn were largely destitute refugees.

Given the chaos, mass violence and administrative collapse that characterized Partition in Pakistan, corruption flourished and firmly established itself within the new state's bureaucratic and political setup. While corruption had been present in the Raj's (the colloquial name for British rule in India) administration, it was almost invariably confined to the lower levels of the vast bureaucracy that governed India. Its higher ranks, both British and (in the twentieth century) Indian, were remarkably free from personal enrichment via corruption, and the presence of this incorrupt top layer kept the middle and lower level excesses in check.

In the sense of building state institutions capable of rooting out bureaucratic and political corruption, Pakistan never recovered from the initial trauma of independence, even if it was successful in rebuilding its bureaucracy and economy in the years following. Since then corruption in the country has grown continuously, despite several major attempted reform efforts, to the point where one can only describe it as an all-pervasive and debilitating disease that affects all levels of the Pakistani administrative and political apparatus.

CORRUPTION DEFINED AND MEASURED

There is no one comprehensive definition of "corruption" that meets all eventualities or covers all possibilities. One could be interested in, for example, public versus private sector corruption, or petty versus grand, or sporadic versus systemic, or explicit versus implicit, or some combination thereof. The definition that best conveys the general sense of "corruption" as it is understood by most observers is the one that Stanislav Andreski employed 50 years ago in the context of personal enrichment by politicians and civil servants: corruption is "using the power of [public] office for making private gains in breach of laws and regulations *nominally* in force" (Andreski 1968, p. 92, emphasis added). Klitgaard's (1991/1988, p. 75) now famous equation of corruption, $C = M + D - A$, captures the essence of this definition quite well: **C**orruption occurs when government officials have **M**onopoly (sole), **D**iscretionary power, and no **A**ccountability, that is, when the rules "nominally in force" can be evaded with impunity by unaccountable officials.

Ever since the international anti-corruption non-governmental organization Transparency International issued its first annual Corruption Perception Index (CPI) ranking of the world's countries by the level of perceived honesty in 1995, Pakistan has regularly placed among the world's most corrupt countries. On a scale from 0 to 100 (where higher scores reflect greater honesty), the first CPI survey put Pakistan 3/54 in the corruption rankings (i.e., the third most corrupt country in the sample) with a score of 22.5; the most honest country then was New Zealand (score 95.5) and the least was Indonesia (19.4). Pakistan reached its corruption nadir the next year when the 1996 CPI put Pakistan as the world's second most corrupt country at 2/54 (score 10). A common joke making the rounds in Pakistan then was that it was actually Pakistan and not Nigeria that was dead last but that the Pakistanis had bribed Transparency International to move it up a notch!

The most recent 2016 survey put Pakistan's honesty at 116/179 (score 32) with a tie for both the most and least honest countries: the most honest were Denmark and New Zealand (score 90 each) and the least honest were Libya and Sudan (score 14 each), which edged out Afghanistan (score 15). Considering that the average CPI score was 43 and the median was 38, Pakistan clearly has a long way to go before moving out of the bottom half of the world corruption rankings even if it has made some progress over the past two decades-plus of the CPI.[1]

Transparency International is not the only organization to rank Pakistan as an extremely corrupt country. The World Bank's "Control of Corruption" subscore of its comprehensive Worldwide Governance Indicators dataset for 1996–2015 put Pakistan's percentile score range from a high of 27.3 to a low of 8.8. That is, in terms of controlling corruption in this period, Pakistan varied from being just outside the worst quarter of countries ranked (27th percentile in 2003) to being in the worst tenth (9th percentile in 1996). The most recent (2015) ranking put Pakistan at the 23.6th percentile. However, rather than a steady albeit slow upwards progression, Pakistan appears to be generally cycling around the 20th percentile mark.

Further survey data from Pakistan confirms the impression conveyed by the Transparency International and World Bank corruption scores. Transparency International's Global Corruption Barometer's 2017 survey of 1,078 Pakistanis indicated that only 28 percent of respondents felt that corruption in Pakistan had decreased over the past year while 35 percent felt that it had increased and very large majorities (75, 68 and 61 percent, respectively) of those respondents who had had any contact with the police, the courts or public utilities reported having to pay a bribe or provide other illegal consideration; this was the highest reported proportion of any of the 16 Asian-Pacific countries included in the survey (Transparency International 2017, p. 15). Furthermore, 40 percent of all respondents who had had any contact with other government entities reported that they had had to pay a bribe or other illegal consideration. Unsurprisingly, for those

---

[1]   CPI data is from Transparency International's website, https://www.transparency.org/news/feature/corruption_perceptions_index_2016#resources (accessed May 20, 2018). While the CPI data and rankings in the early years were criticized for being overly reliant on surveys of expatriate business persons and international consulting firms located in High Income countries that were likely biased against Low and Middle Income countries, rankings from later years explicitly sourced a wider range of surveys. More recent criticisms of the CPI focus on the impossibility of reducing a concept as complex as "corruption" to a single numerical score, that there are no absolute measures of corruption possible and that it is a comparative ranking so even "very honest" countries still have at least some corruption. See Hough (2016) and Thompson and Shah (2005) for more details on criticisms of the CPI.

respondents who gave a specific reason, the modal response (16 percent) for why the corruption incident was not reported to the authorities was that no action would be taken against the corrupt and the second most common reason (11 percent) was fear of the consequences to themselves of reporting corruption; 54 percent of respondents felt that the Pakistani government was handling the fight against corruption "very" or "fairly" badly; a large plurality of respondents felt that "nothing" could be done about corruption in Pakistan and only 12 percent felt that corruption should be reported to the authorities. The most disheartening result was that of the 16 countries surveyed, Pakistanis "felt least empowered with only a third agreeing that [ordinary] people can make a difference" in the fight against corruption (Transparency International 2017, p. 33). The results of the 2013 survey are not markedly different and Pakistani averages are generally below the Asian average in terms of how hopeless the proverbial "person in the street" thinks the corruption problem is and the likelihood of the situation improving in the future.[2]

Experience bears this out. For example, in late 2012 the Auditor General of Pakistan's office, which is responsible for "public accountability and fiscal transparency in governmental operations," with the right to inspect all government financial documents "on demand" (Transparency International Pakistan 2014, p. 145), issued orders to 16 government departments to undergo a full expenditure audit. These orders were ignored with complete impunity and the president and parliament's Public Accounts Committee refused to issue any reprimand or compliance orders to the offending departments (Transparency International Pakistan 2014, p. 145). These orders were perhaps motivated by audits in 2011–2012 that indicated that of a total expenditure of PKR 4.474 billion rupees examined, PKR 1.857 billion was flagged for additional scrutiny due to major procedural irregularities and suspected corruption (Transparency International Pakistan 2014, p. 145).

Unsurprisingly, this rampant corruption seriously impacts the ability of the Pakistani state to generate the funds necessary to carry out even the most basic tasks expected of a functioning government. Pakistan, for example, spent only 1.1 percent of gross domestic product (GDP) in 2013 on 'Social Protection' expenditures (e.g., health care, pensions, unemployment benefits, child welfare, disaster relief, labor force skill enhancement), down from 1.4 percent in 2012 but up from 2011's low 0.8 percent and 2010's even lower 0.4 percent. Consequently, while the average Social

---

[2] The 2015–16 Global Corruption Barometer Survey data is available on the Transparency International website at https://www.transparency.org/files/content/publica tion/2017_GCB_AsiaPacific_RegionalResults.xlsx (accessed August 1, 2017).

Protection Index (SPI) score for Lower Middle Income countries (i.e., Pakistan's group) in 2009 was a 0.096, Pakistan's score was only 0.047; this was even lower than the average score for Low Income Countries: 0.061 (Asian Development Bank 2013, p. 104). While no one would compare Pakistan to High Income or even Upper Middle Income economies, Pakistan underperformed even relative to countries poorer than itself.

A major reason for such low spending on Social Protection is that the Pakistani state is increasingly unable to collect enough revenue to fund such expenditure after it has paid for a massive defense establishment. An internal 2003 Federal Board of Revenue (FBR; the national tax collection service) study estimated that at least one-quarter of all General Sales Tax (GST)[3] refunds were fraudulent (Hoti 2003) and the situation has deteriorated considerably since then. In 2014, internal FBR estimates were that fraudulent refunds cost the country roughly US$2.4 billion annually in lost revenue; that same year the FBR revised its official tax collection estimate down from approximately US$24.75 billion to 23.45 billion while unofficial internal estimates were that it would actually be closer to US$23 billion (Dawn 2014). The reason for the substantial tax shortfalls were not a slowdown in economic activity but fraudulent corporate tax refunds and rampant tax evasion.

The extent of the corruption in the FBR can be gauged by the fact that the usual fee for approving these fraudulent refunds is a 40–50 percent kickback to tax officers, which is then distributed among FBR officials on a sliding scale. The value of these payments, universally called "monthlies" in Pakistan, can be many multiples of the senior officials' salaries.[4] In 2014, in an attempt to improve Pakistan's abysmally low tax-to-GDP ratio (World Bank 2015b) and reduce the extent of rampant tax evasion and fraud, the government reintroduced tax audits and abandoned the failed "self assessment" scheme (Dawn 2016). However, no improvement in the

---

[3] The GST is the Pakistani version of the Value Added Tax (VAT). As with all VATs, the tax due from the manufacturer is only on the value added during the manufacturing process. That is, the value of final products less the value of the raw/intermediate product and the manufacturer is due a refund of the GST levied on raw/intermediate products. Since many manufacturing sectors in Pakistan are favored with lower tax rates than the normal 16 percent GST rate, the value of the refund due to the firm might be extremely large. For example, let's say an exporting firm is levied GST at 4 percent instead of the usual 16 percent. The firm exports PKR (Pakistani rupees) 500 million worth of textiles and purchases PKR 400 million of raw/intermediate goods. In this case the firm is due about a PKR 35 million tax refund from the FBR. Massively overstating the value of exports and/or raw materials will obviously inflate the value of the tax refund. It is also common practice to set up dummy companies to sell non-existent raw materials to firms, purchase fake invoices (called "flying invoices"), and other such methods to claim fraudulent refunds.

[4] Personal knowledge of the author; also cited in Dawn (2014).

efficiency or efficacy of Pakistan's tax authorities is actually likely to occur since news reports indicate that while 93,000 cases have been selected ("at random by computer balloting") for a full tax audit, which statutorily has to be carried out within six months of notification to the taxpayer, there are only 651 qualified FBR personnel who have the authority to audit returns and so each auditor will have to handle 24 audits a month (Rana 2017). It is not clear what the FBR will do about this since the odds of an auditor actually examining 24 income tax returns a month for six months straight *and* doing a careful, thorough and honest job are remote.

The amount of money that senior officials who want to abuse their positions can make is truly astounding. The poster child for corruption among Pakistani officials is Mushtaq Ahmed Raisani, the Secretary Finance for Balochistan (i.e., the senior civil servant in the provincial finance department). A 2016 raid on his residence by officers of the National Accountability Bureau (NAB, Pakistan's main Anti-Corruption Agency, or ACA) recovered approximately seven million US dollars in Pakistani and foreign currencies; video clips of NAB officers literally loading sacks full of dollars, pounds sterling, euros and Pakistani rupees into minivans were played on TV channels for days afterwards (Shah 2016). The reaction of well-informed Pakistanis was blasé. While acknowledging that the amount found in the house was indeed astounding, many said that they could name several other government officials and politicians who had hundreds of thousands of dollars-equivalent in cash in their house and what was extraordinary was that someone that high in the civil service was actually arrested.

The preceding impression of the extent of corruption in Pakistan is reflected in media reports from Pakistan which are full of accounts of corruption ranging from the petty (e.g., traffic policemen shaking down bus and taxi drivers for "tea money") to the serious (e.g., millions of dollars "missing" from the accounts of government agencies ranging from the education department to the ministry of defense) to the embarrassingly humiliating for the nation.[5] Pakistan's rankings in Transparency International surveys are not incorrect.[6]

---

[5] For example, when former Prime Minister Yousaf Raza Gilani admitted in 2015 that a necklace donated by the wife of the Turkish prime minister to the flood victims fund in 2010 and officially listed as untraceable had been in his possession for years; he claimed that he had given its monetary value (~US$3,300) to the fund but that could not be verified (I. Khan 2015).

[6] In the past few years there has been a veritable explosion of Pakistan-based scholars publishing articles on all aspects of Pakistan's economy and society, including naturally corruption, in open access journals. Unfortunately, the vast majority of these publications are in open access "pay to publish" journals listed in Beall's (2012) "List of predatory publishers" and are of very dubious quality. High quality information about the level of corruption on Pakistan is provided by Transparency International's Pakistan's Chapter, which is the

THE MASSIVE GROWTH OF CORRUPTION IN PAKISTAN

It is now a stylized fact in the social sciences that the "resource curse" (the sudden influx of massive wealth arising from the discovery of natural resources such as oil) can be deleterious, even debilitating, for a country (e.g., Collier 2008; Humphreys et al. 2007). While most of the literature on the "resource curse" focuses on commodity exports and the resulting weakening of local institutional quality and increased government corruption, recent studies have found that while the effects on recipient families' welfare may well be generally positive, the negative impact of substantial worker remittances "may have adverse effects on domestic institutional quality, specifically on the quality of domestic governance, that are similar to those of large natural resource flows" (Abdih et al. 2012, p. 663). A different study (Berdiev et al. 2013, p. 185) also found "that remittances increase corruption, especially in non-OECD [Organisation for Economic Co-operation and Development] countries," presumably because of the lower institutional quality therein and their lower ability to contain bureaucratic corruption. The "resource curse" thus acts as a "pull factor" with regards to corruption.

Given that Pakistan is both one of the world's more corrupt countries and a large recipient of worker remittances, the overall effects on Pakistan are undoubtedly worse than on countries less dependent upon remittances. The "resource curse" is indeed just that since what is, in effect, 'unearned' income promotes corruption and retards both the development of robust domestic socio-economic institutions and indigenous institutional capability building. In Pakistan's case, this has allowed its ruling elite to maintain its highly privileged rentier status. Clientelism and corruption go hand-in-hand with rentierism and Pakistan is no exception.

Pakistan has suffered from five separate rounds of "windfall gains." The departure of many of Pakistan's non-Muslim residents in 1947 left behind much valuable property to be redistributed. The country received immense geostrategic rents from the US foreign policy priorities during the Cold War and the global war on terror, and in the 1970s it sent millions

---

largest and most comprehensive source of information about corruption there. It maintains a comprehensive website about corruption, it produces Pakistan-specific corruption reports and surveys, provides a "hot line" for reporting fraud and other form of corruption, helps prepare procurement manuals for large Pakistani state operated enterprises and the Karachi city government in conjunction with the National Accountability Bureau (the national anti-corruption agency), and archives media reports about corruption. See the Transparency International Pakistan website, http://www.transparency.org, for more details on their manifold activities.

of workers to the petroleum exporting countries of the Middle East and elsewhere, which allowed it to benefit from manpower export as a natural resource boom. These three windfalls were:

1. The Evacuee Property Trust Board 1947
2. US Military and Economic Aid
    (a) Round One 1953–65
    (b) Round Two 1982–90
    (c) Round Three 2002–16
3. Expatriate Worker Remittances from 1973 onwards.

This section examines each in turn.

1. The Evacuee Property Trust Board 1947

The mutual exchange of populations between India and Pakistan has been described quite accurately as "one of the ten great tragedies of the 20th century" (Ahmed 2012). While both sets of populations, in total between 14–18 million, often arrived in their new homes with little more than they could carry, they left behind enormous wealth. Thus, when between 7–9 million Hindus and Sikhs left Pakistan for India, their valuable urban and rural property, including houses, factories, commercial enterprises and farms,[7] did not go with them. Although some assets were simply appropriated by local Muslims, they were all (in theory) handed over to the newly created Evacuee Trust Property Board[8] of the Pakistani successor state. The Board soon became a byword for systemic corruption that quickly reached into the highest levels of the new government as the Board attempted to resettle and compensate Muslim refugees for their losses.

Jalal (1991) provides a detailed analysis of how almost immediately after independence the level of corruption at all levels of the Pakistani

---

[7] In West (i.e., Pakistani) Punjab alone an estimated eight million acres of land were abandoned by non-Muslims fleeing to India while some four million acres were abandoned by Muslims in what then became East (Indian) Punjab. In 1950 the Indian government officially estimated the value of immovable property abandoned by Hindus and Sikhs in Pakistan at approximately US$8 billion (over US$75 billion in 2017 purchasing power). While the Pakistani government strongly disputed the accuracy of this valuation, it should suffice to give some idea of the amounts involved. See Schechtman (1951) for more details.

[8] Almost seven decades after its creation, the Board is still the source of legal dispute as the Federal and provincial governments wrangle over who controls its assets since appointments to it are rightly seen as essentially winning the lottery. Almost 70 years after Partition the Board owned 85,500 acres of land in Punjab province alone and was at the center of a dispute between the Federal and Punjab governments over control of the Board's assets. See Aslam (2014) for details.

government skyrocketed and as a 1951 internal government inquiry commission put it, "bribery [was] rampant in every department" and extended into levels of government where it had never existed before (quoted in Jalal 1991, p. 80). Furthermore, official patronage also boomed. The British colonial administration used massive official patronage, including extensive land and cash grants and favored recruitment quotas in government employment for loyal natives and preferred ethnic groups, but now the top layer of British colonial administrators, an extremely well-compensated group with minimal social or other ties to those they ruled, was removed. This meant that there was now no check on those civil servants or their political masters who had the power to dispense patronage; this lack of accountability meant that corruption flourished as patronage was now dispensed on personal whim or a quid pro quo basis and not as official government policy according to set standards designed to further some objectively determined policy criteria (Khan 2007).

There were several unsuccessful attempts to control corruption in the Pakistan civil service. One of the first acts of the new legislature was to pass the Prevention of Corruption Act of 1947, followed by the creation of the Special Anti-Corruption Establishment in 1948 and then the Civil Service (Prevention of Corruption) Rules of 1953; this last set of regulations showed the extent of the problem since it did away with civil service due process and procedural rules and stated that a civil servant was guilty of corruption simply if he had "a general or persistent reputation of being corrupt,"[9] but these attempts were all unsuccessful as the politicians were as, if not more, corrupt than the civil servants and moreover were beholden to them to carry out whatever directives the former issued.[10]

2. US Military and Economic Aid

Round one: 1953–65
Pakistan has been among the world's largest recipients of US military and financial assistance since the early 1950s even though the US-Pakistani relationship has experienced tortuous ups and downs. Pakistan quickly signed on as a major Cold War ally of the USA against the Soviet Union, correctly seeing anti-communism as the surest method of achieving American aid largess. The USA envisaged a series of military treaties encircling ("containing") the Soviet Union and China patterned on the

---

[9] Quoted in Callard (1957, p. 298); see also for details.
[10] See Khan (2007) for more on early attempts by the Pakistani state to control corruption in its civil service.

North Atlantic Treaty Organization (NATO). Accordingly, Pakistan joined the Central Treaty Organization (CENTO) and the Southeast Asia Treaty Organization (SEATO) in 1955 as a founding member and the aid began immediately. The USA had already sweetened the pot by giving a very large amount of economic assistance in 1953 to show what its Cold War allies could expect: US economic aid, which was only US$0.07 billion in 1952 and non-existent before then immediately jumped to US$0.75 billion in 1953 and then immediately declined to US$0.16 billion in 1954 (in constant 2009 dollars).

By 1955, with the signing of the military treaties, total aid jumped again to one billion dollars and military assistance had begun (at just over a quarter of the total). During the period 1953–65, US economic and military assistance to Pakistan totaled US$21.6 billion, or about US$1.67 billion annually. The brief 1965 Indo-Pakistan war greatly reduced US military aid (as the USA embargoed both sides) but economic assistance continued albeit at a reduced rate.

Round two: 1982–90
Pakistan continued to be a major recipient of US economic assistance between 1966 and 1980 (total aid of US$9.6 billion; averaging about US$0.6 billion annually) even if it got virtually no military aid, although even economic assistance virtually dried up after the 1977 military coup. The start of the Reagan administration in 1981 heralded the second coming of US economic and military assistance as the Soviet invasion of Afghanistan in 1979 turned Pakistan from a pariah military regime into (again) a vital frontline ally against the global Communist threat. Between 1982 and 1990, Pakistan received US$9.01 billion in military and economic aid (or about one billion dollars annually). In 1989, almost immediately after the Soviet withdrawal from Afghanistan, the George H.W. Bush administration determined that Pakistan's long-standing nuclear weapons research program would no longer receive a national security waiver and virtually embargoed all future assistance to Pakistan. From 1991 to 2001 US assistance to Pakistan totaled only US$0.84 billion (or about US$0.08 billion annually) but the terrorist attacks of September 11, 2001 changed all that.

Round three: 2002–16
US aid to Pakistan recommenced almost immediately after the 9/11 attacks when, under duress, Pakistan signed on as an ally in the American global war on terror. This last round of US assistance rapidly exceeded all earlier aid. From a negligible US$0.05 billion in 2000, US aid rose to US$0.23 billion in 2001 and to a staggering US$4.06 billion in 2002.

US aid peaked at US$5.6 billion in 2010 before declining to US$0.99 billion in 2016 as successive American administrations grew more and more frustrated with Pakistani policy regarding Afghanistan. Between 2002 and 2016, inclusive, Pakistan received US$42.6 billion (in constant 2009 dollars) in US assistance. To put it in current (2017) buying power terms, it received roughly US$48 billion over this 15-year period. If all US assistance to Pakistan since 1947 is converted into current buying power, it has received approximately US$95 billion since 1947.[11] US aid to Pakistan has made it one of the biggest recipients of US economic and military largess since World War II. Despite this largess, Pakistan has managed to move up only one step in the World Bank's income rankings: from Lower Income to Lower Middle Income status.

3. Expatriate Worker Remittances from 1973 Onwards

Pakistan also started to accrue another financial windfall when remittances from expatriate Pakistanis working abroad (mainly in the Middle East) started increasing tremendously after the 1973 and 1979 oil price hikes. By 1980 remittances constituted a hefty 10.25 percent of Pakistan's GDP, up from barely 3 percent in 1975, before declining to a low of 1.5 percent in 2000 and finally stabilizing to between 4–6.5 percent of GDP from the 2000s onwards.[12] While remittances as a proportion of GDP may have declined as the Pakistani economy grew, in dollar terms they still continue to pump vast sums into the country annually: US$15.8 billion in Fiscal Year (FY) 2013–14, US$18.7 billion in FY 2014–15, US$19.9 billion in FY2015–16 (State Bank of Pakistan 2016, p. 70).

The importance of worker remittances to Pakistan cannot be overstated as the country runs massive Current Account deficits and worker remittances have covered between 80–94 percent of Pakistan's imports in recent years (State Bank of Pakistan 2016, p. 73). While Pakistanis abroad, especially the manual laborers in the Middle East, have worked extremely long and hard under very difficult conditions for the money they remit back to Pakistan, for the domestic elite these funds are easy money, a fantastic

---

[11] Data on US economic and military assistance to Pakistan is calculated from the Guardian (n.d.) and Kronstadt and Epstein (2016). Of the total US assistance to Pakistan since 2002, US$14.56 billion was in Coalition Support Funds (CSF) which is technically not US aid but reimbursement to the Pakistani government for expenditures incurred in support of US troops and objectives in Afghanistan. Where necessary, conversions to 2009 constant dollars and 2017 current dollars were made using the Bureau of Labor Statistics online CPI Inflation Calculator.

[12] Remittance data is from the US Federal Reserve FRED data service (World Bank 2015a).

windfall that supports an extravagant lifestyle that the domestic economy simply cannot sustain.

Thus, Pakistan's series of "windfalls" over its history have resulted in a version of the "resource curse" that has plagued it since independence. The ultimate result of this has been a state structure that is "systemically corrupt" from top to bottom. Despite this, there have been several attempts by various Pakistani administrations to reduce bureaucratic and political corruption and, when that failed to stem the rot, to radically overhaul and reform the system.

While remittances and foreign aid may act as a "pull factor" in increasing corruption in Pakistan, there have also been "push factors" that have furthered the spread of corruption in the country. While possible push factors could include low salaries, lack of effective oversight or monitoring of the bureaucracy, inappropriate recruitment and selection criteria and so on, low salaries have been among the strongest push factors in Pakistan.

Even fairly low-ranking officials of the British Raj had been relatively well paid and high-ranking ones were extremely well paid, and this practice had been continued for the first few decades after independence. However, by the late 1970s inflation had eroded civil service salaries in both real and comparative terms and it had become an unattractive career path . . . at least from the perspective of honest compensation. For example, unskilled construction workers in 1983 made about the same salary as government clerks, not the entry level position in the Pakistani civil service pay scale as it was at Step 5 on the Basic Pay Scale of government service (Chew 1992, p. 110). While the wages of both skilled and unskilled laborers rose steeply due to the massive labor migration to the Middle East in the 1970s and 1980s, it was the Pakistani government's unwillingness to raise civil service salaries in line with inflation that was the real issue here. The effect of this has been that what is often termed "corruption" is simply an "alternative system" that has to be resorted to by extremely badly paid civil servants, especially at the lower (non-gazetted) levels, in order to carry out at least some of their duties or even to just survive (Ali and Nyborg 2010). As the report of the 1971 Hong Kong Salaries Commission, which took cognizance of the extensive corruption at all levels – local and expatriate – of the Hong Kong colonial administration, put it in its recommendations to substantially increase the salaries and allowances of the civil service: "it is not a question of paying sufficient salary to make a man incorruptible, but rather of not paying salary on which a man is encouraged to be corrupt in order to meet his reasonable commitments" (quoted in Lee 1981, p. 190).

The long overdue 2001 salary revision for all Pakistani public sector workers only met 75 percent of the increase in the Consumer Price Index since the last such revision in 1993 (Manning and Rinne 2002, p. 3) and the

next such revision in 2005 was even worse as it met just under half (15 percent) of the Consumer Price Index increase (31.5 percent) since 2001 (Irfan
2008, p. 23). Thus, between 1990 and 2007 the official income of Pakistani
civil servants had dropped substantially in real terms and this drop was
most severe at the highest levels of the service (Irfan 2008, p. 24). While
the salary revisions of 2008, 2011 and 2015 (Government of Pakistan
2015) went a long way towards clawing back the lost buying power of civil
service salaries, as Van Rijckeghem and Weder's (2001) detailed analysis
shows, simply raising civil service salaries will not have any meaningful
impact on the level of corruption in an already corrupt country.[13]

Thus, a combination of "pull" (the lure of evacuee property and the
influx of massive amounts of foreign aid and worker remittances) and
"push" (sustained substantial reduction in the real purchasing power
of civil service salaries over several decades) factors have ensured that
the "steel frame of the Raj" (as the Indian civil service was admiringly
described both before and after independence) inherited by Pakistan
rapidly decayed and almost rusted away into nothingness. However, the
corrosive effects of easy money were recognized early on as the many
attempts to remove the rust and decay from the frame testify. However,
these were all ultimately unsuccessful.

REFORM EFFORTS, MAJOR AND MINOR

As mentioned earlier, attempts were made as soon as Pakistan became
independent to strengthen the laws dealing with bureaucratic corruption (e.g., the Prevention of Corruption Act of 1947) and this was soon
followed by the creation of the Special Anti-Corruption Establishment
(Pakistan's first Anti-Corruption Agency, or ACA) in 1948, and the
Public Representative Offices Disqualification Act (PRODA) of 1953 that
was specifically aimed at elected politicians and was repealed in 1954 after
those charged under it alleged political victimization. The uselessness of
these early efforts can be gauged by the fact that in 1953 the draconian
Civil Service (Prevention of Corruption) Rules were passed that allowed

---

[13]   It should be noted that many high-ranking Pakistani civil servants enjoy considerable perks (official residences, cars, servants, etc.) that if appropriately monetized would
effectively double or triple their salaries. However, these perks are only enjoyed by a small
minority of the highest ranking bureaucrats. Monetary allowances in lieu of actual in-kind
perks are only a small fraction (roughly one-third of the actual salary) of the actual value of
these perks; most middle-rank and higher civil servants would only receive, say, a transportation allowance rather than a car and driver worth much more than the actual allowance
would have been.

for a civil servant to be declared guilty of corruption and dismissed from service simply if he had "a general or persistent reputation of being corrupt" (quoted in Callard 1957, p. 298).

Pakistan's first (of three) military government took over from a discredited civilian establishment in 1958 and immediately enacted new legislation (the Public Offices Disqualification Order, followed by the Elective Bodies Disqualification Order) along with the existing Prevention of Corruption Rules to purge, in effect, anyone suspected of opposing the new regime on (usually legitimate) corruption charges. Gen. Ayub's military successor Yahya Khan (ruled 1969–71) did the same and senior civil servants were again dismissed wholesale, probably on legitimate corruption grounds although their real crime was being out of favor with the new military regime.

The restoration of civilian rule after Pakistan's defeat in the 1971 Indo-Pakistan war also saw the new government of Zulfikar Ali Bhutto (ruled 1971–77) creating a new ACA (Federal Investigation Agency or FIA) that had jurisdiction over national corruption cases. The FIA continued as the premier national ACA until the very end of the 1990s despite another military coup in 1977 and a series of revolving-door civilian governments starting in 1989. Far from being curtailed, corruption increased substantially over this time period.

While some attempts to reduce the worst corruption excesses in Pakistan at the highest levels had limited initial success – for example, the early Ayub regime in 1958 or the 1993 attempt by Moeen Qureishi's interim administration (see Khan 2007 for details) – the speed with which these were abandoned by successor governments which were themselves tainted by serious corruption charges rendered the preliminary success meaningless. Other anti-corruption efforts were an exercise in pure political victimization when, as is too often the case worldwide, corruption charges were used against political opponents. For example, the second Nawaz Sharif administration (1997–99) set up a special "Ehtesab [Accountability] Bureau" ostensibly to combat political corruption among the nation's ruling elite (which the FIA had completely failed to curb) but it was widely viewed as partisan harassment. The only cases pursued by the Ehtesab Bureau were against Sharif's political opponents, the leader of the opposition former Prime Minister Benazir Bhutto and her husband (and future President of Pakistan) Asif Ali Zardari (Hajari 1999). The blatantly political nature of the prosecution garnered Zardari a great deal of popular sympathy despite his well-earned reputation of 'Mr. Ten Percent' and the disclosure that the Bhuttos owned a multi-million dollar estate in the UK and had millions of dollars in Swiss bank accounts.

The most sweeping and most promising attempt to clean the Augean Stables of corruption that Pakistan had become was the one launched by General Pervez Musharraf's (initially at least) reforming military government that took power in the third of Pakistan's bloodless military coups in 1999. Musharraf passed the National Accountability Ordinance 1999 that established the National Accountability Bureau (NAB), Pakistan's most recent and powerful ACA.[14] The NAB remains Pakistan's main anti-corruption agency and as it proclaims on its website: "The National Accountability Bureau is Pakistan's apex anti-corruption organization. It is charged with the responsibility of elimination of corruption through a holistic approach of awareness, prevention and enforcement."[15] The NAB was originally charged with pursuing politicians and public servants guilty of corruption, bribery, extortion and the like (including fraudulent bank loan write-offs and embezzlement[16]). The NAB extended its activities in 2002 with the National Anticorruption Strategy (NACS) project to go beyond "traditional enforcement based routines" to look at international anti-corruption best practices, citizen surveys, local stakeholder participation exercises and a host of other activities designed to "undertake prevention and [raise] awareness." The NAB defines its long-term objective as "The elimination of corruption by engaging all the stakeholders in the fight against corruption through a programme which is holistic, comprehensive and progressive." The NAB is headed at the national level by a Chair and each provincial/federal territory unit is headed by a regional head.

Initially at least, the Musharraf administration seemed to be making a concerted effort to attack high-level, that is, "grand" or "systemic" corruption in Pakistan, but this reforming zeal did not extend much past 2002. General Musharraf removed NAB's first Chair, Lieutenant General Syed Amjad, when he requested permission to go after serving military officers implicated in serious corruption cases (Khan 2007, p. 239).[17] Amjad's brief

---

[14]   See Transparency International Pakistan (2014, p. 27) for more details.
[15]   National Accountability Bureau website, http://www.nab.gov.pk/home/introduction. asp (accessed May 20, 2018).
[16]   A very common practice in Pakistan where the nationalized banking system was used to enrich many politically well-connected persons. This practice drove the entire national banking system to the point of failure in the 1990s.
[17]   There have been only a very few instances that I have been able to find of senior military officers being convicted of corruption. Admiral Mansur ul Haq, former Pakistan naval chief, was convicted in 2001 of accepting kickbacks related to submarine purchases by the Navy, ordered to pay US$7.5 million in restitution and stripped of his rank and all benefits; his sentence of seven years' imprisonment was reduced to time served upon payment of restitution although the 14 houses he, his wife and daughter owned in Pakistan were not included in the restitution decree (see Mohiuddin 2012 for more details). In 2015, of three generals accused of being involved in illegally investing approximately US$43 million in funds from a military-operated firm in the stock exchange and suffering losses of approximately US$18

tenure as Chair of the NAB is perhaps reflected in Pakistan's highest score of 25.76 in 2003 on the World Bank's Worldwide Governance Indicator Control of Corruption variable.

The NAB's relative effectiveness during its first few years can be judged by two indicators: the number of high-profile corruption convictions and the amount of restitution to the national exchequer from those convicted of corruption. In its first full year of operation 2000–01, the NAB launched 935 investigations, convicted 266 persons and acquitted 60 (with 609 investigations pending) for a conviction rate of 28.4 per cent and a completion rate of 34.9 per cent. The list of those convicted reads like a veritable "Who's Who" of Pakistan: prime ministers, chief ministers of provinces, senators, members of parliament and senior civil servants but only one or two (retired) senior military officers (Abbasi 2015, pp. 37–43).[18]

In 2002, General Musharraf, at the insistence of the USA which did not want to deal with a nakedly military government in its global war on terror, decided to replace direct military rule with a civilian facade. Civilian politicians convicted of massive corruption and other ill-deeds under Amjad's tenure at the NAB were allowed to pay token fines, redeem themselves and stand for office again. The predictable effects of this civilianization of the government were that Musharraf's rule rapidly became as corrupt as the civilian administrations that preceded his, and Musharraf's cabinet was soon stocked with politicians convicted by his own regime of corruption.[19]

The culmination of the descent of Musharraf's rule into the typical morass of corruption that has characterized Pakistan was his issuance of the National Reconciliation Ordinance in 2007 that provided a complete amnesty to 8,041 officials (mainly government bureaucrats but also a few politicians, among them ex-Prime Minister Benazir Bhutto, her husband and Musharraf's successor as President of Pakistan, Asif Ali Zardari, and future Prime Minister Yousuf Raza Gillani), who faced corruption charges pending against them, some dating back to 1986. The common view of the Ordinance among most well-informed Pakistanis was that it was a quid pro quo arrangement brokered by the USA that would allow the increasingly unpopular but vital US ally Musharraf to stay on as a "civilian" president and allow Benazir Bhutto to be elected to a third term as prime minister. The annulling of the Ordinance by Pakistan's

---

million which they then tried to hide by embezzling money from a pension fund, a major general was found guilty and "dismissed from service" with full forfeiture of rank, pension, benefits and so on, while one lieutenant general was issued a "severe displeasure" (i.e., a slap on the wrist) and the other found not guilty (see Dawn 2015 for more details).

[18] See Khan (2016) for more details on NAB's setup and initial success and subsequent failure.

[19] See Khan (2007) for more details on this.

Chief Justice and the Chief Justice's subsequent dismissal by Musharraf, followed by Benazir Bhutto's assassination in December 2007, began a mass movement spearheaded by Pakistan's lawyers who took to the streets to oust Musharraf from power. Musharraf's resignation in August 2008 ended the last major attempt to clean up corruption in Pakistan, although Musharraf had given up that task long before his resignation.

Asif Ali Zardari, Benazir Bhutto's widower, who had one of the worst reputations in Pakistan for massive corruption, succeeded Musharraf as president after the latter's resignation in 2008. Zardari proceeded to essentially dismantle the NAB as its budget was slashed almost in half and its staff massively reduced, and for all intents and purposes the NAB ceased to function as an Anti-Corruption Agency. A compromise was reached between an activist Supreme Court and President Zardari when a new director general of NAB was appointed in 2011 who, apparently, agreed to exempt politicians from NAB's scrutiny. Under Zardari, the NAB gave up even the pretense of going after major offenders and contented itself with what were, in effect, penny-ante cases. A change of government in 2013, Pakistan's first transfer of power from one civilian administration to another in its history, reinvigorated NAB slightly and prosecutions and investigations increased somewhat.[20]

When there are actual anti-corruption cases filed by the NAB or the provincial anti-corruption establishments, all too often they tend to sit in the courts for months or years before slipping away into oblivion. For example, for almost a year (May 2011 to April 2012) no judge was appointed to the Federal Anti-Corruption Court in Karachi, Pakistan's largest city and financial center, and over 1,400 high-level corruption cases simply continued, putting the accused out in effect on permanent bail. The situation only came to public attention when the Supreme Court intervened to transfer jurisdiction to itself and rejected all motions to have the cases dismissed (Tanoli 2012). Some of these cases involved embezzlement charges totaling into the hundreds of millions of dollars.

Despite the fact that the Raisani case, involving a senior civil service officer caught red-handed with several million dollars in cash and kilos of gold hidden in his house, and the TV channels endlessly playing video clips showing bags full of cash taken from his house, became the "poster child" for massive corruption at the highest level of Pakistan's governance, how the NAB handled it is illustrative of the current wholly ineffective approach to combating corruption. While accounts of the full extent of

---

[20]   See Khan (2016) for details on NAB's organization and evolution and its overall anti-corruption track-record.

Raisani's corruption (i.e., fraud, kickbacks, embezzlements and so on) are hotly disputed by the NAB itself – it very quickly revised its initial estimate of approximately PKR 40 billion (about US$380 million) down to only PKR two billion (about US$19 million) – the fact that Raisani was allowed a "voluntary return" agreement meant that he was not actually guilty of anything and could, in fact, ask to be reinstated into government service as had previously happened with at least 1,500 other civil servants.

Section 25(a) of the NAB Ordinance (1999) gives the NAB Chairman the power to dismiss all charges against an accused if "the holder of a public office or any other person accused of any offence under this ordinance, returns the assets or gains acquired through corruption or corrupt practices to NAB, then the [NAB] chairman may release him" (quoted in Express Tribune 2015). The unilateral right of the head of NAB to decide on this was reaffirmed in a 2015 High Court (the penultimate appeals court in the Pakistani legal system) ruling in which that court overturned a lower court's order limiting NAB's discretion in this regard (Express Tribune 2015). However, the extreme leniency shown to Raisani – voluntary return is one step below a plea bargain in that the accused does not actually have to admit to any guilt and so is not liable to any legal consequences for his actions – aroused the ire of the activist Pakistani Supreme Court, which in October 2016 issued an injunction preventing the NAB from granting any voluntary return orders without the Court's concurrence. Using extremely harsh language against the NAB, the Supreme Court said that far from being an anti-corruption agency, the NAB had essentially told the corrupt that they can "get away with it" and so "NAB itself is facilitating corruption within the country" (quoted in N. Siddiqui 2017). In earlier hearings, in response to direct orders from the Supreme Court, the NAB Prosecutor General had reluctantly provided the Court with a partial list of 1,548 senior government officials who had made "voluntary returns" totaling approximately PKR two billion (~US$20 million) and been allowed to return to duty (Iqbal 2016). A clearly incensed Chief Justice of the Supreme Court told the NAB Prosecutor General that the NAB's position on voluntary returns was such that it seemed as if after stealing millions, "A person can visit the NAB office, settle the matter over a cup of coffee, and save his job" (quoted in Siddiqui 2016) and that, furthermore, the guilty parties could negotiate with the NAB over how much they had to return and even then they did not have to make the full payment up front but could put only one-third down and the rest in equal monthly installments (Iqbal 2016). The Supreme Court further ordered the Federal and provincial governments to actively pursue for additional punishment all "voluntary return" cases where the official was allowed to return to his employment without any punishment.

The continuation in service of literally thousands of mid-rank and senior civil servants who have "voluntarily returned" some of the billions of rupees they had stolen earlier cannot help but further reduce the minimal amount of competence and probity left in the Pakistani civil service. The NAB's message to the corrupt seems to be "If you're corrupt: heads you win and tails, you get to try again a little later." While the Pakistani government vows to deal with corruption and nominally follows the World Bank's anti-corruption guidelines and so on, it really has no interest in actually doing anything about corruption. During Musharraf's rule and subsequently, Pakistan has ratified several international anti-corruption treaties and codes of conduct, for example, the United Nation's Convention Against Corruption (UNCAC), but there has been no diminishment in the level of corruption there. Quite the contrary, corruption has increased. The issue is not that of inadequate laws but of a simple lack of enforcement: "laws against corruption are comprehensive and strict [but] implementation is very weak. There are many law enforcement institutions, Acts, Orders, Ordinances and a plethora of other legal instruments, yet they so far seem to be insufficient for eliminating corruption" (Transparency International Pakistan 2014, p. 8). The exemplar of this lack of enforcement is that there were no prosecutions for income tax fraud for 25 years. In keeping with the Potemkin village aspect of Pakistan's anti-corruption efforts, Musharraf's infamous National Reconciliation Ordinance was issued 56 days after Pakistan ratified the UNCAC (Transparency International Pakistan 2014, p. 161).

Following the release of the Panama Papers[21] in 2015, in which Prime Minister Nawaz Sharif's family was prominent among the Pakistanis using off-shore tax havens, the Pakistan Supreme Court disqualified him and removed him from office on the highly technical grounds that he had not disclosed a paid directorship held in his son's company in the United Arab Emirates; the Court dismissed his argument that he had not drawn the salary and that the non-disclosure was wholly inadvertent. Sharif's appeal against the judgment was also dismissed and the NAB was ordered to reopen various corruption inquiries into the source of the Sharif family's fortunes. Sharif's finance minister, who had earlier worked for

---

[21]   The "papers" were some roughly 11.5 million documents taken from the internal files of Mossack Fonseca, a Panamanian law firm specializing in off-shore tax havens, and given to the International Consortium of Investigative Journalists who released them in the form of a searchable database online (see their full website https://www.icij.org/investigations/panama-papers/ for details (accessed May 20, 2018)). Many Pakistanis, including ruling Pakistan Muslim League politicians had considerable assets hidden away in tax havens offshore.

various family-owned firms, was also ordered to be investigated, as were several other prominent politicians allied with the Sharifs.[22]

The names of Pakistani politicians and senior government servants continually appear on various lists of those accused of serious graft and corruption charges by the NAB. Starting from the first such list of loan defaulters prepared by General Amjad in 1999–2000[23] to a more recent one delivered to the Pakistan Supreme Court at its order during the summer of 2015, every such list reads like a veritable Who's Who of Pakistan precisely because it *is* a Who's Who. The Summer 2015 list of 150 "influentials" included former President of Pakistan Asif Ali Zardari, former Prime Ministers Nawaz Sharif, Yusuf Raza Gilani, Raja Pervaiz Ashraf, Shujaat Hussain, and current and former cabinet ministers Ishaq Dar and Firdous Ashiq Awan, along with many other current and former politicians and government servants (Ghayur 2015). These "influentials" ran the gamut of political parties and ideologies and showed that corruption played no favorites and shunned no one. The combined value of the graft indulged in ran into the billions of dollars and virtually all of the "influentials" faced multiple charges.

Thus, the Pakistani anti-corruption establishment is not particularly effective since the biggest, most powerful and the most corrupt perpetrators of corruption are never punished. What "punishment" does exist is not even particularly harsh even by "slap on the wrist" standards since "voluntary return" makes a mockery of even the pretense of an anti-corruption strategy. The negative demonstration effect of the highest and most powerful in the land getting away with virtually unbridled greed must be tremendous and it makes any possible countervailing action all the more difficult.

CONCLUSION: POLICY RECOMMENDATIONS

Unsurprisingly, given the extent of the corruption problem, Pakistan's economic growth has been severely impacted in the past two decades. A detailed World Bank study, "What Do We Know about Growth Patterns in Pakistan," concluded that "Pakistan's growth potential is eroding . . . In the last fifty years Pakistan's steady annual [potential] GDP growth

---

[22] See Z. Siddiqui (2017) for a good summary of the Pakistani Supreme Court's reasoning in this case.

[23] Actually, the interim, caretaker government of Prime Minister Moeen Qureishi installed by the Pakistani Army in 1993 also published a list of loan defaulters that featured many names repeated in subsequent lists; this list, however, predated the creation of the NAB by several years.

rate has fallen from 6 percent to about 4.5 percent. This is much lower than the 7 percent the Government estimates is needed to absorb youth unemployment" (Lopez-Calix et al. 2012, p. 6) and, predictably, Pakistani labor is the least productive of South Asia's major economies (Lopez-Calix et al. 2012, p. 6). The World Bank report concludes that deteriorating infrastructure, insufficient investment in workforce education, frequent macroeconomic policy changes, inadequate levels of investment, and so on – in short, the usual suspects – are responsible for this dire state of affairs but in keeping with the Bank's traditional circumspection involving corruption, its recent anti-corruption efforts notwithstanding, there is no mention of the "C-word" in the entire report. However, using sophisticated econometric modeling techniques to analyse Pakistani economic data from 1987 to 2009, Farooq et al. (2013, p. 628) determined that in the long run "a 1% rise in corruption retards economic growth by 0.1489% keeping other things constant" and further econometric analysis (Granger causality testing) showed that the causality arrow ran from corruption to growth retardation and not vice versa, that is, corruption retarded growth and not that economic stagnation caused corruption.

Considering that the most recent International Monetary Fund bailout of Pakistan, agreed upon in 2013, was for a mere US$6.64 billion over three years (F. Khan 2015, p. 187), or about US$2.213 billion annually, plugging just the fraudulent refund hole in the Pakistani tax collection service, the Federal Board of Revenue (FBR), would have given it about US$2.65 billion annually. Plugging all the other holes in the FBR and expanding the ludicrously small Pakistani tax net[24] would have of course generated much greater tax revenue but that is a task that would make Hercules' cleaning of the Augean Stables mere child's play. Thus, Pakistan's educational, physical and institutional infrastructure continues to crumble as the government's budgetary woes preclude sufficient spending on maintenance, much less expansion or improvement.

Pakistan already has a comprehensive National Accountability Ordinance and an anti-corruption body, the National Accountability

---

[24]   For example, in 2013 the total number of registered income taxpayers in the country was 768,000 individuals or roughly 0.57 percent of the population; in other countries at a comparable level of economic development the figure is about 15 percent (BBC 2013). The vast majority, over two-thirds, of Pakistani members of Parliament did not file income tax returns in 2011 or did not even have an income tax number; the then President of Pakistan Asif Ali Zardari, notoriously referred to as "Mr. Ten Percent," had not bothered to file an income tax return that year and no one had been prosecuted for income tax fraud for the past quarter century (F. Khan 2015, p. 189)! Of those lawmakers who did file a tax return, the average tax paid was approximately US$1,000 and the average value of declared assets of the lawmakers was US$800,000 (Houreld 2012).

Bureau, with both expertise and power. The NAB has shown that it has the ability to craft corruption cases against influential and well-connected individuals, even powerful members of the ruling political elite, but this has come to naught in the past due to a complete lack of political will on the part of the Pakistani leadership. As long as the Pakistani leadership, regardless of political party, themselves benefit from corruption, no anti-corruption agency is ever going to prevail in its efforts. In the past, some Pakistani rulers have appointed trusted allies as NAB Chairmen to ensure that its investigations not only do not threaten them but can be channeled along the "right" path. This is of course a recipe for failure as far as NAB's stated mission is concerned.

A successful anti-corruption policy requires a leadership that is committed to it. Singapore's example shows that in this case at least, success flows from the top down. Once the anti-corruption agency's personnel know that their institutional leadership and their leadership's superiors back the official policy and are determined to make it succeed, in effect then the battle is half won already. If the anti-corruption agency's personnel *know* that their leadership and the national leadership is determined to thwart the agency's stated mission, the battle is lost already. So other than a national leadership determined to succeed at reducing if not eliminating corruption in Pakistan, the necessary setup was established by the National Accountability Ordinance in 1999.

Hope apparently springs eternal among some analysts and perhaps the future is not all bleak on the corruption front. Verkaaik (2001), focusing on the Karachi-based Muhajir Qaumi Movement (MQM)[25] political party in Pakistan, extensively documents the interconnections among politics, ethnicity, corruption and violence in Pakistan. Verkaaik (2001, p. 348) encapsulates the people's perception of the Pakistani state around the concept of corruption: "the theme of corruption is the most important argument through which the [people] construct the image of a polluted state or a besieged idea of impartiality and justice." However, presaging the rise to prominence of playboy-cricketer turned political savior Imran Khan's Pakistan Tehrik e Insaaf (Pakistan Justice Movement) anti-corruption political campaign in the 2013 elections, Verkaaik (2001, p. 364) ends on an optimistic note when he concludes that the

---

[25] The Muhajir Qaumi Movement, that is, Refugee's National Movement, which renamed itself as the Muttahida Qaumi Movement (United National Movement), was the nationalist/separatist political party that gained prominence in Karachi, Pakistan's largest city and financial capital, from the 1980s onwards. The renaming was done in 1997 in an unsuccessful attempt to broaden its appeal outside its narrow base of mainly Urdu-speaking refugees and their descendants in urban Sindh province.

growing attention to the theme of corruption must therefore be seen primarily as a sign of emancipation rather than an increase of immoral behavior among state officials. These practices now increasingly labelled as corruption are not new. What is new is the growing sense of indignation about these practices in Pakistani society.

This was written almost two decades ago. Unfortunately for Pakistan, Verkaaik was hopelessly overoptimistic about the MQM, which showed itself to be as bad as, if not worse than, any other Pakistani political party as far as corruption was concerned. While Khan's movement, which aroused great enthusiasm among the relatively young and well-educated urban population, ultimately fizzled out with a whimper, Bashir et al. (2011) found that the very few who would blow the whistle on wrongdoing in government organizations were overwhelmingly young and well educated. Perhaps there might be a brighter future for Pakistan after all.

REFERENCES

Abbasi, M.M. 2015. *National Accountability Bureau: An Independent Review of Structure and Performance.* Islamabad: Pakistan Institute of Legislative Development and Transparency. https://pildat.org/governance1/national-accountability-bureau-an-independent-review-of-structure-and-performance (accessed May 20, 2018).

Abdih, Y., R. Chami, and J. Dagher. 2012. Remittances and institutions: are remittances a curse? *World Development* 40(4), 657–66.

Ahmed, I. 2012. *The Punjab Bloodied, Partitioned and Cleansed: Unravelling the 1947 Tragedy through Secret British Reports and First-person Accounts.* Karachi: Oxford University Press.

Ali, J. and I.L.P. Nyborg. 2010. Corruption or an alternative system? Re-assessing corruption and the role of the Forest Services in the Northern Areas, Pakistan. *International Forestry Review* 12(3), 209–20.

Andreski, S. 1968. *The African Predicament: A Study in the Pathology of Modernisation.* New York: Atherton Press.

Asian Development Bank. 2013. *The Social Protection Index: Assessing Results for Asia and the Pacific*, Manila: Asian Development Bank. https://think-asia.org/bitstream/handle/11540/79/social-protection-index.pdf?sequence=1 (accessed May 20, 2018).

Aslam, S. 2014. Who will own the billionaire Evacuee Trust Property Board? *News International*, February 13. https://www.thenews.com.pk/archive/print/484753 (accessed May 20, 2018).

Bashir, S., H. Khattak, A. Hanif, and S. Chohan. 2011. Whistle-blowing in public sector organizations: evidence from Pakistan. *American Review of Public Administration* 41(3), 285–96.

BBC (British Broadcasting Corporation). 2013. Stop extra UK aid to Pakistan unless taxes increase, urge MPs, 4 April. http://www.bbc.com/news/uk-politics-22017091 (accessed May 20, 2018).

Beall, J. 2012. List of predatory publishers 2013. https://beallslist.weebly.com/ (accessed May 20, 2018).

Berdiev, A., Y. Kim, and C.-P. Park. 2013. Remittances and corruption. *Economics Letters* 118, 182–5.

Callard, K. 1957. *Pakistan: A Political Study.* London: Allen and Unwin.

Chew, D.C.E. 1992. *Civil Service Pay in South Asia.* Geneva: International Labour Office.

Collier, P. 2008. *The Bottom Billion: Why the Poorest Countries are Failing and What Can Be Done About It.* Oxford: Oxford University Press.

Dawn. 2014. Billions lost in bogus sales tax refunds. *Dawn*, March 16. https://www.dawn.com/news/1093513 (accessed May 20, 2018).

Dawn. 2015. Army sentences two former generals in NLC corruption case. *Dawn*, August 6. http://www.dawn.com/news/1198561 (accessed May 20, 2018).

Dawn. 2016. FBR notifies rules to select cases for tax audit. *Dawn*, January 29. https://www.dawn.com/news/1236025 (accessed May 20, 2018).

Express Tribune. 2015. NAB's ability to collect voluntary returns restored. *Express Tribune*, July 15. https://tribune.com.pk/story/920928/nabs-ability-to-collect-voluntary-returns-restored/ (accessed May 20, 2018).

Farooq, A., M. Shahbaz, M. Arouri, and F. Teulon. 2013. Does corruption impede economic growth in Pakistan? *Economic Modelling* 35, 622–33.

Ghayur, Z. 2015. 150 influentials on NAB list. *Daily Times*, July 8. http://www.dailytimes.com.pk/national/08-Jul-2015/150-influentials-on-nab-list (accessed October 26, 2015).

Government of Pakistan. 2015. Revision of basic pay scales, allowances and pension of civil servants of the federal government, July 7. Regulations Wing circular. Ministry of Finance, Islamabad. http://www.finance.gov.pk/circulars/circular_07072015_pay.pdf (accessed May 20, 2018).

Guardian. n.d. Sixty years of US aid to Pakistan: get the data. *Guardian*. https://www.theguardian.com/global-development/poverty-matters/2011/jul/11/us-aid-to-pakistan (accessed May 20, 2018).

Hajari, N. 1999. Bhutto brought to book. *Time*, April 26. http://content.time.com/time/world/article/0,8599,2054147,00.html (accessed May 20, 2018).

Hoti, I. 2003. Only 33pc businessmen paid GST in 2002–03. *Dawn*, December 13. http://forum.pakistanidefence.com/index.php?showtopic=23184 (accessed May 20, 2018).

Hough, D. 2016. Here's this year's (flawed) Corruption Perception Index. Those flaws are useful. *Washington Post*, January 27. https://www.washingtonpost.com/news/monkey-cage/wp/2016/01/27/how-do-you-measure-corruption-transparency-international-does-its-best-and-thats-useful/?utm_term=.2185e1904495 (accessed May 20, 2018).

Houreld, K. 2012. Promoting investigative journalism in Pakistan, one tax return at a time. *Reuters*, December 12. http://blogs.reuters.com/pakistan/2012/12/12/promoting-investigative-journalism-in-pakistan-one-tax-return-at-a-time/ (accessed May 20, 2018).

Humphreys, M., J. Sachs, and J. Stiglitz (eds) 2007. *Escaping the Resource Curse.* New York: Columbia University Press.

Iqbal, N. 2016. NAB chief barred from voluntary return deals. *Dawn*, October 25. https://www.dawn.com/news/1292113 (accessed May 20, 2018).

Irfan, M. 2008. *Pakistan's Wage Structure during 1990/91–2006/07.* Pakistan Institute of Development Economics Working Paper, December. Pakistan Institute of Development Economics, Islamabad. http://www.pide.org.pk/pdf/pws.pdf (accessed May 20, 2018).

Jalal, A. 1991. *The State of Martial Rule: The Origins of Pakistan's Political Economy of Defense.* Lahore: Vanguard Books.

Khan, F. 2007. Corruption and the decline of the state in Pakistan. *Asian Journal of Political Science* 15(2), 219–47.

Khan, F. 2015. Pakistan's self-inflicted economic crises. In C. Fair and S. Watson (eds), *Pakistan's Enduring Challenges.* pp. 178–203. Philadelphia, PA: University of Pennsylvania Press.

Khan, F. 2016. Combating corruption in Pakistan. *Asian Education and Development Studies* 5(2), 195–210. https://doi.org/10.1108/AEDS-01-2016-0006 (accessed May 20, 2018).

Khan, I. 2015. Missing necklace found but plot thickens. *Dawn*, (June 13. http://www.dawn.com/news/1187952/ (accessed May 20, 2018).

Klitgaard, R. 1991/1988. *Controlling Corruption.* Berkeley and Los Angeles, CA: University of California Press.

Kronstadt, K. and S. Epstein. 2016. *Direct Overt U.S. Aid Appropriations for and Military Reimbursements to Pakistan, FY2002–FY2017.* Washington, DC: Congressional Research

Service. https://digital.library.unt.edu/ark:/67531/metadc855748/m2/1/high_res_d/crs_ls_20 16Feb24_pakaid.pdf (accessed May 20, 2018).

Lee, P.N.S. 1981. The causes and effects of police corruption: a case in political moderniza-tion. In R.P.L. Lee (ed.), *Corruption and its Control in Hong Kong*, pp. 167–98. Hong Kong: Chinese University Press.

Lopez-Calix, J.R., T.G. Srinivasan, and M. Waheed. 2012. What do we know about growth patterns in Pakistan? World Bank Policy Paper Series on Pakistan No. PK 05/12. World Bank, Washington, DC. http://documents.worldbank.org/curated/en/746861468090592263/pdf/862540NWP0Worl02000Growth0Patterns.pdf (accessed May 20, 2018).

Manning, N. and J. Rinne. 2002. Pakistan civil service brief. World Bank, Washington, DC. http:// www1.worldbank.org/publicsector/civilservice/PayEmpModels/ pakbrief.pdf (accessed May 20, 2018).

Mohiuddin, S. 2012. NAB agrees to free Admiral Mansur for Rs 457.5 million. *News International*, January 12. https://www.thenews.com.pk/archive/print/340478 (accessed May 20, 2018).

Rana, S. 2017. FBR faces mammoth task of auditing over 93,000 cases within six months. *Express Tribune*, January 7. https://tribune.com.pk/story/1287565/fbr-faces-mammoth-task-auditing-93000-cases-within-six-months/ (accessed May 20, 2018).

Schechtman, J. 1951. Evacuee property in India and Pakistan. *Pacific Affairs* **24**, 406–13.

Shah, S. 2016. NAB recovers Rs730m from Balochistan finance secretary's residence. *Dawn*, May 6. https://www.dawn.com/news/1256686 (accessed May 20, 2018).

Siddiqui, A. 2016. SC orders actions against beneficiaries of NAB "voluntary return." GeoTV, October 26. https://www.geo.tv/latest/118841-SC-orders-action-against-officials-who-benefited-under-NAB-voluntary-return (accessed May 20, 2018).

Siddiqui, N. 2017. NAB is facilitating corruption in country, says Supreme Court. *Dawn*, January 2. https://www.dawn.com/news/1305972 (accessed May 20, 2018).

Siddiqui, Z. 2017. The Supreme Court still has some pretty damning things to say regarding Panamagate. *Dawn*, November 8. https://www.dawn.com/news/1369017/the-supreme-cou rt-still-has-some-pretty-damning-things-to-say-regarding-panamagate (accessed May 20, 2018).

State Bank of Pakistan. 2016. *Annual Report 2015–2016: The State of the Economy*, Chapter 6 External Sector. Karachi: State Bank of Pakistan. http://www.sbp.org.pk/reports/an nual/arFY16/Chapter-06.pdf (accessed May 20, 2018).

Tanoli, I. 2012. Graft cases pile as anti-corruption court vacant for a year. *Dawn*, April 1. http://www.dawn.com/news/707221/graft-cases-pile-as-anti-corruption-court-vacant-for-a-year (accessed May 20, 2018).

Thompson, T. and A. Shah. 2005. Transparency International's Corruption Perceptions Index: whose perceptions are they anyway? http://siteresources.worldbank.org/PSGLP/Resources/ShahThompsonTransparencyinternationalCPI.pdf (accessed May 20, 2018).

Transparency International. 2017. Global Corruption Barometer (People and Corruption: Asia Pacific). https://www.transparency.org/whatwedo/publication/people_and_corruptio n_asia_pacific_global_corruption_barometer (accessed May 20, 2018).

Transparency International. Various years. Corruption Perception Index. http://www. transparency.org/research/cpi/ (accessed May 20, 2018).

Transparency International Pakistan. 2014. National integrity system country report 2014. https://www.transparency.org/whatwedo/nisarticle/pakistan_2014 (accessed May 20, 2018).

Van Rijckeghem, C. and B. Weder. 2001. Bureaucratic corruption and the rate of temptation: do wages in the civil service affect corruption, and by how much? *Journal of Development Economics* **65**(2), 307–31.

Verkaiik, O. 2001. The captive state: corruption, intelligence agencies, and ethnicity in Pakistan. In T. Hansen and F. Stepputat (eds), *States of Imagination: Ethnographic Explorations of the Postcolonial State*, pp. 345–63. Durham, NC: Duke University Press.

World Bank. 2014. *Worldwide Governance Indicators: Country Data Report Pakistan, 1996–2014*. Washington, DC: World Bank. .http://documents.worldbank.org/curated/en/410231467996713318/pdf/105541-WP-PUBLIC-Pakistan.pdf (accessed May 20, 2018).

World Bank. 2015a. Remittance inflows to GDP for Pakistan. Washington, DC. https://research.stlouisfed.org/fred2/series/DDOI11PKA156NWDB/

World Bank 2015b. Tax revenue (% of GDP). Washington, DC. http://data.worldbank.org/indicator/GC.TAX.TOTL.GD.ZS (accessed May 20, 2018).

18. Corruption in Bangladesh: insights from the financial sector
Mohammad Nurunnabi

Corruption undermines development by weakening the institutional foundation on which economic growth depends. Corruption is almost universal across the planet but varies widely in severity, type and consequences. There is a growing literature that suggests corruption is an epidemic globally and focuses on the determinants of corruption (Weyland 1998; Easterly 2001; Becker et al. 2009; Ortega et al. 2010). Corruption occurs when the expected benefits exceed the costs in the form of rent-seeking behavior (Klitgaard 1988). The literature classifies the primary sources of corruption into various determinants, including social, economic, political and geographical.

The objective of this chapter is to provide economic determinants of corruption taking into account financial sectors in Bangladesh. The banking sector dominates the financial markets in Bangladesh (Asian Development Bank 2009). The financial crisis has had little impact on Bangladesh's credit markets since loan disbursements grew by 43 percent in the quarter ending September 2008.

This study investigates corruption in the financial sector, which is the backbone of the Bangladeshi economy. The findings contribute to literature that shows that corruption in the financial sector is highly politicized, which endangers its long-term development. The findings of the study offer an agenda for policy makers. The next section briefly discusses an overview of corruption in Bangladesh. A case study is then presented of the Hallmark Group and, finally, chapter conclusions.

CORRUPTION AND FINANCE IN BANGLADESH

There is ongoing political instability throughout Bangladesh history due to the failure of democratic politics and the return of military rule. Zafarullah and Akhter (2001) explore military rule, civilianization and electoral corruption in Bangladesh. According to Transparency International's Corruption Perceptions Index (CPI), Bangladesh was the 145th least corrupt nation out of 175 countries in 2016, or the 30th most

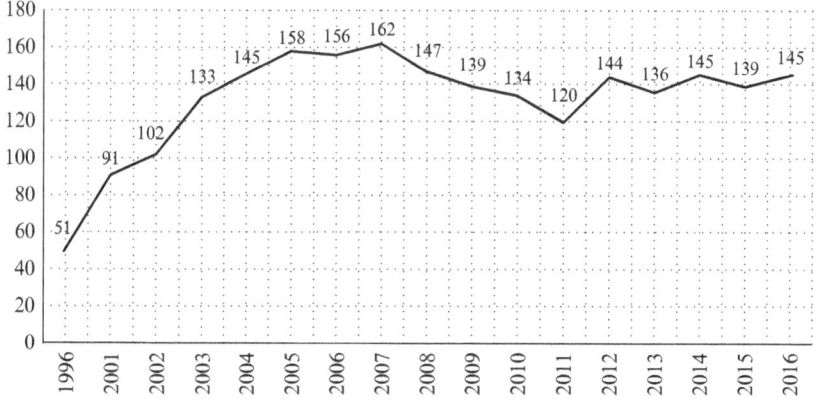

Figure 18.1   Corruption perceptions rank of Bangladesh, 1996–2016

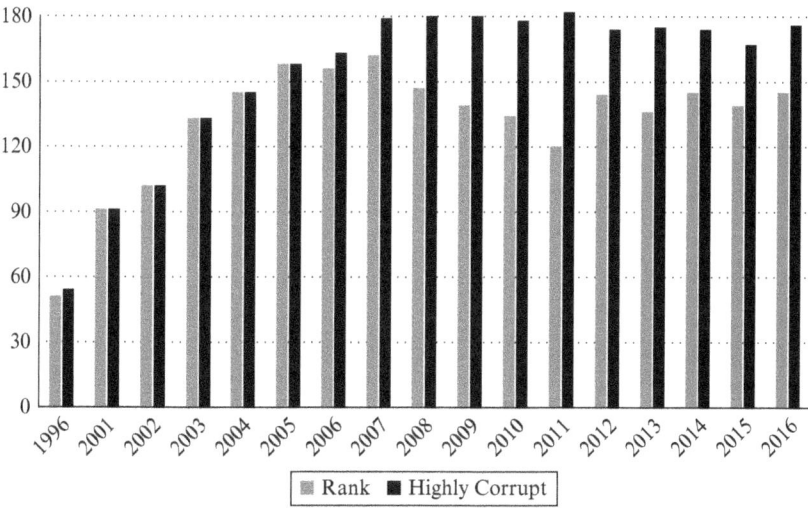

*Figure 18.2   Corruption rank of Bangladesh and the highly corrupt
countries rank, 1996–2016*

corrupt. Transparency International's CPI in Bangladesh averaged 132.18
from 1996 to 2016, reaching an all-time high of 162 in 2007 and a record
low of 51 in 1996 (Figures 18.1–18.4).

Prior research has identified that there is a positive association between
financial development and economic growth (Ahmed and Ansari 1998;
Treisman 2000; Rahman 2004; Khan 2008; Monnin and Jokipii 2010). On

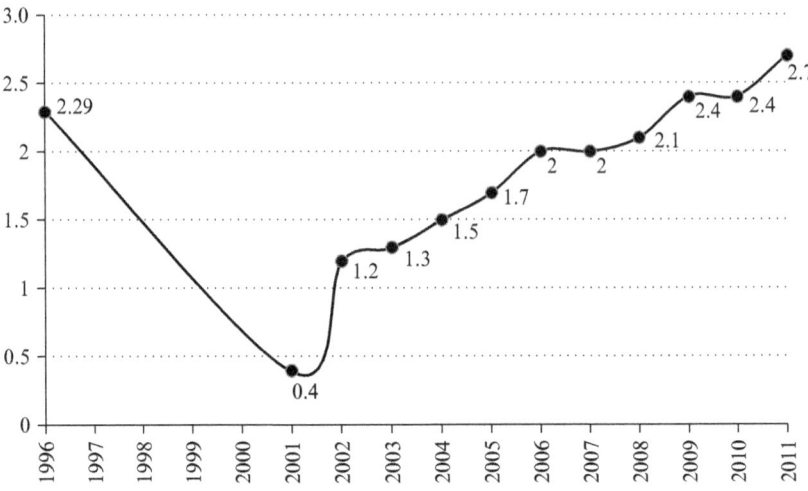

Figure 18.3   Corruption score of Bangladesh, 1996–2011

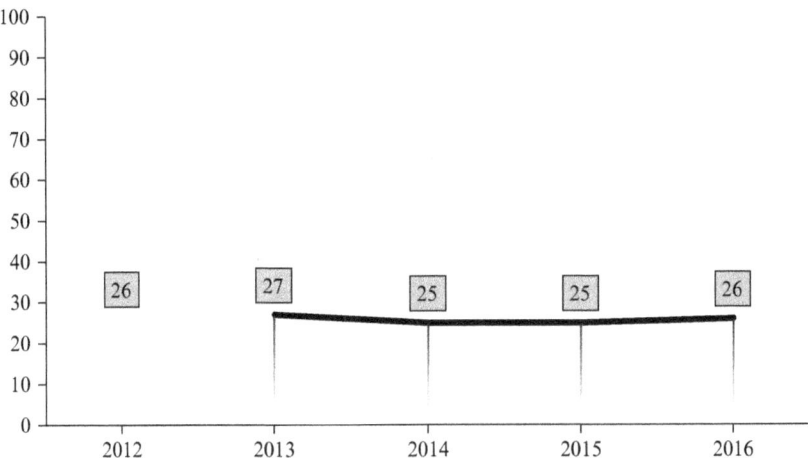

Figure 18.4   Corruption score of Bangladesh, 2012–16

the other hand, crisis, volatility and corruption in the banking sector have negative implications for the growth of the banking industry (Lin and Huang 2012; Moshirian and Wu 2012; Park 2012).

Zafarullah and Siddiquee (2001) strongly stress that corruption in the Bangladeshi public sector is rampant. They outline various factors including bribery, rent-seeking and misappropriation of funds, excessive lobbying, delays in service provision, pilferage and larceny, irresponsible conduct of officials, bureaucratic intemperance, patronage and clientelism. Golub (2007) discusses the top-down, state-centered paradigm ("rule of law orthodoxy") and legal empowerment (the use of legal services, legal capacity-building and legal reform by and for disadvantaged populations). The study finds that legal empowerment varies considerably; for example, gender-biased, non-state justice systems exist in Bangladesh.

Knox (2009a, p. 449) reports that "The commitment of the democratically elected government to tackle corruption at all levels will be a key determinant of whether trust and stability can emerge from the volatility of Bangladesh's politics ... holding public bodies to account for corrupt practice in delivering key services." Knox (2009b) explains that the anti-corruption efforts of Transparency International Bangladesh produced significant results in terms of education and health services. He also discusses case studies by Transparency International Bangladesh and the results of a large-scale survey on government services. Effective anti-corruption policies in Bangladesh suffer from a history of political instability.

Anik et al. (2013) explore the household-level determinants of corruption and its different forms in Bangladesh. They find that the households with higher expenditures are more likely to face high levels of corruption and bribery. However, these households are less likely to face nepotism or favoritism. They argue that relationships with different power entities of the government reduce households' probability of facing corruption. Using the 2007 household survey data collected by Transparency International Bangladesh, Choe et al. (2013) investigate corruption in the education sector in Bangladesh. The key findings of the study were:

- The incidence of corruption and the amount of bribes increase with the level of red tape.
- Poorer households, households with a less educated household head, and households with girls studying in school are more likely to be victims of corruption.
- Households with higher social status are more likely to use informal networks to bypass the red tape or pay less amounts in bribes.
- As a result, corruption is likely to be regressive.

Islam (2013) argues that since independence, successive governments and political leaders have consistently attempted to monopolize state power in various ways in Bangladesh. In addition, a society structured around patron-client relations helps to personalize the state power and elites use elements of the state and political system to engage in corruption.

Waheduzzaman and As-Saber (2015) evaluate the participation of citizens in implementing locally based development initiatives by international aid agencies. Using six aid-assisted project-based case studies, they explore the reasons for failure and find that the dysfunctional political system and corruption in Bangladesh have compromised the role of the state in ensuring local-level development activities. Regards the political and economic state of the countries of Central and South Asia, De Pedro and Barbero (2016) discuss the financial policies implemented by the Narendra Modi government of India and examine the extent of corruption there.

Using household surveys, focus groups and key informant interviews, Islam et al. (2017, p. 361) find that

> local government provides important support, for example relief distribution, livelihood assistance, and reconstruction of major community services. However, patronage relationships (notably favouring political supporters) and bribery play a substantial role in how those responsibilities are discharged. The equity and efficiency of these contributions to recovery are markedly diminished by corruption . . . however, general reform of governance in Bangladesh would be needed to bring this about.

According to Transparency International Bangladesh (2006), the total amount of financial loss from the 423 reported cases for which data were available was more than US\$67.43 million. However, the amount does not represent total financial loss due to corruption in the country as a whole. Transparency International Bangladesh also noted two factors responsible for corruption, a monopoly of discretionary power and a lack of accountability and the authority's failure to take steps against corruption and absence of deterrence.

Beck and Rahman (2006) argue that while Bangladesh has made progress in its contractual framework over the years, it is still ranked as one of the countries with the highest degree of corruption. The creditor and minority shareholder rights are relatively poor and enforcement is weak. The Asian Development Bank (2009, p. 9) reports that

> The financial market suffers from low investor confidence. Debt and equity markets have yet to fully recover from the domestic capital market crisis of 1996 and the departure of foreign investors. Despite allegations of irregularities among stock market brokers, not a single case has been successfully prosecuted. Prevalent and widely accepted rent seeking behavior is the major barrier to improving the financial sector. Foreign aid and investment accomplish little if

assistance does not flow directly to the targeted beneficiaries. Corruption is a financial mechanism and the financial sector is a key test area for establishing and implementing law and order. The financial sector must adopt market-based and transparent decision making to build confidence. Furthermore, institutions need to be set up to ensure that corruption is reduced.

As shown in Table 18.1, business owners and top managers in 1,442 firms were interviewed from April 2013 through September 2013 (Enterprise Surveys 2013). The key findings were:

- percent of firms expected to give gifts to secure government contract: 50.8
- percent of firms expected to give gifts to get an operating license: 57.0
- percent of firms expected to give gifts to get an import license: 76.8
- percent of firms expected to give gifts to get a construction permit: 54.3
- percent of firms expected to give gifts to get a water connection: 54.7.

In the case of the financial sector, powerful bank supervisors might be associated with higher levels of corruption. The dominant government-owned banks and politically connected private banks are reluctant to establish good governance (Khatun 2012). As shown in Figure 18.5, corruption in the public sector is a concerning issue too. Although the government has taken an encouraging step in this regard by revitalizing the Anti-Corruption Commission, which was financed with US$150 million from the Good Governance Program.

CASE STUDY: THE HALLMARK GROUP

The fraudulent misappropriation by the Hallmark Group is the biggest scandal in the banking industry in Bangladesh. In May 2012, Bangladesh Bank revealed that the Ruposhi Bangla Hotel Branch of the state-owned Sonali Bank illegally distributed Tk 36.48 billion (US$460 million) in loans between 2010 and 2012. The largest share, Tk 26.86 billion (US$340 million), went to the Hallmark Group (Farashuddin 2012). Khatun (2012) argues that despite the positive results from earlier reforms in 2007, the banking industry could not sustain its growth due to a weak institutional framework and corruption that threatened the stability of the banking industry.

Table 18.1   Corruption encountered by private sector firms in Bangladesh

	Bangladesh		South Asia	All countries
	2013	2007	2013	2013
Bribery incidence (percent of firms experiencing at least one bribe payment request)	46.9	63.1	24.8	18.0
Bribery depth (percent of public transactions where a gift or informal payment was requested)	41.9	43.3	21.0	14.0
Percent of firms expected to give gifts in meetings with tax officials	37.8	53.0	19.6	13.3
Percent of firms expected to give gifts to secure government contract	50.8	38.6	45.5	29.5
Value of gift expected to secure a government contract (percent of contract value)	2.8	1.4	2.9	1.8
Percent of firms expected to give gifts to get an operating license	57.0	37.0	25.3	14.3
Percent of firms expected to give gifts to get an import license	76.8	46.3	27.4	14.4
Percent of firms expected to give gifts to get a construction permit	54.3	22.5	30.9	23.3
Percent of firms expected to give gifts to get an electrical connection	46.7	46.8	37.3	16.2
Percent of firms expected to give gifts to get a water connection	54.7	58.6	36.7	16.2
Percent of firms expected to give gifts to public officials "to get things done"	48.2	84.5	25.5	22.5
Percent of firms identifying corruption as a major constraint	44.8	54.9	40.1	32.7
Percent of firms identifying the courts system as a major constraint	9.5	19.9	16.8	14.5

Source: Enterprise Surveys (2013).

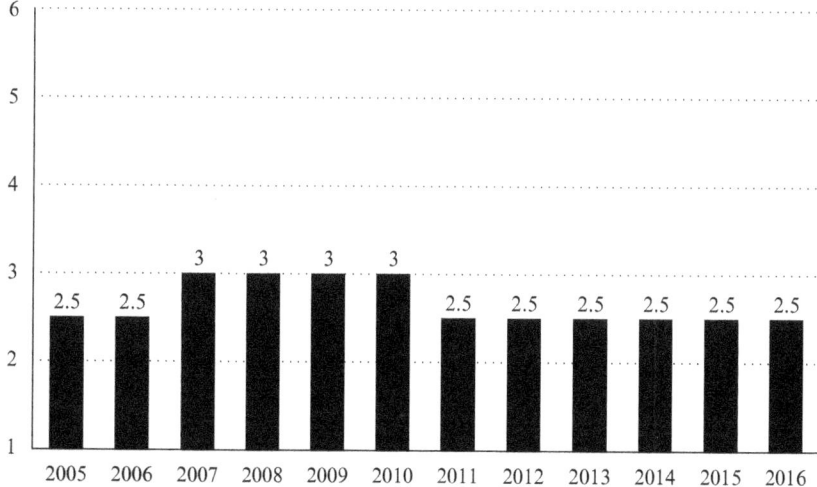

Note: * Transparency, accountability and corruption in the public sector assess the extent to which the executive can be held accountable for its use of funds and for the results of its actions by the electorate and by the legislature and judiciary, and the extent to which public employees within the executive are required to account for administrative decisions, use of resources and results obtained. The three main dimensions assessed here are the accountability of the executive to oversight institutions and of public employees for their performance, access of civil society to information on public affairs, and state capture by narrow vested interests.

Figure 18.5   Country Policy and Institutional Assessment (CPIA)* transparency, accountability and corruption in the public sector (1 = low to 6 = high)

The Hallmark Group case of forgery through inland bills is a case of poor management, weak internal control and risk management, and above all, poor governance. The Hallmark Group established fictitious companies, such as Anwara Spinning Mills, Max Spinning Mills and Star Spinning Mills, which were shown as recipients of letters of credit (LCs). These fictitious companies submitted falsified paperwork that were then paid for by the LCs from Sonali Bank.

According to Khatun (2012, p.107), the value of Tk 36.07 billion is equivalent to:

- 320.6 percent of Sonali Bank's capital!
- 6.6 percent of the Annual Development Program (ADP) of FY2012–13
- 15.9 percent of allocation for the social safety net program in FY2012–13

- 38.6 percent of allocation for health in FY2012–13
- 16.8 percent of allocation for education in FY2012–13
- 0.3 percent of projected gross domestic product (GDP) of FY2012–13
- 15.0 percent of the finance requirement of the Padma Bridge
- 42.9 percent of the envisaged support by the World Bank for the Padma Bridge.

The most important measure of asset quality is the non-performing loans (NPLs) ratio. According to Bangladesh Bank (2017, p. 32), at the end of 2015, private commercial banks (PCBs) had the lowest and state-owned development financial institutions (DFIs) had the highest ratio of gross NPLs to total loans. Private commercial banks' (PCBs) gross NPLs to total loans ratio was 4.9 percent, whereas that of state-owned commercial banks (SCBs), foreign commercial banks (FCBs) and DFIs were 21.5, 7.8 and 23.2 percent, respectively. The gross NPL ratios to total loans for the SCBs, PCBs, FCBs and DFIs were recorded as 25.7, 5.4, 8.3 and 26.1 percent, respectively, at the end of June 2016 (Table 8.2). The increasing trend of NPLs also indicates the lack of effective governance in the financial industry in Bangladesh.

CONCLUSIONS AND POLICY AGENDA

This chapter reviewed two cases of corruption in the financial sector in Bangladesh. The findings of the case study suggest that corruption slows down economic development. The findings also suggest the corruption is an ongoing major obstacle impeding the Bangladeshi economy. As suggested by Montinola and Jackman (2002) and Ortega et al. (2010), the chapter also finds that authoritarian regimes have lower corruption levels than democratic regimes in Bangladesh. Notably, it can be argued that both economic and political actions may have a significant impact on corruption levels in the Bangladeshi financial sector. It is noteworthy to mention that despite several attempts by the government (Figure 18.6), the good governance framework is still largely missing.

Some policy prescriptions can be drawn that may help policy makers to tackle the levels of corruption in Bangladesh:

- Corporate governance in Bangladesh is still weak. This is because of the political nature of corporate governance. The government should enforce good governance in the financial sector.
- The political appointment of board directors is common in Bangladesh. This appointment is based on political loyalty and

BOX 18.1 RESULTS OF THE FINANCIAL AUDIT OF
RUPOSHI BANGLA BRANCH OF SONALI BANK

The financial irregularities at the Ruposhi Bangla branch of Sonali Bank Limited reveal that as on May, 31, 2012 total outstanding loans and advances related to international trade was Tk 3,699.53 crore, of which funded and non-funded unauthorized loans and advances was Tk 3,606.48 crore. These unauthorized bank loan facilities were given by the general manager and assistant manager of the branch to Hallmark Group (Tk 2,667.45 crore), T and Brothers Group (Tk 685.63 crore), Paragon Group (Tk 144.44 crore), DN Sports Group (Tk 28.54 crore), Nakshi Knit Group (Tk 65.3 crore), and others (Tk 15.12 crore).

These loans and advances were given by disregarding the rules and regulations of the bank. Also, branch officials did not maintain relevant documentation properly on purpose. A significant fund was misappropriated through the inland bills trading. The branch opened inland letters of credit (LCs) in favor of three fictitious spinning mills which were customers of the said branch on account of another company, a concern of the Hallmark Group. The branch transferred the money to the accounts of the three companies, which after a few days advised the bank to transfer the money to the account of Hallmark Group.

The branch manager resorted to a number of unauthorized ways to disburse a huge amount of money to Hallmark Group and other customers. Some of these are as follows:

- Provided credit facilities without any sanction.
- Allowed credit facilities after the expiry of the sanctioned period.
- Illegally opened local back-to-back LCs and provided acceptance for documents raised by different banks in favor of non-existent spinning and textile mills on account of Hallmark Group, T and Brothers Group, Paragon Group, DN Sports Group and Nakshi Knit Group.
- Illegally opened local back-to-back LCs and provided acceptance for documents raised by inter-/intra-branch customers, namely, Star Spinning Mills, Max Spinning Mills, Anwara Spinning Mills and Master Cotton Yarn Limited on account of Hallmark Group and T and Brothers Group.
- Created unauthorized fresh loans to adjust unauthorized overdue loans.
- Provided loans against fake export documents of Hallmark Group, T and Brothers Group, Paragon Knit Composite Limited and Nakshi Knit Composite Limited.
- Provided excess pre-shipment credits (PSC) over the approved limit.
- Allowed PSC without LCs.
- Allowed PSC after shipment date.
- Opened cash LCs without customers' existing liabilities and limits.
- Opened cash LCs for capital machinery at zero margin without the approval of the Head Office of the bank.
- Foreign Currency Management Division (FCMD) debited Head Office NOSTRO account without obtaining reimbursement from the branch.
- Reported irregular Payment Against Documents loans as regular in classified loan reports.

- Opened foreign back-to-back LCs without office note and realizing commission.
- Transferred fund illegally by debiting sundry deposit accounts.
- Purchased export bills before completion of exports.
- Applied incorrect exchange rates for purchasing deferred payment export bills.
- Overdue export bills were not reported to the central bank.
- Allowed temporary overdraft illegally against cash incentives.

Source: Report on Financial Audit of Ruposhi Bangla Branch of Sonali Bank Limited, August 2012; see also Khatun (2012, pp. 106–7).

Table 18.2   Non-Performing Loans (NPL) ratios by banks

Bank types	2008	2009	2010	2011	2012	2013	2014	2015	2016
SCBs	25.4	21.4	15.7	11.3	23.9	19.8	22.2	21.5	25.7
DFIs	25.5	25.9	24.2	24.6	26.8	26.8	32.8	23.2	26.1
PCBs	4.4	3.9	3.2	2.9	4.6	4.5	4.9	4.9	5.4
FCBs	19	2.3	3.0	3.0	3.5	5.5	7.3	7.8	8.3
Total	10.8	9.2	7.3	6.1	10.0	8.9	10.0	8.8	10.1

Note: State-owned commercial banks (SCBs), state-owned development financial institutions (DFIs), private commercial banks (PCBs) and foreign commercial banks (FCBs).

party preferences. An independent institution should appoint directors based on fair selection criteria. According to Khatun (2012), the board directors should be appointed based on "Fit and Proper Test Criteria" to meet their responsibilities as outlined in the Bangladesh Bank Guidelines 2010.

- The Anti-Corruption Commission must be proactive. It should be fully operational, independent and effective. In some cases, the Commission is politically biased and unable to take effective action. The Commission should bring corrupt directors to justice (Transparency International Bangladesh 2006).
- The Anti-Corruption Commission should monitor newspaper reports on corruption in the financial sector and take appropriate legal action.
- There should also be a central corruption monitoring unit of the Anti-Corruption Commission for the financial sector in Bangladesh. This special unit will ensure that enforcement functions effectively.

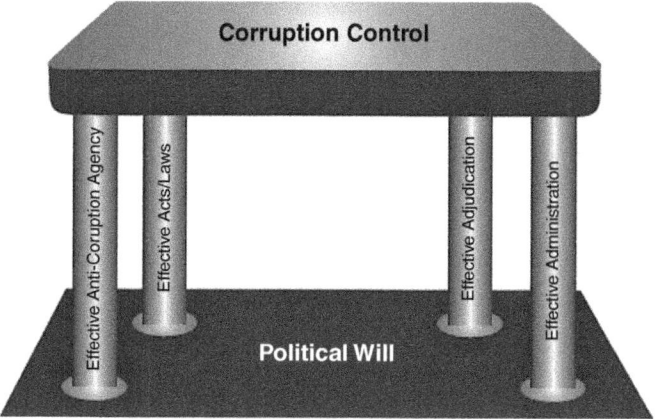

Source: Anti-Corruption Commission Bangladesh (2015, p.12).

Figure 18.6   The four-legged table of corruption control

REFERENCES

Ahmed, S. and Ansari, M. 1998. Financial sector development and economic growth: the South Asian experience. *Journal of Asian Economics* **9**(3), 503–17.

Anik, A., Bauer, S., and Alam, M. 2013. Why farm households have differences in corruption experiences? Evidences from Bangladesh. *Agricultural Economics* **59**(10), 478–88.

Anti-Corruption Commission Bangladesh. 2015. *Annual Report 2015.* Segun Bagicha, Dhaka, Bangladesh.

Asian Development Bank. 2009. *Bangladesh Financial Sector: An Agenda for Further Reforms.* Manila: Asian Development Bank.

Bangladesh Bank. 2017. *Annual Report 2015–2016.* Motijheel C/A, Dhaka, Bangladesh.

Beck, T. and Rahman, M. 2006. Creating a more efficient financial system: challenges for Bangladesh. Policy Research Working Paper 3938. World Bank, Washington, DC.

Becker, S., Egger, P., and Seidel, T. 2009. Common political culture: evidence of regional corruption contagion. *European Journal of Political Economy* **25**(3), 300–10.

Choe, C., Dzhumashev, R., Islam, A., and Khan, Z. 2013. The effect of informal networks on corruption in education: evidence from the Household Survey data in Bangladesh. *Journal of Development Studies* **49**(2), 238–50.

De Pedro, N. and Barbero, I. 2016. Asia central y meridional: Las dinamicas domesticas y regionales se aceleran. *Anuario Internacional* **1**, 178–86.

Easterly, W. 2001. *The Elusive Quest for Growth: Economist's Adventures and Misadventures in the Tropics.* Cambridge, MA: MIT Press.

Enterprise Surveys. 2013. Corruption: http://www.enterprisesurveys.org; The World Bank: http://www.enterprisesurveys.org/Custom-Query/bangladesh (accessed May 14, 2017).

Farashuddin, M. 2012. Hall-mark fraud: lessons for the banking system. *Daily Star* (September 9). http://www.thedailystar.net/news-detail-248925 (accessed May 14, 2017).

Golub, S. 2007. The rule of law and the UN peacebuilding commission: a social development approach. *Cambridge Review of International Affairs* **20**(1), 47–67.

Islam, M. 2013. The toxic politics of Bangladesh: a bipolar competitive neopatrimonial state? *Asian Journal of Political Science* **21**(2), 148–68.

Islam, R., Walkerden, G., and Amati, M. 2017. Households' experience of local government during recovery from cyclones in coastal Bangladesh: resilience, equity, and corruption. *Natural Hazards* **85**(1), 361–78.

Khan, M. 2008. Financial development and economic growth in Pakistan: evidence based on autoregressive distributed lag (ARDL) approach. *South Asia Economic Journal* **9**(2), 375–91.

Khatun, F. 2012. State of governance in the banking sector: dealing with the recent shocks. Paper presented at the Centre for Policy Dialogue (CPD), Dhaka, Bangladesh, November 5.

Klitgaard, R. 1988. *Controlling Corruption.* Berkeley, CA: University of California Press.

Knox, C. 2009a. Building trust amidst corruption in Bangladesh. *Round Table* **98**(403), 449–59.

Knox, C. 2009b. Dealing with sectoral corruption in Bangladesh: developing citizen involvement. *Public Administration and Development* **29**(2), 117–32.

Lin, P.C. and Huang, H.C. 2012. Banking industry volatility and growth. *Journal of Macroeconomics* **34**(4), 1007–19.

Monnin, P. and Jokipii, T. 2010. *The Impact of Banking Sector Stability on the Real Economy.* Swiss National Bank Working Papers (2010-5), Zurich.

Montinola, G. and Jackman, R. 2002. Sources of corruption: a cross-country study. *British Journal of Political Science* **32**(1), 147–70.

Moshirian, F. and Wu, Q. 2012. Banking industry volatility and economic growth. *Research in International Business and Finance* **26**(3), 428–42.

Ortega, D., Florax, R., and Delbecq, B. 2010. Primary determinants and the spatial distribution of corruption. Working Paper No. 10-6. Department of Agricultural Economics, Purdue University.

Park, J. 2012. Corruption, soundness of the banking sector, and economic growth: a cross-country study. *Journal of International Money and Finance* **31**(5), 907–29.

Rahman, M. 2004. Financial development-economic growth nexus: a case study of Bangladesh. *Bangladesh Development Studies*, **XXX**(3–4), 113–28.

Transparency International Bangladesh. 2006. *Corruption Database Report 2005.* Dhaka, Bangladesh.

Treisman, D. 2000. The causes of corruption: a cross-national study. *Journal of Public Economics* **76**, 399–457.

Waheduzzaman, W. and As-Saber, S. 2015. Politics and policy in achieving participatory governance in a developing country context. *Politics and Policy* **43**(4), 474–501.

Weyland, K. 1998. The politics of corruption in Latin America. *Journal of Democracy* **9**(2), 108–21.

Zafarullah, H. and Akhter, M. 2001. Military rule, civilianisation and electoral corruption: Pakistan and Bangladesh in perspective. *Asian Studies Review* **25**(1), 73.

Zafarullah, H. and Siddiquee, N. 2001. Dissecting public sector corruption in Bangladesh: issues and problems of control. *Public Organization Review* **1**(4), 465–86.

19. Corruption in China
Vanesa Pesqué-Cela

Soon after coming to power in 2012, President Xi Jinping launched the most aggressive anti-corruption campaign seen since the days of Mao Zedong (Saich 2017). According to the Central Commission for Discipline Inspection, the Communist Party's anti-corruption agency, more than 1.3 million officials have been punished for graft since 2013 (Reuters 2017). In spite (or because) of this, perceptions of corruption have worsened across the country. Transparency International's 2015/16 Global Corruption Barometer survey found that almost 75 percent of the respondents believed corruption had increased over the previous three years (Transparency International 2017a).

One of the main criticisms made of the current anti-corruption campaign is that with its focus on catching corrupt "tigers and flies" (i.e., high- and low-level officials), it fails to address the institutional roots of corruption (Quah 2015). At the same time, there is growing concern that failure to combat corruption will hinder the Party's ability to deliver on its commitment to "build a well-off society", "deepen economic reform", "promote the (socialist) rule of law", and "strengthen Party discipline" (the so-called "Four Comprehensives") (Wei 2017). Corruption, and the failure to bring it down, is one of the greatest threats the Communist Party currently faces, as President Xi Jinping admitted in his speech at the recent 19th Party Congress (*Guardian* 2017).

Against this background, this chapter examines the changing forms, causes and consequences of corruption in post-reform China from a historical and comparative perspective.[1] To this end, it addresses five main questions: (1) What have been the major trends and patterns of corruption in post-1978 China? (2) What explains differences in corruption levels over time and across regions? (3) What are the prospects for reducing corruption in China under the current institutional framework? (4) To

---

[1] Corruption is defined here as the use of public office for private gain. A key challenge in research on corruption in China and elsewhere is how to measure it. Unlike most cross-national studies, which have relied on "subjective" measures of corrupt activity, studies focusing on China have mostly used "objective" measures of corruption, whether "demand-" or "supply-side" measures, such as the number of public officials convicted for corruption or the amount of money spent on bribes by firms. For a discussion of the measurement of corruption in China, see Wedeman (2012) and X. Li (2016).

what extent, and in what ways, does corruption pose a threat to economic growth and regime stability? (5) Do the scale, causes and consequences of corruption in China differ from those found in other developing and transitional countries? In what ways, and why?

The chapter is structured as follows. The first section examines the quantitative and qualitative changes in corruption patterns in China over the last four decades. The second section provides an analysis of the institutional factors underlying variation in corruption levels over time and across regions. In the third section, I discuss the past and potential consequences of corruption for economic development and regime stability. The last section concludes.

THE DYNAMICS OF CORRUPTION IN REFORM-ERA CHINA

There has been an "explosion" of corruption in China during the reform period (Manion 2004). Corruption has worsened in terms of both its incidence and severity, especially since the 1990s. This growth is reflected in a dramatic increase in the number of corrupt officials prosecuted and convicted, and in a parallel increase in the seniority of the officials, and the amount of money, involved in corrupt activities[2] (Kwong 1997; Sun 2004; Guo 2008; Wedeman 2012; Manion 2014).

This dramatic increase in corruption levels has been associated with equally important changes in corruption patterns. The first change concerns the types of corruption. Since the 1990s, there has been a shift in the most common forms of corruption: from *non-transactive* to *transactive* corruption (e.g., Ko and Weng 2012; Wedeman 2012), and from *individual* to *collusive* corruption (e.g., Gong 2002; Pei 2016). There has been a rise in the number of bribery cases (i.e., transactive corruption) and a decrease in corruption related to embezzlement and misappropriation of public funds (i.e., non-transactive corruption). At the same time, the number of corruption cases involving three or more officials has increased (i.e., collusive corruption).[3] As a result, the theft or misuse of public assets by individual officials is no

---

[2]   The number of cases filed by the Procuratorate (see endnote 13) increased from 9,000 in 1980 to 41,500 in 2014 (Wedeman 2012, 2017). The number of cases involving senior officials (i.e., holding positions at or above the county level) jumped from 190 in 1988 to 2,380 in 2007. The amount of money recovered per case also grew significantly during this period from RMB 1,200 in 1979 to RMB 273,000 in 2007 (Wedeman 2012).

[3]   In recent years, the cases of bribery have accounted for about 60 per cent of all cases filed by the Procuratorate (Wedeman 2012), while those of collusive corruption for around 50 per cent (Pei 2016).

longer the most common form of corruption in China. Instead, bribery between public officials and private actors has become the most prevalent form, and a large proportion of bribery and other corruption cases involves groups of officials cooperating and protecting each other.

A second change in corruption patterns is related to the arenas of corruption. While officials working in the Party and government institutions have accounted for the largest proportion of officials charged with corruption since the early 1980s, there has recently been a significant increase in corruption cases involving officials working in judicial and law enforcement institutions, as well as in financial institutions, such as state-owned banks. The proportion of corrupt officials working in the business sector has declined since the 1990s, although more as a result of the privatization of small- and medium-sized state-owned enterprises than the elimination of corruption in large state-controlled firms[4] (Wedeman 2012).

In view of these trends, a central issue of scholarly debate is whether corruption is "under control" in China; and whether the "new" patterns of corruption are more or less destructive than the "old" ones. At one end of the debate, some scholars view the growth of corruption as a by-product of market reforms, and argue that the regime's anti-corruption efforts have been effective in preventing corruption from growing out of control (e.g., Ko and Weng 2012; Wedeman 2012; Ramírez 2014). The corruption situation in China is not "catastrophic" and will likely improve as the economy continues to reform and grow. Indeed, China's corruption level in the late 2000s was similar to that of the US in the late 1920s, when both countries were at an equivalent stage of development (Ramírez 2014). Equally important is that as market reforms have progressed, non-transactive corruption, considered the most economically detrimental type of corruption, has become less prevalent than transactive corruption, except in poor regions and rural areas, where "highly predatory, non-transactive" forms of corruption still tend to prevail[5] (Wedeman 2012, p. 6). In most of the country, however, the majority of corruption cases no longer involve officials preying on public resources, but rather officials

---

[4] Between 1978 and 1984, officials working in government and Party institutions accounted for 31 per cent of all public officials charged with corruption, while those working in the business, judicial and financial sectors represented 48 per cent, 5 per cent, and 5 per cent of the total, respectively. In the period 2002–09, the percentage of officials charged with corruption who worked in government and Party institutions still amounted to 27 per cent. This percentage, however, was 14 per cent for those officials working in the business sector, 17 per cent for those employed in the judiciary, and 10 per cent for those working in the financial sector (Wedeman 2012).

[5] See Bernstein and Lü (2003) and Guo (2001) for an analysis of the problems of predatory taxation and land expropriation in rural China.

exchanging favours for money with private actors. Overall, corruption has become less "predatory" and more "developmental" over time, and this is a reason for optimism (Guo 2006).

At the other end of the debate are those scholars who point to the progressive and unstoppable worsening of corruption in China, both in terms of levels and patterns (e.g., Gong 2002; Sun 2004; Pei 2008, 2016). They call into question the effectiveness, and even the sincerity, of the Party's efforts to keep corruption under control, while drawing attention to the dramatic increase in collusive corruption among officials within the Party-state, between officials and business, among state-owned enterprise managers, and between law enforcement agencies and organized crime.

Their research suggests that this type of corruption is equally pervasive across regions and levels of government,[6] and that it manifests itself in a multiplicity of ways, such as: officials buying and selling appointments and promotions; businessmen bribing officials in exchange either for obtaining undervalued state assets (e.g., enterprises and land) and large contracts (e.g., infrastructure projects), or for not complying with taxes and regulations; state-controlled enterprise managers colluding to loot their enterprises or to extract kickbacks from suppliers and contractors; or criminal groups using bribes to obtain protection from the police and the courts. They argue that collusive corruption is not only more complex and difficult to detect than individual corruption, but it is also much more destructive (Pei 2016). It provides the basis for the emergence of various forms of institutional corruption,[7] including patrimonialism (Zhu 2017), cronyism (Pei 2016), and state capture (Shieh 2005), which erode the Party-state's capacity to govern.

THE CAUSES OF CORRUPTION: AN INSTITUTIONAL ANALYSIS

In comparison with other countries, China is perceived as highly corrupt. It sits in the bottom third of Transparency International's 2016 Corruption Perceptions Index, with a score of 40 (out of 100) and a rank of 79 (out of 176 countries) (Transparency International 2017b). This is perhaps not surprising in light of the findings of the cross-country

---

[6]   See, for example, recent research by Pei (2016) and Zhu (2017).

[7]   Institutional corruption refers to "organizational situations in which officials and others collectively adapt state institutions, procedures, and rules for their own ends" (Scott 2017, p.13). See Scott (2017) also for an analysis of the various forms that institutional corruption takes in Asia.

literature on the determinants of corruption (Lambsdorff 2006; Treisman 2007; Pellegrini and Gerlagh 2008; Warf 2016).[8] As Manion (2014, p. 128) pointed out, "we should not be surprised to find widespread corruption in a non-democracy where the state continues to play an important role in the economy, where market competition is tempered by political intervention, and where press freedoms are greatly limited".

Nonetheless, corruption is more serious now than at the beginning of the reform period, and it is more serious in some regions than in others. Studies examining the reasons for these differences suggest that the causes of corruption in China at the national and local level are largely similar to those found in other countries. Corruption in China is also positively correlated with higher rents from natural resources, lower trade openness, lower regulatory quality, greater fiscal centralization,[9] weaker law enforcement, higher income inequality, lower educational levels, and greater ethnic diversity (Chen 2004; Dong and Torgler 2013; Ko and Zhi 2013).

One important difference, however, between the within-country evidence from China on the causes of corruption and the cross-country evidence concerns the relationship between corruption and economic development. As noted by Treisman (2007, p. 223), the most consistent finding in the cross-national literature is that higher (perceived) corruption is correlated with lower economic development. In China, by contrast, higher levels of (actual) corruption seem to be associated with higher levels of economic development (Dong and Torgler 2013). While this positive relationship between development and corruption is unusual, it is not unique to China and has also been found in other East Asian countries, such as Japan and South Korea (Wedeman 2012).[10]

---

[8] The cross-national empirical literature on corruption suggests that it is caused by a range of economic, political, social, and geographic factors. There is strong evidence that a country's corruption is positively associated with: lower economic development, lower trade openness, more intrusive business regulations, less intense market competition, the absence or weakness of democratic institutions, lower representation of women in government, the lack of an independent judiciary and a free press, lower social trust, and an abundance of natural resources (Lambsdorff 2006; Treisman 2007; Pellegrini and Gerlagh 2008).

[9] The increase in corruption in China during the reform period has often been attributed to decentralization (Gong 1997; Sun 2004; Guo 2008; Pei 2016). However, recent empirical studies using different measures of both decentralization and corruption have found that (fiscal) decentralization actually reduces corruption (Dong and Torgler 2013), particularly in jurisdictions with strong law enforcement (Ko and Zhi 2013). This finding would explain why the intensification of corruption has coincided with the implementation of re-centralization reforms since the mid-1990s. More research needs to be conducted on how the evolving features of the intergovernmental and local fiscal systems have affected the relationship between (de-)centralization and corruption over time, across regions, and across levels of government.

[10] The reasons underlying the possible two-way positive relationship between corruption and economic development in China will be discussed in detail in the third section.

Economic development and the other socio-economic factors previously mentioned are among the most critical determinants of corruption in China and elsewhere and have attracted much attention in the cross-national literature. However, addressing the questions of why corruption in China has worsened over time, and why it varies across regions, also requires a detailed analysis of its institutional causes at the national and local levels.[11]

Explaining the Growth of Corruption during the Reform Period

Institutional explanations of corruption in China have focused on the analysis of the opportunities and incentives for corrupt behaviour[12] created by the interplay of several key factors, including: (1) the gradual yet incomplete implementation of market and state-owned enterprise reforms (e.g., Chow 2005; Wedeman 2012); (2) persistent weaknesses in the country's intergovernmental fiscal and administrative systems (e.g., Wong and Bird 2008; Wong 2009; Birney 2014); (3) the "organizational involution" of the Chinese Communist Party and its failure to develop into a Weberian bureaucracy (Bernstein and Lü 2003); (4) the authoritarian nature of the regime and the stalling or reversal of reforms aimed at building "checks and balances" or at promoting limited political competition within it (e.g., Pei 2016); (5) the lack of a free press and civil society (e.g., Saich 2016); and, last but not least, (6) the lack of effective anti-corruption agencies and an independent judiciary[13] (e.g., Manion 2004).

---

[11] The focus here will be on formal institutions rather than on informal institutions or cultural factors.

[12] See Rose-Ackerman (2010) for a review of the institutional economics framework and its relevance to the study of corruption.

[13] China's anti-corruption system consists of four agencies: (1) the Party's Discipline Inspection Commission; (2) the State's Ministry of Supervision; (3) the Supreme People's Procuratorate; and (4) the National Corruption Prevention Bureau. The Discipline Inspection Commission is responsible for investigating corruption involving Party members, while the Ministry of Supervision is responsible for investigating corruption cases involving government officials. The two agencies merged in 1993. The Procuratorate is part of the judicial system and is responsible for investigating and prosecuting serious crimes committed by Party and government officials. Finally, the National Corruption Prevention Bureau is responsible for corruption prevention. In theory, the agencies responsible for corruption investigation and prosecution should work in a coordinated manner: the Discipline Inspection Commission and the Ministry of Supervision should conduct preliminary investigations of corruption cases, impose Party and/or administrative sanctions on individuals found guilty of non-criminal offenses and pass individuals suspected of more serious offenses on to the Procuratorate for criminal investigation and prosecution by the courts (Sun 2004; Wedeman 2012; Quah 2015). In practice, one of the main problems with this system is the Party's "appropriation of cases and the substitution of party disciplinary actions for criminal punishments" (Manion 2004, pp. 135–6).

Each of these institutional factors has fostered corruption, either by increasing its potential benefits or by decreasing the potential risk of detection and punishment. Yet, the extent to which, and ways in which, they have shaped corruption levels and patterns have changed over time. In the 1980s and 1990s, the major institutional sources of corruption in China were:

1. The coexistence of the "plan" and the "market" as complementary mechanisms for allocating and pricing resources in the economy under the "dual-track system" (Guo 2008).
2. The presence of high tariff and non-tariff barriers to imports (Gong 1997).
3. The implementation of management reforms in state-owned enterprises, which increased managerial autonomy without a corresponding increase in managerial accountability (Sun 2004).
4. The existence of an unfavourable business environment for non-state enterprises, in general, and private enterprises, in particular (Sun 2004).
5. The absence of a meritocratic civil service (Ko and Weng 2012), and the relatively low wages of civil servants (He 2000).
6. The poor design and implementation of decentralization reforms, which transferred fiscal and administrative decision-making power to local governments – from the province to the village level – but failed to provide the central government and, more importantly, citizens with effective mechanisms to hold local governments to account for the exercise of such power (Gong 2006).
7. The underdevelopment of anti-corruption laws and agencies (Manion 2004).
8. The lack of independence of the judiciary from the government and the Party, especially (but not only) at the local level (Ko and Weng 2012).
9. The reversal after the 1989 Tiananmen protests of the timid political reforms initiated in the 1980s, such as those aimed at separating the Party from the state (Pei 2006).

All these factors combined to increase the opportunities and incentives for corrupt officials to engage in: (1) profiteering activities, by purchasing goods at (low) plan prices and selling them at (higher) market prices; (2) the smuggling of foreign goods into the country; (3) the stripping of state-owned enterprise assets; (4) the taking of bribes from collective and private businesses in exchange for support and protection; as well as (5) the levying of illegal taxes and the embezzlement and misuse of local public funds, among other illicit activities.

Since the mid-1990s, the problem of corruption has worsened, and the main factors contributing to it have been:

1. A continued high degree of state intervention in the economy (Manion 2014).
2. The failure of the banking, securities, and insurance regulators to reduce corruption in the financial sector (Chow 2005).
3. The poor design and implementation of the privatization of small and medium state-owned enterprises (Sun 2004).
4. The limited success in improving the governance of large state-controlled firms following their corporatization and stock market listing[14] (Naughton 2015).
5. The failure of civil service reforms to eliminate agency problems – such as conflicts of interest and asymmetric information – between superiors and subordinates within the administrative hierarchy (Burns and Xiaoqi 2010).
6. A drastic re-centralization of fiscal revenues, which has made local governments heavily dependent on (often, discretionary) transfers from the central government, as well as on non-tax revenues from local land sales, that is, on less transparent sources of revenues compared to local tax revenues (Shue and Wong 2007; Man and Hong 2011).
7. The re-centralization of administrative authority within the local Party-state (Oi et al. 2012), as well as within bureaucracies (Mertha 2005), and the resulting development of "local or vertical kingdoms" at the county and provincial levels, which are subject to little monitoring from above or below (Zhu 2017).
8. The Party's inclination to discipline corrupt officials internally, instead of transferring their cases to the Procuratorate for criminal investigation and prosecution,[15] as well as the lack of political independence[16] and resources of its discipline inspection committees (Manion 2004; Quah 2015).
9. The lack of an independent judiciary and the lax enforcement of existing anti-corruption laws (Pei 2008, 2016).

---

[14] The People's Bank of China reported in 2011 that from the mid-1990s to 2008, thousands of government officials and state-owned enterprises' executives moved a total of US$126 billion overseas (Li 2012, p. 617).

[15] The main consequence of this is that "milder party penalties often substitute for harsher punishment (including the death penalty) for criminal prosecution" (Manion 2014, p. 135).

[16] The Chinese Communist Party Charter stipulates that "when disciplinary agencies at various levels discover violations by members of the Party committees at that level, they should report to the Party committees at that level and seek approval before proceeding to open investigations" (Sun 2004, p. 169).

10. An over-concentration of political power in the hands of the top leadership, particularly since the 18th Party Congress (Lee 2017).
11. The Party's abandonment of the 1980s attempts to "separate the party and state", as well as its weakening commitment to carry on with the reforms initiated in the 1990s and 2000s to promote "grassroots democracy", "inner-party democracy", and "extra-party consultation and supervision", among others (Shambaugh 2008; O'Brien and Zhao 2011).
12. The tightening of control over the media and civil society (Saich 2016).

These factors have combined to create a context in which the potential benefits of engaging in corruption continue to outweigh the potential costs, and in which new types of corruption have emerged and become widespread, including: land, construction, and real estate-related corruption, financial corruption, judicial corruption, as well as corruption related to the sale of official positions[17] (Manion 2014). This latter type of corruption is actually one of the most destructive sources of corruption, as it facilitates the development of corruption networks in Party and government institutions, as well as in judicial, regulatory, and law enforcement agencies (Pei 2016).

Looking back, we can see that the evolution of corruption in China over the last four decades has been intrinsically linked to the evolution (or involution) of its economic reforms, intergovernmental institutions, and political and legal system in various ways. On the one hand, we can see how different economic reforms have created different opportunities for corruption among officials at each stage of the economic transition process,[18] and how the move from a decentralized to a centralized system of governance has, in turn, shifted the locus of corruption from lower to upper levels of the Party-state. On the other hand, we can also see that the country's anti-corruption agencies, laws, and campaigns have thus far been relatively ineffective in deterring corruption, given the low probability of detection and punishment faced by corrupt officials. In recent years, only 3 per cent of all officials punished by the Party for corruption have been criminally prosecuted[19] (Pei 2008, pp. 235–6).

---

[17] A recent survey of corruption covering all major Chinese cities found that the areas perceived as being plagued by corruption are: "construction", "organizational and personnel management" (i.e., sale of public office), "police", "public finance and taxation", "courts", "SOE [state-owned enterprise]management", and "financial investment" (Song and Cheng 2012).

[18] For a detailed analysis of how different economic reforms have given rise to different opportunities for corruption, see Gong (1997), He (2000), Sun (2004), Guo (2008), and Wedeman (2012).

[19] This percentage was 6 per cent in the 1980s and 1990s (Pei 2008, pp. 235–6).

Exploring Differences in Corruption across Regions

Differences in corruption levels across and within regions are significant, and can be partly explained by differences in the quality of local institutions, whether economic, political or legal.[20] Firstly, regions and localities with weak market-supporting institutions appear to be more corrupt. Firm-level corruption (proxied by the time spent by a firm interacting with government officials) is greater in provinces with intrusive business regulations and poorly functioning markets (Jiang and Nie 2014; You and Nie 2017). Likewise, bribery-related corruption (proxied by firms' entertainment and travel expenses) is more widespread in cities where property rights are less secure, the quality of public services poorer, and tax burdens higher (Cai et al. 2011; Wang 2014).

Secondly, the problem of corruption seems to be particularly serious in provinces, municipalities, and counties where political power has become concentrated in the hands of the Party secretary (i.e., the so-called first-in-command or *yibashou*) (Ren and Du 2008; Ko and Zhi 2013; Pei 2016). The phenomenon of "first-in-command-corruption" is typically associated with the most damaging abuses of power, such as the sale of public offices (Pei 2016), the embezzlement of public funds (Ren and Du 2008) or the illegal taking of land (Ko and Zhi 2013). More worryingly, it is at the basis of the emergence of "local mafia states" in parts of China, where a large number of senior local officials are involved in collusive corruption with subordinates, private businessmen, and organized crime groups (Pei 2016, pp. 243–55).

Lastly, inter- and intra-regional differences in corruption levels also seem to be explained by differences across jurisdictions in the level of law enforcement and, to a lesser extent, of media exposure and civil society activism. Corruption tends to be higher in provinces with a smaller budget

---

[20]   Consistent with the findings of the cross-national literature on the causes of corruption, Dong and Torgler (2013) have found that provinces with abundant natural resources, higher levels of income inequality and ethnic diversity, and with lower levels of educational attainment, tend to be more corrupt. However, in contrast with the cross-national literature, they have found that provinces with higher levels of economic development (as proxied by their per capita real gross regional product) have higher levels of corruption (as proxied by the number of registered cases of corruption per 100,000 people). They attribute this finding to the transitional nature of the Chinese economy. While there seems to be a positive relationship between the levels of provincial gross domestic product (GDP) and corruption, different forms of corruption may be more prevalent in some Chinese regions than in others. For example, the sale of public offices seems to be more common in inland provinces, while land-related corruption appears to be more prevalent in coastal ones (Yep 2013; Pei 2016). Future research on the relationship between corruption and economic development in China would benefit from using multiple measures of both variables.

for law enforcement (i.e., the police, procuratorates and courts), lower newspaper circulation, and less social participation (Dong and Torgler 2013; Ko and Zhi 2013).

THE CONSEQUENCES OF CORRUPTION: A REASSESSMENT

Corruption in China, unlike in other developing and transitional countries, has not led to economic stagnation and crisis, nor has it resulted in political instability and regime change – at least, not yet. Despite its high levels of corruption, China has experienced unprecedented high rates of economic growth since the early 1980s. This growth has contributed critically to the legitimacy and stability of the regime.[21]

Given the centrality of growth for economic development and political stability, much of the literature on corruption in China has focused on explaining the seemingly paradoxical coexistence of high corruption and high growth during the reform period. Five major arguments have been made as to why corruption has not been an obstacle to growth, which relate to: (1) the sequencing of corruption and growth in China; (2) its model of public administration; (3) its transitional economy; (4) the nature of corruption networks in the country; as well as (5) the prevailing patterns of corruption.

One reason behind China's "corruption paradox" is the sequencing of corruption and growth (Wedeman 2012). Corruption was limited before the introduction of market reforms, and its increase has thus been an outcome of the country's gradual economic reforms and high growth, rather than an obstacle. According to Wedeman (2012, p. 81), "because corruption was not a serious problem initially, when reform stimulated the Chinese economy it had a chance to enter dynamic growth *before* corruption reached significant levels" (emphasis in the original).

Another argument for why China has been able to sustain high levels of economic growth despite having high levels of corruption focuses on the Chinese model of public administration (Birney 2014; Rothstein 2015). While China lacks a Weberian bureaucracy, it has made use of its cadre management system to make local governments accountable to the central government for the implementation of the latter's policy priorities. This mechanism of top-down control has proved to be "selectively effective"

---

[21] Nonetheless, the 1989 Tiananmen pro-democracy protests were to a great extent fuelled by popular discontent with corruption.

(Birney 2014), in that it has created strong incentives for local officials to promote growth but has failed to create equally strong incentives for them not to engage in corruption.

A third reason why high corruption has not prevented, and may have facilitated, high growth in the Chinese case is related to the efficiency-enhancing effects of corruption in the context of a transitional economy, where bribes can help firms cut red tape and access key resources, such as credit. In this view, corruption has "greased the wheels" of the Chinese economy in its transition from plan to market, by eliminating institutional barriers to entrepreneurship (e.g., Sun 1999; Dong and Torgler 2010; Cai et al. 2011; Wang and You 2012; X. Li 2016).

Another explanation for the coexistence of high corruption and high growth in China lies in the "industrial organization of corruption" (Shleifer and Vishny 1993). It is believed that the more centralized corruption networks are, the less detrimental is their impact on the economy: "when those (corruption) networks are organized and managed by a strong centralized state, as in a monopoly industry, corruption is likely to be less corrosive to investment and growth than when it is organized by numerous government officials acting as independent monopolists" (Rock and Bonnet 2004, p. 1003). The fact that corruption in China is more centralized and less chaotic and unpredictable than in other developing and transitional economies might thus explain why it has not stifled growth[22] (e.g., Rock and Bonnett 2004; Sun and Johnston 2009).

A final way of explaining China's corruption-growth paradox is by distinguishing between more and less damaging forms of corruption, and examining the relative prevalence of "predatory", "parasitic", and "developmental" forms of corruption both across jurisdictions and over time. According to Sun (2004), the coexistence of corruption and growth in reform era China can be explained by the prevalence of less predatory or more developmental forms of corruption in most parts of the country. Wedeman (2012, p. 141) likewise argues that corruption has been "more parasitic than predatory" in China over the reform period, as it has "fed off the growing economy rather than on the economy vitals".

An important point in the discussion about the seemingly paradoxical relationship between corruption and rapid development in post-1978 China that is rarely made in the literature is the issue of how we define "development".[23] If development is narrowly defined as GDP growth,

---

[22]   In this respect, Sun and Johnston (2009, p. 10) refer to the "localized centralization" of corruption in China, where "local chief executives" have become the main locus of "favour exchanges" (i.e., joint monopoly of corruption).

[23]   One of the most notable exceptions is Pei (2006).

corruption has clearly not impeded development in China, which is somewhat puzzling from a comparative perspective. However, if we define development more broadly and consider its economic, social, environmental, and institutional dimensions, there is little doubt that corruption has been as detrimental to development in China as anywhere else. When viewed from this perspective, the relationship between corruption and development in the Chinese case does not seem paradoxical.

Corruption in China has had a profound negative impact on the quality of economic growth, government performance, and Party/state-society relations during the reform period. Firstly, it has been, and remains, a major obstacle for China to develop a model of inclusive and sustainable growth in its transition towards a "socialist market economy". It has weakened the developmental role of the state by reducing its spending on education, health, and research and development,[24] as well as by undermining its ability to implement laws and regulations effectively (Dong and Torgler 2010). Partly as a result of this, it has been one of the major drivers of the growing inequality and environmental problems facing the country[25] (Dong and Torgler 2010).

Secondly, corruption has damaged Party/state-society relations by increasing distrust in government (L. Li 2004, 2016; Chen 2017). Of particular concern is the growing perception that anti-corruption laws are not effectively implemented, and that corruption is neither sufficiently investigated nor punished (Song and Cheng 2012). The growth of corruption has resulted in an increase in social unrest, as indicated, for example, by the fact that 65 per cent of the 180,000 "mass incidents" registered in 2010[26] were caused by land-related corruption (Göbel and Ong 2012). It has likewise resulted in sporadic episodes of political unrest, in particular the 1989 and 2011 pro-democracy protests.

If not curbed, corruption will be more damaging to the economy and the regime than it has already been. This is because it may now be sufficiently entrenched to delay or even block the economic reforms needed to maintain high growth, and to make such growth socially and

---

[24]   A conservative estimate of the direct costs of corruption in the areas of government procurement and administration, infrastructure investment, and land transactions amounts to 3 per cent of GDP (Pei 2008, p. 237).

[25]   See the empirical findings of Dong and Torgler (2013) showing that higher levels of corruption in China are associated, on the one hand, with lower levels of inward foreign direct investment, tax revenues, and government spending on education, health and research and development and, on the other hand, with higher levels of income inequality and environmental pollution.

[26]   In 1993, this figure was 8,700. Land disputes are often caused by the illegal expropriation of land, inadequate compensation to farmers, and forced land seizures (Göbel and Ong 2012).

environmentally sustainable (Chow 2005). At the same time, continued high levels of corruption may also have a more corrosive effect on trust in, and support for, the regime. Under the current centralized system, citizens may not only blame local governments but also, and increasingly, the central government and the Party for the country's corruption problems,[27] as recent research suggests (Lewis-Beck et al. 2014; L. Li 2016; Chen 2017).[28] If this happens, the combination of a declining economy, systemic corruption, and growing distrust of both local and central leaders would have a devastating effect on the regime's legitimacy and stability.

CONCLUSION

Corruption in China has grown to alarming levels over the last decades and become the top concern among its citizens in recent years (Pew Research Center 2015). Institutional explanations of corruption emphasize the role of China's economic, fiscal, administrative, and politico-legal institutions in shaping opportunities and incentives for corruption among officials at the central and local levels. For some scholars, the dramatic rise in corruption is a side-effect of the country's gradual market reforms and high economic growth (Wedeman 2012); for others, a symptom of the failure to get intergovernmental relations "right" (Wong and Bird 2008); and yet for others, the unavoidable consequence of the Communist Party's monopoly on political power and absence of rule of law (Pei 2016).

Taken together, these and other studies suggest that (1) different economic reforms have created opportunities for corruption in different areas of economic activity throughout the reform period; (2) the recent centralization of China's fiscal and administrative systems has merely shifted discretionary power over the allocation of resources (and the opportunities for corruption that go with it) from lower to higher levels of the Party-state, without improving accountability;[29] and (3) China's

---

[27]   Paradoxically, Xi Jinping's anti-corruption campaign and the high-profile corruption cases it has thus far uncovered, including those of Bo Xilai, Zhou Yongkang, and other central leaders, may further contribute to changing the public perception of corruption as a national, rather than as an exclusively local, problem.

[28]   Recent research suggests that: (1) trust in the central government is weaker than it looks (L. Li 2016); and that (2) it is contingent not only on evaluations of the national economy and perceptions of corruption at the central level (Chen 2017), but also on levels of dissatisfaction with the performance of local governments (Lewis-Beck et al. 2014).

[29]   For instance, since the tax-sharing system was introduced in 1994, no significant steps have been taken to establish: (1) a clear and appropriate division of expenditure responsibilities and tax revenues among *all* levels of government; (2) a transparent system of intergovernmental transfers; or (3) an effective local tax system.

anti-corruption agencies have been ineffective in preventing, detecting, and punishing corruption.

While it may be too soon to tell, there are few reasons to be optimistic about the effectiveness of the current anti-corruption campaign in combating the country's deep-rooted corruption problem. In China, like elsewhere, "effective (anti-corruption) policy cannot just concentrate on catching and punishing the 'rotten apples'. Policy must address the underlying conditions that create corrupt incentives, or it will have no long-lasting effect" (Rose-Ackerman 2010, p. 63).

REFERENCES

Bernstein, T.P. and Lü, X. 2003. *Taxation without Representation in Contemporary Rural China*. Cambridge and New York: Cambridge University Press.

Birney, M. 2014. Decentralization and veiled corruption under China's "Rule of Mandates". *World Development* **53**, 55–67.

Burns, J. and Xiaoqi, W. 2010. Civil service reform in China: impacts on civil servants' behaviour. *China Quarterly* **201**, 58–78.

Cai, H., Fang, H., and Xu, L.C. 2011. Eat, drink, firms, government: an investigation of corruption from the entertainment and travel costs of Chinese firms. *Journal of Law and Economics* **54**, 55–78.

Chen, D. 2017. Local distrust and regime support: sources and effects of political trust in China. *Political Research Quarterly* **70**, 314–26.

Chen, K. 2004. Fiscal centralization and the form of corruption in China. *European Journal of Political Economy* **20**, 1001–9.

Chow, G. 2005. Corruption and China's economic reform in the early 21st century. CEPS Working Paper No. 116. Center for Economic Policy Studies, Princeton, NJ.

Dong, B. and Torgler, B. 2010. The consequences of corruption: evidence from China. CREMA Working Paper No. 73.2010. Center for Research in Economics (CREMA), Basel.

Dong, B. and Torgler, B. 2013. Causes of corruption: evidence from China. *China Economic Review* **26**, 152–69.

Göbel, C. and Ong, L.H. 2012. Social unrest in China. Europe China Research and Academic Network (ECRAN). https://ssrn.com/abstract=2173073 (accessed 23 October 2017).

Gong, T. 1997. Forms and characteristics of China's corruption in the 1990s: change with continuity. *Communist and Post-Communist Studies* **30**, 277–88.

Gong, T. 2002. Dangerous collusion: corruption as a collective venture in contemporary China. *Communist and Post-Communist Studies* **35**, 85–103.

Gong, T. 2006. Corruption and local governance: the double identity of Chinese local governments in market reform. *Pacific Review* **19**, 85–102.

Guardian. 2017. *Xi* Jinping heralds "new era" of Chinese power at Communist Party congress. *Guardian*, 18 October. https://www.theguardian.com/world/2017/oct/18/xi-jinping-speech-new-era-chinese-power-party-congress (accessed 18 October 2017).

Guo, S. (ed.) 2006. *China's Peaceful Rise in the 21st Century: Domestic and International Conditions.* Aldershot: Ashgate.

Guo, X. 2001. Land expropriation and rural conflicts in China. *China Quarterly* **166**, 422–39.

Guo, Y. 2008. Corruption in transitional China: an empirical analysis. *China Quarterly* **194**, 349–64.

He, Z. 2000. Corruption and anti-corruption in reform China. *Communist and Post-Communist Studies* **33**, 243–70.

Jiang, T. and Nie, H. 2014. The stained China miracle: corruption, regulation, and firm performance. *Economics Letters* **123**, 366–9.

Ko, K. and Weng, C. 2012. Structural changes in Chinese corruption. *China Quarterly* **211**, 718–40.

Ko, K. and Zhi, H. 2013. Fiscal decentralization: guilty of aggravating corruption in China? *Journal of Contemporary China* **22**, 35–55.

Kwong, J. 1997. *The Political Economy of Corruption in China.* Armonk, NY: M.E. Sharpe.

Lambsdorff, J. 2006. Causes and consequences of corruption: what do we know from a cross-section of countries? In S. Rose-Ackerman (ed.), *International Handbook on the Economics of Corruption*, pp. 3–51. Cheltenham, UK and Northampton, MA, USA: Edward Elgar.

Lee, S. 2017. An institutional analysis of Xi Jinping's centralization of power. *Journal of Contemporary China* **26**, 325–36.

Lewis-Beck, M., Tang, W., and Martini, N. 2014. A Chinese popularity function: sources of government support. *Political Research Quarterly* **67**, 16–25.

Li, C. 2012. The end of the CCP's resilient authoritarianism? A tripartite assessment of shifting power in China. *China Quarterly* **211**, 595–623.

Li, L. 2004. Political trust in rural China. *Modern China* **30**(2), 228–58.

Li, L. 2016. Reassessing trust in the central government: evidence from five national surveys. *China Quarterly* **225**, 100–21.

Li, X. 2016. Measuring local corruption in China: a cautionary tale. *Journal of Chinese Political Science* **21**, 21–38.

Man, J.Y. and Hong, Y.-H. (eds) 2011. *China's Local Public Finance in Transition.* Cambridge, MA: Lincoln Institute of Land Policy.

Manion, M. 2004. *Corruption by Design: Building Clean Government in Mainland China and Hong Kong.* Cambridge, MA: Harvard University Press.

Manion, M. 2014. The challenge of corruption. In A. Goldstein and J. Lisle (eds), *China's Challenges: The Road Ahead*, pp. 125–38. Philadelphia, PA: University of Pennsylvania Press.

Mertha, A.C. 2005. China's "soft" centralization: shifting *tiao/kuai* authority relations. *China Quarterly* **184**, 791–810.

Naughton, B. 2015. The transformation of the state sector: SASAC, the market economy, and the new national champions. In B. Naughton and K. Tsai (eds), *State Capitalism, Institutional Adaptation and the Chinese Miracle*, pp. 46–74. New York: Cambridge University Press.

O'Brien, K. and Zhao, S. (eds) 2011. *Grassroots Elections in China.* New York: Routledge.

Oi, J.C., Babiarz, K., Zhang, L., Luo, R., and Rozelle, S. 2012. Shifting fiscal control to limit cadre power in China's townships and villages. *China Quarterly* **211**, 649–75.

Pei, M. 2006. *China's Trapped Transition: The Limits of Developmental Autocracy.* Cambridge, MA and London: Cambridge University Press.

Pei, M. 2008. Fighting corruption: a difficult challenge for Chinese leaders. In C. Li (ed.), *China's Changing Political Landscape: Prospects for Democracy*, pp. 229–50. New York: Brookings Institution Press.

Pei, M. 2016. *China's Crony Capitalism: The Dynamics of Regime Decay.* Cambridge, MA and London: Harvard University Press.

Pellegrini, L. and Gerlagh, R. 2008. Causes of corruption: a survey of cross-country analyses and extended results. *Economics of Governance* **9**, 245–63.

Pew Research Center. 2015. Corruption, pollution, inequality are top concerns in China. http://www.pewglobal.org/2015/09/24/corruption-pollution-inequality-are-top-concerns-in-china/ (accessed 19 October 2017).

Quah, J.S.T. 2015. *Hunting the Corrupt: "Tigers" and "Flies" in China: An Evaluation of Xi Jinping's Anti-corruption Campaign.* Maryland Series in Contemporary Asian Studies 2015. Baltimore, MD: Carey School of Law, University of Maryland.

Ramírez, C. 2014. Is corruption in China "out of control"? A comparison with the US in historical perspective. *Journal of Comparative Economics* **42**, 76–91.

Ren, J. and Du, Z. 2008. Institutionalized corruption: power overconcentration of the first-in-command in China. *Crime, Law and Social Change* **49**, 45–59.

Reuters. 2017. Chinese watchdog says 1.34 million officials punished for graft since 2013. https://www.reuters.com/article/us-china-corruption/chinese-watchdog-says-1-34-million-officials-punished-for-graft-since-2013-idUSKBN1CD04B (accessed 18 October 2017).

Rock, M. and Bonnett, H. 2004. The comparative politics of corruption: accounting for the East Asian paradox in empirical studies of corruption, growth and investment. *World Development* **32**, 999–1017.

Rose-Ackerman, S. 2010. The institutional economics of corruption. In G. Graaf, P. Maravic, and F. Wagenaar (eds), *The Good Cause: Theoretical Perspectives on Corruption*, pp. 47–63. Opladen: Barbara Budrich.

Rothstein, B. 2015. The Chinese paradox of high growth and low quality of government: the cadre organization meets Max Weber. *Governance* **28**, 533–48.

Saich, T. 2016. Controlling political communication and civil society under Xi Jinping. In S. Heilmann and M. Stepan (eds), *China's Core Executive: Leadership Style, Structures and Processes*, pp. 22–5. Berlin: Mercator Institute for China Studies.

Saich, A. 2017. What does General Secretary Xi Jinping dream about? Harvard Kennedy School Faculty Research Working Paper No. RWP17-038. John F. Kennedy School of Government, Cambridge, MA.

Scott, I. 2017. Institutional corruption and the state in Asia. In T. Gong and I. Scott (eds), *Routledge Handbook of Corruption in Asia*, pp. 13–26. New York: Routledge.

Shambaugh, D. 2008. *China's Communist Party: Atrophy and Adaptation.* Berkeley, CA: University of California Press.

Shieh, S. 2005. The rise of collective corruption in China: the Xiamen smuggling case. *Journal of Contemporary China* **14**, 67–91.

Shleifer, A. and Vishny, R. 1993. Corruption. *Quarterly Journal of Economics* **108**(3), 599–617.

Shue, V. and Wong, C. (eds) 2007. *Paying for Progress in China: Public Finance, Human Welfare and Changing Patterns of Inequality.* London and New York: Routledge.

Song, X. and Cheng, W. 2012. Perception of corruption in 36 major Chinese cities: based on survey of 1,642 experts. *Social Indicators Research* **109**, 211–21.

Sun, Y. 1999. Reform, state, and corruption: is corruption less destructive in China than in Russia? *Comparative Politics* **32**(1), 1–20.

Sun, Y. 2004. *Corruption and Market in Contemporary China.* Ithaca, NY: Cornell University Press.

Sun, Y. and Johnston, M. 2009. Does democracy check corruption? Insights from China and India. *Comparative Politics* **42**, 1–19.

Transparency International. 2017a. Corruption in Asia Pacific: what 20,000 people told us. https://www.transparency.org/news/feature/corruption_in_asia_pacific_what_20000_people_told_us (accessed 4 September 2017).

Transparency International. 2017b. Corruption Perceptions Index 2016. https://www.transparency.org/news/feature/corruption_perceptions_index_2016 (accessed 4 September 2017).

Treisman, D. 2007. What have we learned about the causes of corruption from ten years of cross-national empirical research? *Annual Review of Political Science* **10**, 211–44.

Wang, Y. 2014. Institutions and bribery in an authoritarian state. *Studies in Comparative International Development* **49**, 217–41.

Wang, Y. and You, J. 2012. Corruption and firm growth: evidence from China. *China Economic Review* **23**, 415–33.

Warf, B. 2016. Geographically uneven landscapes of Asian corruption. *Asian Geographer* **33**(1), 57–76.

Wedeman, A. 2012. *Double Paradox: Rapid Growth and Rising Corruption in China.* Ithaca, NY: Cornell University Press.

Wedeman, A. 2017. Xi Jinping's tiger hunt: anti-corruption campaign or factional purge? *Modern China Studies* **24**(2), 35–94.

Wei, S. 2017. "Four comprehensives" light up the future. *China Daily* http://www.chinadaily. com.cn/opinion/2017-07/10/content_30050292.htm (accessed 19 October 2017).

Wong, C. 2009. Rebuilding government for the 21st century: can China incrementally reform the public sector? *China Quarterly* **200**, 929–52.

Wong, C. and Bird, R. 2008. China's fiscal system: a work in progress. In L. Brandt and T. Rawski (eds), *China's Great Economic Transformation*, pp.429–66. Cambridge: Cambridge University Press.

Yep, R. 2013. Containing land grabs: a misguided response to rural conflicts over land. *Journal of Contemporary China* **22**(80), 273–91.

You, J. and Nie, H. 2017. Who determines Chinese firms' engagement in corruption: themselves or neighbors? *China Economic Review* **43**, 29–46.

Zhu, J. 2017. Corruption networks in China: an institutional analysis. In T. Gong and I. Scott (eds), *Routledge Handbook of Corruption in Asia*, pp.27–41. New York: Routledge.

20. An ambivalent state: the crossover of corruption and violence in the Philippines
Cleo Calimbahin

On January 2017, six months into his office as President of the Philippines, Rodrigo Duterte called all incumbent mayors for a meeting in the Presidential Palace. Cursing and showing little effort to hide his anger, Duterte made it clear that whoever was involved in the illegal drug trade would be killed.[1] Seven months since that meeting, three mayors from Duterte's "narco-list," that is, public officials reportedly involved in the drug trade, are dead. One died in a shootout with policemen in a checkpoint.[2] One mayor died inside a provincial jail under police custody. Another, notorious for drug dealing, was killed along with family members inside his home while a warrant of arrest was being served.

The campaign against illegal drugs of the Duterte administration has resulted in more than 7,000 deaths.[3] Often the murders are carried out execution-style and the deaths are disavowed by law enforcement agencies. The deaths have been called extra-judicial killings. In the 14-month period since the Philippine National Police launched Operation Tokhang (Operation Knock and Plead) in 2016, 29 children died.[4] One such death was that of Kian Loyd Delos Santos, seen on CCTV cameras being forcibly taken by identified policemen. His face was covered while being beaten. Kian was heard begging them to stop and to let him go home because he still had school the next day. When the policemen did stop, they gave Kian a gun and told him to run. Three gunshots later, the 17-year old Kian was down on the floor with three bullets in his ear and back of the head. The

---

[1] Nestor Corales, 2017. Duterte threatens, curses mayors at palace meeting. http://newsinfo.inquirer.net/861517/duterte-curses-threatens-mayors-during-palace-meeting. Accessed August 22, 2017.

[2] Ina Andolong, 2017. Duterte to mayors: "I will kill you if your name is in the narco-list." http://cnnphilippines.com/news/2017/01/10/Duterte-to-mayors-I-will-kill-you-if-your-name-is-in-the-narco-list.html. Accessed June 26, 2017.

[3] Michael Bueza, 2016. IN NUMBERS: the Philippiness "war on drugs." https://www.rappler.com/newsbreak/iq/145814-numbers-statistics-philippines-war-drugs. Accessed June 15, 2017.

[4] Amanda Lingao, 2017. CNN special report: at least 29 minors killed. http://cnnphilippines.com/news/2017/08/21/Minors-killed-in-war-on-drugs.html. Accessed August 20, 2017.

policemen, much larger than the teenage boy, claimed that Kian was shot because *nanlaban* ("he fought back").

While the crackdown on illegal drugs has received wide support, the public finds worrisome a growing sense of impunity among some law enforcement agencies and personnel. Perceived by the public as institutions with serious issues of credibility and integrity, it is sometimes unclear who is the criminal and who is the crime fighter.[5] President Duterte is aware of this issue. After the killing of a South Korean businessman inside the headquarters of the Philippine National Police, the president warned policemen who were taking advantage of the war on drugs to extort money, "you policemen are the most corrupt. You are corrupt to the core. It's in your system."[6] Duterte counts on the Philippine National Police to implement his campaign promise of ridding the country of drugs, crime and corruption, yet it is an institution that can only be characterized as weak and compromised. Prone to corruption, the police are often used by national and local politicians to exert control and compliance. As historian Alfred McCoy (2009a, p. 54) wrote in his book that traced the origins of the Philippine National Police, "throughout the twentieth century, all the crises serious enough to threaten the legitimacy of Philippine governments or regimes have somehow revolved around policing, either corruption from syndicated vice or the illegitimate use of legal force." But the Philippine National Police is just one, among many, institutions in the country that has little autonomy from external pressures.

This chapter focuses on the Philippines, one of the oldest democracies in Asia. Philippine state formation is often described as having created weak institutions and strong elites (Hutchcroft 2000; McCoy 2009b; Sidel 1999). Since 1946, the Philippines has been through periods of democracy building, dictatorial control and the re-establishment of democratic institutions (Landé, 1965; Hutchcroft and Rocamora, 2003). Throughout these periods, systemic corruption cut across institutions, players and processes in a society where wealth and power, public and private interests are controlled and defined by a few beneficiaries (Johnston 2005). Two words sum up the corruption issues in the Philippines: impunity and entitlement. The chapter will discuss the inherent weakness of Philippine institutions, the formation of a corrupt elite, and the vulnerabilities of the electoral process

---

    [5]   Maricar Cinco and Madonna Virola, 2016. Two cops kill anti-crime crusader in Mindoro. http://newsinfo.inquirer.net/825201/2-cops-kill-anticrime-crusader-in-mindoro. Accessed May 15, 2017.

    [6]   Karl Malakunas, 2017. You are corrupt to the core, Duterte tells cops. http://news.abs-cbn.com/focus/01/30/17/you-are-corrupt-to-the-core-duterte-tells-cops. Accessed May 15, 2017.

that make corruption deeply embedded in the Philippines. Because of the systemic nature of corruption in the Philippines, it will not be easy to give a clean picture of corruption and abuse. Instead, one will see "the riddle that is the Philippine state, weak yet strong, centralized yet localized" (McCoy 2009b, p. 296). Using the sociological concept of ambivalence (Merton 1976), the Philippines is an ambivalent state because of the conflicting orientation and actions from institutions. The ambivalence in the relationship between central and local institutions has created a type of structural corruption that is now crossing over to a level of violence the Philippines has not yet seen. What exacerbates the situation is the clientelistic relationship between and among national and local elites. Using the lens of syndromes of corruption by Johnston (2005), the Philippines is an example of an Oligarch and Clan type. To further explain the prevalence and persistence of corruption, the chapter looks at state formation, the players involved and the processes that make it difficult to dislodge oligarchs and clans. The chapter not only examines the past but examines contemporary manifestations of the strategy of oligarchs and clans towards weak state institutions.

SYSTEMIC AND SYMPTOMATIC: FALLING WITHIN THE SPECTRUM OF OLIGARCHS AND CLANS

According to Johnston (2005), the Philippines falls within the same class of corruption as Russia and Mexico. The "Oligarch and Clan Symptom" is a combination of a weak state apparatus, the vulnerability to exploitation of these public institutions, the use of violence, and unpredictability because the rule of law is not upheld or is weakened by players who want to maintain power and wealth. As such, oligarchs and clans are about contending interests of a few elite families with far-reaching influence to protect their interests involving the public and private sector. Johnston makes it clear that the Philippines is not a "clone" of the Russia case but exhibits similarities. I would like to raise the possibility that the Philippines is moving closer to the Russian model in so far as the escalating use and level of violence by a state enforcement agency is concerned, the co-mingling of organized crime and syndicates with state actors, and the changing hands of government licenses from one oligarch to a more favored one. There is consistency in the Philippine case in the continuing issues of electoral integrity in the form of electoral fraud, conflict of interest issues, and poor enforcement of campaign finance laws that threaten the credibility of elections.

The rising level of violence currently evident in the Philippines is happening at an alarming pace. According to Human Rights Watch

Table 20.1   Official numbers of deaths related to the drug war, July 1,
*   2016–March 31, 2017*

Number of anti-illegal drug operations	53 509
Number of homicide cases	9 431
Official number of deaths related to drug operations	1 847
Deaths not related to drug war	1 894
Under investigation	5 691

Source: Philippine Drug Enforcement Agency.

and Amnesty International, the number of extra-judicial killings since the Duterte administration took office is now close to 9,000 in 2017.[7] In one day alone, an operation in Bulacan, a suburb near Metro Manila, resulted in the deaths of 32 people.[8] In the 2017 Freedom House report, a downward trend in the freedom index is expected because of "the rule of law and application of justice are haphazard and heavily favor ruling dynastic elites."[9] It likewise mentioned that the protracted and violent war against insurgents will continue as it has for decades and "impunity remains the norm for crimes against activists and journalists." Violence is not new in the Philippines. Over the years, there have been multiple fronts where a generalized type of violence is present. Even in the capital, firearms are highly visible in business and commercial areas not just for uniformed policemen but for private security hired by offices and stores. The expected decline in the freedom index is due to the violent war on drugs by the administration, which has resulted in an alarming rate of extra-judicial killings and vigilante-style justice. The government has provided figures showing that the deaths were not part of legitimate police operations (Table 20.1). While that may be enough for the Philippine National Police to consider themselves absolved from the deaths, the public is left to wonder who is responsible for the killings and why is it that law enforcers are unable to solve these crimes?

Many observers are aware of the corruption that the Philippines has

---

[7] Human Rights Watch Report 2017; CHR backs Human Rights Watch and Amnesty International on war on drugs. CNN Philippines. July 4, 2017. cnnphilippines.com/news/2017/07/04/chr-backs-hrw-amnesty-intl Accessed May 18, 2018.

[8] Frances Mangosing, 2017. Bulacan PNP war on drugs: 32 suspects killed in 24 hours. http://newsinfo.inquirer.net/923117/drug-suspects-bulacan-drug-bust-war-on-drugs-president-rodrigo-duterte-anti-illegal-drugs-pnp. Accessed August 16, 2017.

[9] Freedom House Country Report 2017. https://freedomhouse.org/report/freedom-world/2017/philippines. Accessed August 20, 2017.

experienced over the years. Ferdinand Marcos and his wife Imelda ruled for 20 years under what was called "the conjugal dictatorship." Years of pursuing their unexplained and not-so-hidden wealth have resulted in the sequestration of some properties and accounts. But the lion's share of the $10 billion ill-gotten wealth remains elusive.[10] The World Bank's Stolen Asset Recovery Initiative reports securing $683 million from Swiss banks.[11] The Philippine government's Presidential Commission for Good Government (PCGG), tasked to recover the Marcos's ill-gotten wealth, only recovered $3.7 billion and has not successfully prosecuted any members of the Marcos family or their associates in the 30 years since the agency was created. President Joseph Estrada was embroiled in a corruption scandal which involved receiving payoffs of millions of dollars from the illegal numbers game *jueteng* and concealed business interests amounting to hundreds of millions of US dollars (Johnston 2005). President Gloria Macapagal-Arroyo had her own corruption scandal involving bribery from the Chinese telecommunications firm ZTE for a $300 million project.[12] In all these presidencies, the Philippine National Police or members of the law enforcement agencies were implicated. It was, however, Ferdinand Marcos who controlled and manipulated the Philippine National Police. Marcos used the police for the execution of Chinese drug trafficker Lim Seng, deployed them as anti-subversion squads to arrest student or pro-democracy groups, and used extra-judicial operations for those who organized against his regime (McCoy 2009a). Marcos justified the need for strongman rule by saying that the country needed discipline, yet what he did was to give more power to the police by "arming them with formal decrees and informal impunity" (McCoy 2009a). The post-Marcos years were a period of continued impunity but this time with some senior police officers protecting syndicates or are themselves involved in kidnap for ransom targeting wealthy Chinese families.

The Duterte administration's virulent anti-drug campaign is putting the spotlight on the unholy merger of state actors and organized crime. The Philippines, at this critical juncture, is seeing a new kind of corruption in

---

[10]   Nick Davies, 2016. The 10bn dollar question. What happened to the Marcos millions? https://www.theguardian.com/world/2016/may/07/10bn-dollar-question-marcos-millions-nick-davies. Accessed February 7, 2017.

[11]   Case ARW 61 Ferdinand and Imelda Marcos. World Bank and UNODC Stolen Asset Recovery Initiative. http://star.worldbank.org/corruption-cases/node/18497. Accessed February 7, 2017.

[12]   GMA Network News, 2008. IBON: corruption scandals under Arroyo cost Filipinos P7.3B.   http://www.gmanetwork.com/news/news/nation/83278/ibon-corruption-scandals-under-arroyo-cost-filipinos-p7-3b/story/. Accessed February 7, 2017.

the form of violence that is either tolerated by some state actors and law enforcers or the state actors are simply too weak to address the problem effectively. There is a deadly combination of weak capacity and compromised autonomy within the Philippine National Police, the agency tasked with law enforcement. With the declaration of Duterte as the winner in the 2016 Philippine presidential elections, there was a rise in vigilante-style killings. Death was both brutal and turned into a deliberate spectacle. The victim is often left hogtied with a handwritten sign "I am a drug user/pusher."[13] The initial spate of violence was considered by some as perhaps the "cleansing" or killing of runners and middlemen by elements of the police who live a double life as crime fighter and criminal.[14] They were killing individuals who might whistleblow on them regarding their participation in the illicit trade. President Duterte has offered a reward for the death of policemen involved in drugs.[15] The Philippine National Police suffers from having both loyal policemen and some members in their ranks who have close links with organized crime. The contradiction makes apparent the tragedy of the Philippine state and society.

What makes the Duterte administration's war on drugs a critical juncture in Philippine history is the increased level of violence that is felt in the communities of the urban poor. While smugglers, Bureau of Customs' officials and warehouse owners who have been caught red-handed with $120 million worth of *shabu* or methamphetamine from China are able to defend themselves in open public hearings and in court, the deaths of users and small-time dealers are now in the thousands.[16] The term "extrajudicial killing" or "ejk" has become common in the Philippines in only a few months. The use of violence is just one more manifestation, albeit a new one for the Philippines, of corruption. What makes the situation problematic for the state and society is that this involves cops against cops.[17] A stronger internal affairs unit within the police force should

---

[13] Clair Baldwin, Andrew Marshall, and Damir Sagoji. 2016. Police rack up an almost perfectly deadly record in Philippine drug war. http://www.reuters.com/investigates/special-report/philippines-duterte-police/. Accessed February, 7, 2017.

[14] See Ai Balagtas. 2016. Cop keeps P7M, "shabu" in home fit for general. http://newsinfo.inquirer.net/787531/cop-keeps-p7m-shabu-in-home-fit-for-general. Accessed May 15, 2017.

[15] Allan Nawal, 2017. Kill cops into drugs and get a P2-M reward – Duterte. http://newsinfo.inquirer.net/923565/rodrigo-duterte-crime-drugs-scalawag-policemen-war-on-drugs-drug-killings. Accessed August 20, 2017.

[16] Jodesz Gavilan, 2017. TIMELINE: how P6.4-B worth of shabu was smuggled into PH from China. https://www.rappler.com/newsbreak/iq/178667-timeline-smuggled-shabu-china-customs. Accessed August 20, 2017.

[17] Mike Frialde, 2016. Cops vs cops in drug war probe. http://www.philstar.com/headlines/2016/06/25/1596515/cops-vs-cops-drug-war-probe. Accessed May 20, 2017.

have enough capacity and autonomy to hold rogue cops accountable. On October 13, 2016, the head of the Citizen's Crime Watch was shot by two men on a motorcycle in front of her house. When policemen were alerted to the shooting, a chase ensued and shots were fired. The gunmen lost control of their motorcycle and when the police caught up with them, they were surprised to find the Chief of Police of Socorro, Oriental Mindoro, and another policeman wearing a ski mask and a wig surrendering to them.[18] While the two policemen were relieved from their posts, they were eventually allowed to post bail ten months after the incident.[19] A second incident in which corruption involved violence by state actors happened on August 21, 2016 in Payatas, Quezon City, a poor urban community in metro Manila. Five men who worked as garbage collectors were playing pool when four armed policemen asked them to kneel inside a hut, tied them with electrical wires and accused them of using illegal drugs. One survivor gave this testimony:

> The armed man made Daa sit on a wooden chair and Morillo on the armrest thereof. Then, without warning, he pointed his firearm at Morillo and shot him on the chest. Morillo fell to the ground bleeding, but he did not lose consciousness. Next, the armed man shot Daa, who fell to the ground beside Morillo. Daa was shot a second time on the head as he lay on the ground. He died.
> They made the three kneel on the ground at the back of the house and shot them to death. Jessie Cule was the last of the three to be killed. He begged to be spared, hugging the legs of one of the armed men and sobbing. As he would not let go of his hold, the man shot him on the nape.[20]

The four policemen responsible were identified by the dense community of urban informal settlers. They were transferred to another area and an internal affairs investigation was launched. This operation resulted in the suspension of Operation Tokhang (Operation Knock and Plead) or the administration's anti-drug war. In February, 2017, the only survivor among the five men shot filed a petition in the Court of Appeals for grave misconduct, multiple murder, attempted murder, robbery and planting of evidence against four Quezon City policemen.[21] After several weeks,

---

[18] Cinco and Virola, Two cops kill anti-crime crusader in Mindoro.
[19] Maricar Cinco, 2017. Cops in murder of Mindoro anticrime crusader out on bail. http://newsinfo.inquirer.net/921881/cops-in-murder-of-mindoro-anticrime-crusader-out-on-bail. Accessed August 20, 2017.
[20] Lian Buan, 2017. A gruesome tale of TokHang: "Sir, may humihinga pa." https://www.rappler.com/nation/160014-gruesome-tokhang-payatas-quezon-city-petition. Accessed May 20, 2017.
[21] Rappler News Team, 2017. Petitioner asks CA to order probe into Payatas body dumping. https://www.rappler.com/nation/165479-petition-ca-payatas-body-dumping-probe. Accessed May 20, 2017.

Operation Tokhang was relaunched as "Operation Tokhang: Double Barrel Reloaded." The Chief of the Philippine National Police, Ronaldo De La Rosa, claimed that "This time we will make sure it will be less bloody if not bloodless, and target high-value drug suspects."[22]

Apart from violence, the Philippines continues to manifest signs that once there is a change in power, there is likewise changing hands of ownership and access to state licenses and franchises among oligarchs and clans. Because "official powers and institutions are ineffective, unpredictable and often up for rent," there is an increase in unpredictability that can only be addressed through an increase in clientelistic relationships (Johnston 2005, p.121). A former Marcos crony, Roberto Ongpin, found himself singled out and mentioned as an example of an oligarch by President Duterte. In at least two public speeches, Duterte mentioned and named former Trade Secretary of Ferdinand Marcos, Roberto Ongpin, as an oligarch who has taken advantage of his influence to gain wealth.[23] With the threat to make online gambling illegal, Roberto Ongpin quickly sold all his stakes of his online gambling business Philweb. It is not usual to see a former Marcos crony scramble to divest himself of his financial interests. Ongpin, after all, survived and thrived through the years with his deft use of clientelism and building close relationships with those in power. This time, though, Ongpin had to sell his shares at bargain prices to Gregorio Araneta,[24] a member of the Araneta clan who has considerable economic and political clout; he is also son-in-law of former president Ferdinand Marcos and wife, Imelda. It was a case of one oligarch losing out to another.

Between the violence and the use of public office for rent-seeking, the Oligarch Clan symptom of corruption breeds a climate of impunity and a strong sense of entitlement to what should be public goods and resources. The beneficiaries of rent-seeking activities are limited to a very few families. But the threat of violence and impunity by state forces, including collusion with organized crime, victimizes a lot of ordinary families and citizens. Another symptomatic feature of Oligarchs and Clans is electoral fraud, discussed in the fourth section, which shows how the process of electoral shortcomings is part of a legitimization and delegitimization of

---

[22]   Philippine Star News Team. Oplan Tokhang double barrel reloaded. 2017. http://www.philstar.com/headlines/2017/03/07/1678731/double-barrel-reloaded. Accessed May 20 2017.

[23]   Nestor Corrales, 2016. Duterte blasts Ongpin, targets oligarchs in drive to cleanse government. http://newsinfo.inquirer.net/802692/duterte-blasts-ongpin-targets-oligarchs-in-drive-to-cleanse-govt. Accessed May 20, 2017.

[24]   Doris Dumlao, 2016. Ongpin sells Philweb to Araneta for P2B. http://business.inquirer.net/216108/ongpin-sells-philweb-stake-to-araneta-for-p2b. Accessed May 20, 2017.

state institutions. This corruption dissuades new players, especially those from civil society, to participate in the electoral competition. Instead, the use of money and violence, vote buying and intimidation are part of the election cycle.

POLITICAL INSTITUTIONS: STATE FORMATION AND WEAK BUREAUCRACY

The Philippines' colonial past plays a major role in the kind of state and society it has today. The colonial legacy of the Spanish and Americans in their attempt to control one of the world's largest island groups, with 7,107 islands, 78 languages and 500 dialects, had an uneven impact across time and space. Today, the Philippines is composed of 17 regions, 79 provinces, 115 cities, 1,499 municipalities and 41,969 *barangays*, a Filipino term for the smallest political unit. The challenge is how to hold the Philippines' ethno-linguistic diversity together in a common territory and with a common identity. How can the reach of the state be effective?

Abinales and Amoroso (2005) trace the origins of the weak state from the Spanish colonial period. Spain controlled the Philippines through both clerical and state officials. As the farthest colonial outpost of Spain, the

> secular state was weak in personnel, its power did not flow evenly through the territory it claimed, and it remained extremely dependent upon the friars for its most basic functions . . . part of the reason for this weakness was the shortage of lay Spaniards willing to serve in the new colony. (pp. 66–7)

Poor administrative control by Spain meant that the structures were not fully developed and instead became heavily dependent on clerical officials – "in much of the Philippines, the friar was the state" (p. 67). Abinales and Amoroso are not incorrect but recent scholarship suggest secular power and the influence of liberalism during the nineteenth-century Spanish colonial project had an impact not just on the revolution and independence but on the political configuration of the weak central state and strong local autonomy often traced to the American colonial period (1898–1946) (de Llobet 2014). In particular, the introduction of Spain's 1812 constitution had an impact in the Tagalog region and southern Luzon. The constitution

> hoped to transform Spain from a patrimonial to a Liberal empire. However, the Liberal tenets that made citizens out of all Spanish subjects, redefined their relationship with the state, also entailed the centralization and strengthening of government by divesting governmental and economic control from local power holders and redirecting it to the *Cortes*. Regardless of its centralizing

intentions, the *Cortes* ended up reinforcing communal autonomy. (de Llobet 2014, p. 218).

When the Americans took over the Philippine islands for $20 million, President William McKinley signed the Benevolent Assimilation Proclamation on December 10, 1898. This took place even as Philippine revolutionary forces declared independence from Spain six months earlier, on June 12, 1898. To pacify the revolution and control the Philippines as a colonial state, Daniel Williams, Secretary of the Philippine Commission, set out to construct "a new government . . . created from the ground up, piece being added to piece as the days and weeks go by" (Kramer et al. 2003, p. 10). Sociologist Julian Go (2008, p. 3) pointed out this is "a claim [that] is somewhat exaggerated. The Spanish had already constructed a colonial state in the Philippines before the United States." Building a representative government and considering it a working democratic process within a colonial state has also been called "anomalous" by historian Ruby Paredes (1989, p. 41). Inheriting institutions left by the Spanish empire, the Americans continued the colonial project with more aggressive secularization but reinforced the autonomy of local elites.

Philippine state formation is linked to the colonial state formation that the Americans built after Spain. In the Letter of Instructions of President McKinley to the Second Philippine Commission under Taft transmitted through the Secretary of War, the priorities were clearly stated that there was a need for the Commission to "devote its attention to the establishment of an educational system and an efficient civil service system, the organization of the judiciary and of municipal and provincial governments" (Corpuz 1957, p. 164). This instruction from Washington shows the responsibilities of and expectations from American administrators. The expectation was both a project of colonial state building and colonial governance. It was an ambitious project in which American administrators had no experience. Making this more difficult to achieve was the existing political, economic and cultural structures built by 300 years of Spanish colonial rule that did not result in the creation of institutions that bear the imprint of modernity or development. Instead, the "colony was lifted from a bureaucratic world that was as late as 1898, almost an intact survival of the medieval era" (Corpuz 1957, p. 164).

Colonial state building was the creation of basic structures that the Americans needed for political control: it was not an easy task given the spatial challenges of pacifying and containing an archipelago and turning it into a modern state. This was state building that included the creation of formal institutions such as administrative structures for revenue collection and a codified legal system (Go 2008). In many ways, the project of colonial state building was a straightforward endeavor that meant recreating the tools of state rationalization. This included the creation of a police force and a bureau of census to have a better sense of the population that was now under the colonial administrators. The establishment of a wide-reaching public educational system is commonly hailed as an American colonial legacy. This was not the case for the creation of a professional and effective civil service and bureaucracy.

On the other hand, colonial governance was a more difficult task. It eventually led to unintended consequences and the emergence of various institutions, players and processes that were mastered by Filipinos during the American colonial period and carried over through independence. Governing the archipelago meant the implementation of the vision of the rule of colonial administrators. Facing resistance from Filipinos meant that collaboration and compromises were needed for immediate traction in the implementation of governance. Identifying elite Filipinos in various parts of the archipelago was the continuity of patronage that Filipinos had already learned under Spain. It was this game of patronage and clientelism that made it impossible to have a strong, central administrative structure. Identifying the Taft era from 1900 to 1913 as crucial to understanding the Philippine state, Hutchcroft (2000) challenges arguments that the Philippines is overly centralized. Instead, Hutchcroft (2000, p. 278) illustrates that given the

> territorial character of the modern Philippine state ... Manila has long displayed a notably weak capacity for sustained administrative supervision over provincial and local officials. National politicians moreover must commonly rely heavily on local power (and the brokering of arrangements with local bosses and their private armies) to succeed in local contests.

The Taft era in the Philippines showed both colonial state building and colonial governance at play. It was a period that brought about the establishment of democratic institutions such as the creation of a civil service based on merit, an electoral process, an extensive judicial mechanism, a bill of rights, a functioning system of municipal, provincial and national governments, a Filipino legislature and a political party system.

The kind of institutions built and the relationship between central authority and local leaders made this period of state formation vulnerable to misuse of public funds and left a lasting imprint. The systemic corruption and continuing clientelism can be traced to the colonial project of both Spain and the US. The American colonial period provided discretionary powers to provincial and local elites in exchange for political control, yet at the expense of building institutions that were accountable and effective. It was a period of "subordination of an underdeveloped state apparatus to elected municipal, provincial, and national officials in the American colonial era" (Sidel 1999, p. 12). With independence after World War II, the Philippines was a liberal democracy that held much promise. But the seeds of ambivalence in the new republic remained with the "persistence of coercive pressures, local power monopolies in electoral politics and social relations" (Sidel, 1999, p. 9). Corruption in the Philippines is both deeply and widely embedded at the political, economic and familial levels in the form of dynasties and bailiwicks that compete with national institutions for influence to govern.[25] This also explains why the high rates of re-election make local elites into national ones. Marcos was a local elite and his bailiwick, Ilocos Norte, is still controlled by the family. Both Joseph Estrada and Rodrigo Duterte share the same background of moving from local mayor to the presidency.

THE PLAYERS: LOCAL AND NATIONAL ELITES

Various political scientists have written about the corruption of former presidents of the Philippines such as Ferdinand Marcos (Bonner 1987; Johnston 2005; Manapat 1991), Fidel Ramos, Joseph Estrada and Gloria Macapagal-Arroyo (Coronel 2004; Quah 2006) within the political history of the Philippines. In this section, I focus on the kind of relationship that national and local elites have using the lens of the 2016 Philippine presidential elections. It is important to note that all the presidential candidates claimed to be reformists, blurring ideological distinctions, especially with the kind of weak political party system found in the Philippines. All the candidates faced allegations of corruption or issues related to campaign finance at some point in their political careers. In this section, I illustrate the dynastic identities and issues that beset the local and national elites in the 2016 campaign.

Candidates for both the presidential and vice-presidential elections mobilized resources and supporters to persuade voters that they could address the problems of the Philippines. Rodrigo Duterte ran under a banner of change. His loyal supporters, especially in social media, promoted and hinted that "Change is Coming," as a foreshadowing of the Duterte presidency. In various ways, the candidates all hinted at the rhetoric of change. The initial front runners Vice President Jejomar Binay and Senator Grace Poe, an early favorite of voters albeit saddled

---

[25]   I am grateful to Michael Johnston for emphasizing this point and for providing insightful comments.

with citizenship issues, dominated the television and radio airwaves with talk of change. Binay spent $7 million in political advertisements alone.[26] Only incumbent-endorsed Roxas ran under the rhetoric of continuity. Throughout his campaign Roxas constantly echoed Noynoy Aquino's *daang matuwid* (the straight path). This was Aquino's successful 2010 presidential campaign slogan, winning – with 42 percent – the second highest number of votes received by a presidential candidate since his mother Corazon Aquino won in 1986.

Yet, despite the rhetoric of change, leading candidates for the presidency and vice-presidency represent continuity. A quick glance at the leading presidential candidates shows that they are part of dynastic families deeply embedded in Philippine politics and society. Former Davao Mayor Rodrigo Duterte comes from an elite family that can be traced to Cebu's Veloso-Durano-Osmena families from the 1950s.[27] Incumbent Vice-President Jejomar Binay's family ruled Makati continuously for 30 years. In the years 2013–16, Binay and his children simultaneously held office as Vice-President, Senator, Congresswoman of Makati and Mayor of Makati. Manuel Roxas is from the Roxas-Araneta clan that occupied various national and local positions throughout the political history of the republic, as high as the presidency to the Senate and the House of Representatives. Roxas, the grandson of the Philippines' post-war president, is a member of the national elite and part of the leadership of the Liberal Party, the same party as former President Benigno "Noynoy" Aquino III.

Six years of economic growth and the stability that the Aquino administration brought after the consecutive plunder charges of residents Estrada and Arroyo made the incumbent Noynoy Aquino and his party mate Mar Roxas confident that the people would want more of the same. Agencies and commissions such as the Department of Public Works and Highways (DPWH) under Secretary Singson, the Commission on Audit (COA), the Office of the Ombudsman under former Justice Conchita Carpio-Morales, the Supreme Court and the Commission on Elections (COMELEC) have reform-minded leadership and on occasion have shown independence from the executive. The COA, for instance, started issuing the notice of disallowance requiring those who do not pass their audit to return state funds. But Aquino had many missed opportunities

---

[26]   Menchu Macapagal, 2016. Who has spent the most money on campaign ads? http://cnnphilippines.com/news/2016/04/27/biggest-campaign-ad-spenders-pcij-nielsen.html. Accessed May 20, 2017.
[27]   Conversations with historian Michael Cullinane provided me with information regarding the long lineage of dynastic ties that the Duterte family has in Cebu.

as well. These were either corruption scandals, allegations of protection and patronage, or simply weak institutions unable to bear the weight of a crisis. The Aquino administration was highly criticized for the poor disaster response and management during Typhoon Haiyan in 2013. The pork barrel scam involving discretionary funds for development projects revealed collusion between high-level politicians and non-governmental organizations. The president's Disbursement Acceleration Program (DAP) fund, with 144 billion pesos ($3.3 billion), was used to support and win new allies.[28] The Mamasapano incident that killed 44 members of the special forces would not have happened if Aquino had not let his trusted friend General Purisima usurp the chain of command while under suspension for charges of graft. The $81 million money laundering scandal brought suspicion to many Filipino banks and remittance companies that service the 10 million overseas Filipino workers who remit $2 billion a month. The failure to pass the Freedom of Information bill or the whistleblower protection bill has been a sore point for those who supported Aquino. It was a case of national elites failing to deliver campaign promises.

On the other hand, the local elites had a track record of delivering public services. This might involve corruption or violence, but they delivered and remained popular among voters, especially the poor. Duterte did not run on a platform advocating for the poor. His singular and successful campaign message was to curb drugs and corruption. Duterte comes from a long line of provincial elites going back to the 1950s. His daughter and son replaced him as mayor and vice-mayor of Davao, respectively. Various controversies have hounded Duterte's public life. Critics of Duterte were quick to point out the human rights abuses that happened under his watch as mayor of Davao, including tolerating vigilantes such as the Davao Death Squad. Duterte's response was to face these criticisms head on with candor and punctuating his statements with curses, which delighted the crowds that gathered to listen to him. Addressing a crowd of mostly young people, the lone presidential candidate who joined one forum said, "If I become president, there is no such thing as bloodless cleansing. I propose to get rid of the drugs between 3 to 6 months." Political scientist John Sidel (1999) describes local elites as using "coercive forms of control over local populations." Violence and coercion is a strategy more widely used by local strongmen to control their territory. In the end, the voters chose Duterte, who won

---

The Economist, 2014. DAP dancing: the president vs the Supreme Court. http://www. economist.com/news/asia/21610286-president-versus-supreme-court-dap-dancing. Accessed May 20, 2017.

with a relative majority in a five-way race that was both divisive, costly and with a lot of controversies. Duterte's win has been described as the end of a 30-year liberal democracy in the Philippines (Teehankee and Thompson 2016). But it might be more of the same rather than a break. It is strongman rule with the promise to rid the country of drugs and corruption.

THE PROCESS: ELECTORAL SHORTCOMINGS AND THE COST OF ELECTIONS

A democracy is about popular representation and ideally leadership options should come from all sectors of society. In the case of the Philippines, one of the oldest democracies in Asia, the options remain limited to dynastic families. According to the Center for People Empowerment and Governance (CenPeg), 70 percent of the House of Representatives are from political dynasties while the upper house has a higher figure of 80 percent. According to Representative and Liberal Party spokesperson Romero Quimbo, "by the end of the filing, almost 50 percent of the posts are uncontested. This is the first time that a substantial number of congressional seats were uncontested."[29] The rise in the number of uncontested posts illustrates incumbency advantage of deeply entrenched dynasties. By running unopposed, voters have very little or no option on who represents them or who will be the executive for their local government unit. Uncontested elections deprive voters of a choice. This trend, along with the high cost of campaigning and dynasties, dissuades new players from challenging incumbents.

As one of Asia's oldest democracies, is the Philippines truly representative when most of the candidates come from prominent political families? When the number of uncontested elections continues to rise? Political dynasties are not unique to the Philippines. Political dynasties symbolize elite power politics. The high cost of running for public office, the staggering amounts of public money lost to corruption, and the obvious desire to stay in power to avoid accountability and wield greater economic-political influence all add to the frustration of those critical of dynasties. Dynasties symbolize the lack of change and persistence of unaddressed political, economic and social problems. And there is also the seemingly undue advantage of name recall, unparalleled resources and a well-oiled political

---

[29] Gil Cabacungan, 2015. Half of governors running unopposed. *Philippine Daily Inquirer*. http://newsinfo.inquirer.net/749091/half-of-governors-running-unopposed. Accessed May 20, 2017.

Table 20.2   Filipino election expenditures (pesos)

Presidential candidate	Election expenditures
Grace Poe	510 845 262.56
Jejomar Binay	463 375 216.37
Rodrigo Duterte	371 461 480.23

Source: rappler.com; $1 = 47.09 Philippine pesos.

machinery ready and adept to carry a winning candidate to his or her desired public office.

Apart from these issues, there are almost always allegations of electoral fraud in the Philippines. In the 2016 elections, the son of former dictator Ferdinand Marcos ran as vice-president and lost; he claimed electoral fraud and filed an electoral protest in the courts. Electoral integrity remains an issue in the Philippines because the COMELEC had a history of partiality and some of its leadership were involved in election fraud (Calimbahin 2009). Part of the problem is that the COMELEC suffers from a three-pronged pathology that involves externally and internally motivated clientelism and bureaucratic inefficiency. The continued rising cost of running for public office (Table 20.2) is a campaign finance issue that can be addressed with better enforcement and increased capacity. And while the COMELEC in the last seven years has initiated reforms and improved its credibility, it continues to be hounded by issues of fraud and collusion. Although certainly not at levels as in the past such as under Marcos and Arroyo. The current chair of the COMELEC Andres Bautista, an appointee of Aquino, is embroiled in allegations of ill-gotten wealth amounting to 1 billion pesos ($19 million). His official declaration of assets is 170 million pesos ($3.3 million).[30] As a constitutional commission tasked and mandated to handle the election administration in the country, the COMELEC should lead in innovation and create an election environment that is fair and honest. Since 2010, since the automation of elections in the Philippines, the COMELEC's performance continues to improve and voters perceive both the elections and the COMELEC as credible. There are many ways by which COMELEC capacity, including modernization programs, can still improve and the commission continues to have a key role that can level the playing field and re-define the rules

---

[30]   Rappler News Team, 2017. Wife accuses Comelec chair of "unexplained wealth."https://www.rappler.com/nation/177902-wife-comelec-chairman-andy-bautista-accused-unexplained-wealth. Accessed May 20, 2017.

so new players can challenge dynasties and voters can have more choices (Calimbahin 2011).

CONCLUSION

After the first year of the Duterte administration in 2016–17, no one can question the determined political will of the leadership to rid the country of drugs and corruption. The president's strong words akin to an order of battle for the Philippine National Police have been both acclaimed and criticized. But the strong words are often unmatched by the enforcement agencies' capability and capacity. The Philippines has an ambivalent state apparatus. Instead of a synergy of institutions, one finds institutions that are vulnerable to corruption and clientelism. The current climate of increased violence illustrates the contradictions present within the relevant institutions, players and processes. Political will alone is not enough. The strong words and directives of the president mean little if the executing agencies are crumbling due to their lack of capacity and autonomy. A strong leader who is genuinely for reform and development will need strong institutions and a strong bureaucracy. The strongman rule of Duterte will not curb corruption. More than a strong hand, the country needs strong institutions that will not crumble in the face of oligarchs, clans and families with embedded interests to keep themselves in power. Accountability is one way to curb the impunity and entitlement that is bred by the syndromes of corruption coming from oligarchs and clans.

President Duterte himself said,

> corruption persists like a fishbone stuck in the throat. It pains and it is disconcerting. We need to pry corruption from the government corpus [where it] is deeply embedded. We also need to put an end to squabbles and bickering within agencies [and] focus fully on the speedy provision of quality public services to our people. Believe me, it is easier to build from scratch than to dismantle the rotten and rebuild upon its rubble."[31]

State building and governance happen simultaneously and with preexisting conditions, dynamics, players and plenty of compromises. The Philippines will see more of the same until key democratic institutions gain capacity and autonomy from embedded elites and interests.

---

[31] From the transcript of the Second State of the Nation Address, President Rodrigo Duterte, p. 17. http://news.abs-cbn.com/focus/07/24/17/read-transcript-of-president-dutertes-2nd-sona. Accessed May 20, 2017.

The persistence of corruption in the Philippines needs a more developed prescription. The need for strong leadership and the reform of key anti-corruption institutions is necessary but will not be enough. Corruption in the Philippines seems more resistant to reform because of the persistence of strong elites and weak institutions that create contrasting orientations and functions. Corruption in the Philippines is likely to continue for as long as it serves the interests of multiple segments of society and strong elites.

REFERENCES

Abinales, P. and D. Amoroso. 2005. *State and Society in the Philippines*. Latham, MD: Rowman & Littlefield.

Bonner, R. 1987. *Waltzing with a Dictator: The Marcoses and the Making of American Policy*. New York: Random House.

Calimbahin, C. 2009. *The Promise and Pathology of Democracy: The Commission on Elections of the Philippines*. Madison, WI: University of Wisconsin Press.

Calimbahin, C. 2011. Exceeding (low) expectations: autonomy, bureaucratic integrity, and capacity in the 2010 elections. *Philippine Political Science Journal* **32**(55), 103–26.

Coronel, S. 2004. *The Rulemakers: How the Wealthy and Well-born Dominate Congress*. Manila: Philippine Center for Investigative Journalism.

Corpuz, O. 1957. *The Bureaucracy in the Philippines*. Manila: Institute of Public Administration, University of the Philippines.

de Llobet, R. 2014. Chinese mestizo and natives' disputes in Manila and the 1812 Constitution: pld privileges and new political realities (1813–15). *Journal of Southeast Asian Studies* **45**(2), 214–35.

Go, J. 2008. *American Empire and the Politics of Meaning: Elite Political Cultures in the Philippines and Puerto Rico during US Colonialism*. Durham, NC: Duke University Press.

Hutchcroft, P. 2000. Colonial masters, national politicos, and provincial lords: central authority and local autonomy in the American Philippines, 1900–1913. *Journal of Asian Studies* **59**(2), 277–306.

Hutchcroft, P. and J. Rocamora. 2003. Strong demands and weak institutions: the origins and evolution of the democratic deficit in the Philippines. *Journal of East Asian Studies* **3**(2), 259–92.

Johnston, M. 2005. *Syndromes of Corruption: Wealth, Power, and Democracy*. Cambridge: Cambridge University Press.

Kramer, P., J. Go, and A. Foster. 2003. *The American Colonial State in the Philippines: Global Perspectives*. Durham, NC: Duke University Press.

Landé, C. 1965. *Leaders, Factions, and Parties: The Structure of Philippine Politics*. New Haven, CT: Yale University Press.

Manapat, R. 1991. *Some are Smarter than Others: The History of Marcos' Crony Capitalism*. Haydenville, MA: Aletheia Press.

McCoy, A. 2009a. *Policing America's Empire: The United States, the Philippines, and the Rise of the Surveillance State*. Madison, WI: University of Wisconsin Press.

McCoy, A. (ed.) 2009b. *An Anarchy of Families: State and Family in the Philippines*. Madison, WI: University of Wisconsin Press.

Merton, R. 1976. *Sociological Ambivalence and Other Essays*. New York: Free Press.

Paredes, R. (ed.) 1989. *Philippine Colonial Democracy*. Quezon City: Ateneo de Manila University Press.

Quah, J. 2006. Curbing Asian corruption: an impossible dream? *Current History* **690**, 176–9.

Sidel, J. 1999. *Capital, Coercion, and Crime: Bossism in the Philippines.* Stanford, CA: Stanford University Press.

Teehankee, J. and M. Thompson. 2016. Electing a strongman. *Journal of Democracy* **27**(4), 125–34.

21. Indonesia: why democratization has not reduced corruption

Marcus Mietzner

In 1995, as the authoritarian rule of Suharto's New Order regime was approaching its end, Transparency International (TI) issued its first Corruption Perceptions Index (CPI). In that list, which comprised 41 countries, Indonesia came last, making it the most corrupt country surveyed by TI at that time. As the CPI gradually included more surveyed countries in following years, Indonesia initially stayed in the bottom group of the world's most corrupt countries. But by the early 2010s, with the CPI covering the vast majority of countries and territories in the world, Indonesia had moved up to a middle-ranked position. In 2016, it was ranked 90th out of 176 surveyed states. This move upwards was not only a result of the increased number of participants in the CPI survey, however. While still viewed as having high levels of corruption, Indonesia had recorded significant increases in its CPI score, climbing from 32 in 2013 to 37 in 2016. Overall, then, Indonesia's CPI assessment had improved considerably since the onset of democratization in 1998.

In explaining this improved score, TI and corruption specialists focusing on Indonesia have typically pointed to post-1998 political and bureaucratic reforms, greater transparency and, in particular, the successes of the Corruption Eradication Commission (Komisi Pemberantasan Korupsi, KPK), established in 2003 (Bolongaita 2010; Choi 2011; Olken 2007; Schütte 2012). Indeed, the KPK has imprisoned hundreds of parliamentarians, government officials, party leaders, judges, governors, mayors, district heads and entrepreneurs, sending shock waves through the political establishment. But we must ask: have the actions of the KPK – which is only based in the capital Jakarta, not in the regions – really led to a decline in overall political and societal corruption, or have they only created a (misplaced) *perception* of reduced corruption?

In this chapter, therefore, I question claims of reduced levels of corruption in post-authoritarian Indonesia. Borrowing from Warf (2016, p. 659), I instead show that "the benefits of the misuse of public power [still] exceed the expected costs (i.e., the probability of being apprehended and the penalties that might follow)". As a result, corruption remains deeply engrained in Indonesian politics and society; indeed, it is instrumental to their functioning.

This chapter develops its arguments in five analytical steps. The first section gives an overview of the history of corruption in Indonesia, covering its development from the colonial period to Suharto's autocracy. The second section discusses the decentralization and fragmentation of corruption as a result of the post-1998 democratization process. It lays out the logic of corruption in the current democratic polity, and identifies the rising costs of politics – in the absence of a credible political finance system – as a main driver of contemporary corruption. In the third section, I assess why the operations of the KPK – despite the organization's impressive record of arrests – have arguably not led to a significant reduction in the overall levels of corruption. The fourth section discusses the spatiality and geography of corruption in Indonesia, with remote areas having almost no control mechanisms in place while urban centres face higher scrutiny but also enjoy stronger capital flows. Finally, the conclusion highlights some institutional lessons to be learned from the Indonesian case for anti-corruption campaigns in new democracies.

CORRUPTION IN INDONESIA: FROM COLONIALISM TO SUHARTO

The patterns of contemporary corruption in Indonesia developed as a result of a complex interplay between feudal local traditions, colonial exploitation and the creation of modern political institutions. In pre-modern times, the territory now known as Indonesia consisted of a system of local empires, kingdoms, sultanates and smaller aristocratic units. These feudal entities drew their authority from a mixture of coercive capacity, self-crafted spiritual-religious status and regional negotiations (Ricklefs 1981, p. 16). The most successful of these states obtained privileged access to trade and, if particularly powerful, received loyalty payments from vassals (Cribb 2000, p. 87). The funds raised through such efforts were at the exclusive discretion of the rulers, who often used them to enhance their image of splendour and personal wealth. Such blending of public and private interests, which under modern definitions would qualify as corruption, was not only seen as acceptable but as inherent to the natural and God-given concept of power. This attitude did not change much with the arrival of the Spanish and the Portuguese in the sixteenth century and of the Dutch in the early seventeenth century. Indeed, the Dutch East India Company (Vereenigde Oostindische Compagnie, VOC), which dominated the key trading posts of the Indonesian archipelago between 1602 and 1799, introduced new forms of corruption. Many VOC officials viewed the offices they held as a mandate to extract a maximum amount of rents, contributing to the financial collapse of the VOC at the end of the eighteenth century.

The Dutch state replaced the VOC in 1800, establishing a gradually expanding colonial state apparatus. While the Netherlands East Indies tried to run a more professional administration than its private predecessor, vast opportunities for corruption and patronage remained. This was partly due to the cultivation system, which was in place between 1830 and 1870. Under this system, the indigenous peasantry had to reserve a fifth of their land for export crops – mostly indigo, sugar and coffee. The local aristocracy was put in charge of overseeing the implementation of the system on the ground, extracting rents from peasants and – in turn – handing the largest share of the profits to the colonial power. Most importantly, Dutch officials were paid percentages of the produce (Angelino 1931, p. 45) in order to keep them from siphoning off money informally. This, in essence, institutionalized – rather than overcame – the idea that public office should grant its holder a share of the generated profits. Although this practice ended in 1866, and the cultivation system itself in 1870, many Indonesian aristocrats – now firmly part of the lower levels of colonial administration – and Dutch officials maintained their rent-seeking attitude when working in the civil service.

The newly independent Indonesian state – emerging from a guerrilla war against the Dutch between 1945 and 1949 – inherited the corrupt patterns of the colonial state and produced some new ones. For instance, the armed struggle was fought by a myriad of self-funding militias, which only slowly came together as the Indonesian army. The practice of raising funds from the population rather than from an allocated budget would survive the war, and led to decades of military business activity and an associated culture of corruption (Mietzner and Misol 2012). Civilian ministries established similar off-budget and self-funding mechanisms, citing the chaos surrounding the war and the scarcity of official government funds as justification. Democracy, which took hold in 1950, did not change this regime but aggravated it. Political parties viewed themselves largely as representatives of a particular socio-religious constituency, and were intent on channelling state resources to those groups. The Indonesian National Party (Partai Nasional Indonesia, PNI), for example, was mostly supported by *priyayi*, descendants of aristocratic families that had served the Dutch (Rocamora 1975, p. 44). They still dominated the bureaucracy, mixing a traditional sense of feudal rent-seeking entitlement with a newly developed self-confidence of state control. In the same vein, Nahdlatul Ulama, a party of traditional Muslims, used its frequent control over the ministry of religion to shift money to its boarding schools and government jobs to its members (Fealy 1998).

The democratic era (1950–57) also saw the first systematic attempts at anti-corruption campaigns – and their immediate politicization. Corruption charges were primarily laid on political opponents – one foreign minister, for example, was temporarily arrested for corruption by a regional faction of the military in 1956 as part of an intense civil-military power struggle (Ricklefs 1981, p. 241). The image – and reality – of widespread corruption contributed to the breakdown of democracy by 1957 and the establishment of Sukarno's left-leaning, autocratic Guided Democracy regime (1959–65). Under Sukarno, corruption worsened further, infecting especially the judiciary – which had been one of the lesser corrupt bodies in the late colonial regime and the early 1950s. Now, according to Dan Lev (2000, p. 8), judges were "pressured to do the government's bidding, [suffered] from declining real income and loss of status [and thus] were easily enticed into corruption". The collapse of the formal economy in the mid-1960s – partly caused by isolationism and overregulation – accelerated this trend in other areas of the administration as well, with cash-strapped bureaucrats trying to stay afloat by extracting bribes from an already desperate population (Feith 1967).

The poor state of the economy under Sukarno, as well as the rampant corruption, convinced many Indonesians to tolerate – or even support – the military takeover of 1965. Initially presiding over a classic military dictatorship, Suharto established a much more personalized autocracy from the late 1970s onwards. This personalization of presidential rule was accompanied by the centralization of corruption in himself and his family. As McLeod (2011, p. 49) explained, Suharto sat at the apex of a franchise system of corruption, "the fundamental purpose of which was to use the coercive power of government privately to tax the general public and redistribute the revenue to a small elite". The key public institutions forced into this franchise regime were the legislature, the judiciary, the military and police, the bureaucracy and state-owned enterprises. The main function of these bodies was to "to implement policies that would generate rents on behalf of the beneficiaries of the franchise system" (McLeod 2011, p. 50). But according to McLeod (2011, p. 53), Suharto understood "that maximizing tax revenue (whether public or private tax) involves keeping the rate of taxation modest, since high tax rates have the effect of encouraging tax avoidance and evasion, or even driving the taxed activity out of existence". In other words, while corruption under Suharto was extensive and institutionalized, it was also predictable (and bearable) for its "clients".

It was no coincidence, then, that this franchise system broke down when the Asian Financial Crisis of 1997–98 made it unaffordable for the vast majority of Indonesians. The rich, the middle class and the poor all suffered a significant loss of purchasing power as a result of the crisis, and corruption – which previously had been seen as a bearable cost – now seemed like the cause of their economic distress. After unprecedented mass demonstrations, Suharto resigned in May 1998, paving the way for Indonesia's democratization – and a new form of corruption.

POST-1998: CORRUPTION AND THE FUNDING OF POLITICS

After Suharto's resignation, Indonesia completed a messy but eventually successful democratic transition between 1998 and 2004. During that transition, the previously autocratic and centralized polity was democratized and decentralized. Power was diffused across many different institutions and levels of government administration. Consequently, corruption was diffused and decentralized as well – but it was not significantly reduced, as many authors writing about the link between corruption and democracy would postulate (Moreno 2002). Rather than being diminished, corruption shifted from the presidency – where it was previously concentrated and coordinated – to include the other power centres of the new democratic Indonesia as well: parliament, political parties, the judiciary and local governments (Mietzner 2015; Sherlock 2010). In the process, the previously predictable extent and mechanism of corruption became erratic and multi-layered: those requiring a bureaucratic service (such as a business licence) now have to negotiate with a wide range of national and local actors, paying bribes to many of them without the guarantee of a satisfactory outcome. In short, post-Suharto corruption has adjusted to its new institutional framework of multi-party and decentralized democracy while still bearing many of the trademarks of its colonial and pre-1998 manifestations (Hadiz 2010).

What are the reasons for the persistence of corruption despite democratization – and despite upward movements in the CPI? One of the key answers to this question is the exponentially increased cost of politics as a result of the competitiveness of the new democratic system. Democratic Indonesia has one of the most competitive electoral systems in the world. In a five-year cycle, Indonesians vote for the presidency; national, provincial and district parliaments; the "senate"; governors; district heads or mayors; and village heads. Importantly, since 2009 Indonesia has been using an open party list, proportional representation system for its legislative elections – meaning that hundreds of thousands of party candidates compete against nominees from their own party for seats allocated to that party. One effect of this has been a dramatic increase in vote buying (Aspinall and Sukmajati 2016), the levels of which are now the third-highest in the world.[1] Combined with other cost factors created by or increased through democratization – among others, the need for consultants, pollsters, campaign staff, television ads and social media assistants – the escalating vote buying costs have made political candidacies exceedingly expensive. According to the Minister of Home Affairs, some candidates have spent Rp 43 billion (US$ 3.3 million) for their 2014 national legislative campaign, and Rp 75 billion (US$ 5.8 billion) to run for the headship of a medium-sized district (Triyono 2015).

In the absence of a functional party and campaign financing system, these costs need to be recouped by successful candidates once elected to public office. Given the miniscule state funding for parties and extremely low levels of official donations, many politicians raise funds by abusing their public office to shift state contracts to donors or produce policy regulations that favour them. Alternatively, they mark up large infrastructure projects and pocket the excess funds for themselves. In the largest such case to surface in the post-Suharto period, Rp 2.3 trillion (US$ 177 million) was lost to corruption in the early 2010s from the budget for the production of electronic identity cards. The case involved the later Speaker of the House, Setya Novanto (Topsfield 2017). Not only did the scandal highlight the large amounts of money being siphoned off, it also left Indonesia's home ministry with virtually unusable identity cards and delayed the long-developed plan for a digitized citizen registry by many years. While extraordinary in size, the case represented many others in which similar mechanisms of illicit fundraising were used and caused a comparable extent of damage.

Another reason for the continued prevalence of corruption in Indonesia is that despite significant reforms to the country's political and electoral institutions, two key actors remain largely unreformed: the judiciary and the bureaucracy. The judiciary, while no longer doing "the government's bidding", remains highly corrupt, with judicial clerks often acting as middlemen and negotiators between judges and case participants over the size of the payment that could swing a verdict (Butt and Lindsey 2011). Even the Constitutional Court, which for some time was thought to be immune towards these patterns, eventually got affected – its chief justice was sentenced to life in prison in 2014 for manipulating verdicts on local electoral disputes in exchange for bribes. The bureaucracy, on the other hand, has

---

[1] This latter finding is based on an upcoming doctoral dissertation by Burhanuddin Muhtadi at the Australian National University.

resisted pressure to professionalize it (McLeod 2014). The lucrative business of selling positions within the civil service continues, with its overall value estimated to be around US\$ 1 billion a year.[2] Indeed, politicians have actively participated in this trade – local government heads have sold off positions to recoup their expenditure in electoral campaigns.[3]

Furthermore, tolerance of – and indeed, active engagement in – corruption continues to be deeply embedded in society. While surveys consistently indicate a rhetorical rejection of corruption, more detailed questions reveal habitual tolerance towards it. When Indonesians were asked in a 2017 nationwide survey what they thought of using personal connections to "smoothen" a bureaucratic process, only 8 percent considered it a crime; 30 per cent thought it was normal, and 13 percent even endorsed it as "necessary" (ICW and Polling Center 2017, p. 12); 43 percent viewed it as "not ethical" – a much weaker rejection than viewing it as illegal. Unsurprisingly, then, many Indonesian voters have continued to vote for candidates already convicted of corruption, both at the legislative and local executive level.

Ultimately, the reason why democratization has not reduced corruption in Indonesia is linked to Klitgaard's proposition that "corruption occurs when the gains (monetary or political) exceed the expected costs, i.e., the penalties multiplied by the probability of being caught" (Warf 2016, p. 659, referring to Klitgaard 1998). With the exploding costs of politics creating new needs for politicians to raise funds; the enforcement of existing regulations by the judiciary and other agencies weak; and tolerance of corrupt practices in society still strong, the cost-benefit logic of corruption in Indonesia remains tilted in favour of the latter. Against this background, the next section explains why even the KPK has been unable to significantly alter this equation.

THE KPK: PEBBLE IN THE ELITE'S SHOE

On the surface, the record of the KPK in arresting hundreds of state and party officials since 2004 is impressive. Its "victims" have included ministers, national parliamentarians, party chairmen and a chief justice; thus, it cannot be accused of having focused on the small rather than the big fish, which is a criticism often launched against anti-corruption commissions

---

[2]   Interview with Kuntoro Mangkusubroto, the then head of the Presidential Working Unit for Development Monitoring and Control (UKP4), Jakarta, 5 December 2013.

[3]   Interview with Taufiq Effendi, former minister of civil service empowerment and bureaucratic reform, Depok, 30 November 2013.

around the world (Human Rights Watch 2013). Rather, its inability to meaningfully increase the cost of corruption is related to other factors, which are laid out below.

First, while the KPK law of 2002 gave the agency wide-ranging powers (including wiretapping, investigation and prosecution), executive and legislative elites subsequently ensured that its technical capacity remained limited. Since the KPK's inception, the budget annually granted to it has only been sufficient to cover one Jakarta-based office – despite the agency's frequent requests to open branches in the regions. Similarly, in 2016 the KPK had only 96 investigators, who worked on 140 cases. For a country of 260 million people with a long tradition of extensive corruption, this small number of investigators and cases means that the likelihood of corruption being uncovered and prosecuted continues to be tiny and thus tolerable. Furthermore, at the insistence of other law enforcement agencies and affiliated political elites, half of the KPK investigators in 2016 were police officials. Those investigators have the reputation of being much less aggressive in the pursuit of corruption cases than their non-police colleagues. Indeed, one senior KPK investigator was reported to have tried to prevent the indictment of the House speaker, Setya Novanto, over the electronic identity card case (Sitompul 2017).

Second, while a handful of cases processed by the KPK led to high-profile arrests and long prison sentences, the average cases produced much less dramatic outcomes. In 2016, the average sentence handed down for corruption cases in Indonesia was two years and two months (this includes cases handled by the attorney general's office as well) (Sohuturon 2017). Given that Indonesian prisoners often get generous sentence reductions on national and religious holidays, the actual time spent in prison is on average only about half of the given sentence. Put differently, the average corruptor in Indonesia – if caught – spends only a little over a year behind bars. When in detention, convicts can exploit the corruption in the prison system to purchase privileges for themselves: better food, private space, air conditioners, unlimited visiting times and so forth. Indeed, one supposedly incarcerated convict was caught on camera in 2010 while watching an international tennis competition in Bali. Thus, the costs of committing corrupt acts – compared to the huge benefits – remain acceptable.

Third, very few of the legal proceedings against corruptors have led to the confiscation of their ill-gotten wealth. In most cases, corruptors have been sentenced to paying relatively small fines and repaying nominal damages to the state. Accordingly, after having spent a short time in prison, corruptors often can still enjoy the spoils of their transgressions – stashed away in the accounts of relatives and friends, invested in property or hidden in bank deposit boxes. Only in a small number of cases – such as in the corruption trials against the chairman of an Islamic party in 2013 and a district head from Madura in 2015 – have judges endorsed money laundering charges and ordered the confiscation of much of their private property. Once again, this pattern of rare confiscations has reduced the risks and costs of corruption.

Fourth, elite intervention has repeatedly weakened the KPK from within. To begin, KPK commissioners are elected through a long process initiated by the president, but the final selection is made by parliament – whose members are among the main targets of KPK investigations. Naturally, then, many legislators prefer commissioners who "go soft" on corruptors. This selection bias is typically cloaked in euphemistic language, with members of parliament often supporting nominees who agree with them that the KPK should focus on "corruption prevention measures" rather than on prosecution. Similarly, the police have frequently interfered with the KPK's work, especially when one of their senior leaders is investigated for corruption by the agency. In 2009, the police fabricated cases against two KPK commissioners, temporarily removing them from office (Butt 2011). The same approach was used again in 2015, but this time two commissioners were replaced permanently. One of the replacements installed was a former police general. He subsequently became chairman of the KPK and appointed other police officers to key positions in the organization. As indicated above, these officers were often seen hindering rather than advancing the KPK's anti-corruption campaign.

The budgetary, institutional and judicial constraints imposed on the KPK's ability to net more corruptors and deliver more substantive sanctions have served to keep the costs of corruption in Indonesia low. Applying Klitgaard's paradigm, the gains of corruption continue to exceed the expected costs. To be sure, the existence of the KPK has been a nuisance to the political elite, which had agreed to its establishment largely because of popular and foreign donor pressure after the old regime's breakdown in 1998 (Schütte 2012, p. 42). It is also true that some elite politicians had to pay a high price for their corrupt practices. But on average, the probability of being caught is still small, and possible sanctions manageable – indeed, the re-election of many former corruption convicts shows that potential capture does not end one's political career. Thus, the KPK has been more of a pebble in the elite's shoes rather than a game changer. The elite has stayed on the same patronage-defined pathways that were laid in colonial times as well as under Indonesia's pre-1998 political regimes. With the KPK pebble in their shoes, the elite is forced to walk with more care, but it keeps walking nevertheless.

CORRUPTION IN INDONESIA: SPATIAL AND GEOGRAPHICAL DIMENSIONS

The patterns outlined above – that is, the continued prevalence of corruption despite democratization – are also caused by spatial and geographical factors. To begin with, Indonesia likes to compare itself to its regional neighbours in Southeast Asia, including in the field of corruption. In addition to the pressures mentioned earlier, Indonesian politicians also agreed to establish the KPK because of the perception – confirmed by the CPI – that Indonesia was at the bottom of the regional and global corruption scale. But by 2016, Indonesia had taken a middle position on that CPI scale. It ranked behind Singapore, Malaysia and Brunei, but was placed ahead of Thailand, the Philippines, Vietnam, Myanmar and Cambodia. This improvement in the CPI scale has been a double-edged sword: while Indonesia's previous ranking below regional competitors had been a significant motivation to initiate anti-corruption measures (most importantly, the creation of the KPK in 2003), its subsequent middle ranking reduced the pressure to pursue more substantial reforms. Thus, while there may have initially been a positive contagion effect from Indonesia's comparison with its less corrupt neighbours, subsequently this comparison produced complacency and stagnation. This constellation further complicates the already difficult question of whether a rise in CPI rankings indicates reduced corruption or simply increased anti-corruption measures (more on this in the conclusion below).

However, the most important spatial and geographical dimensions of Indonesian corruption are not regional, but domestic in nature. As indicated above, the KPK's exclusive location in the capital Jakarta has predetermined its limited scope and effectiveness. While the Jakarta-based KPK theoretically has authority for local cases as well, in practice its activity radius is restricted to a handful of the major urban centres. Consequently, most local cases are handled by the police and the attorney general's office; of course, it was the structural embeddedness of these two bodies in long-standing cultures of corruption that had led to the establishment of the KPK in the first place. Even today, they are much less aggressive about – and much less interested in – corruption investigations than the KPK. This pattern has produced Indonesia's specific geography of corruption. In the internal logic of this geography, anti-corruption measures are strongest in Jakarta, where much of the national budget flows and private enterprise expenditure converge. This makes Jakarta the space where corruption is most profitable and, at the same time, most risky for its perpetrators (as explained above, even this "highest" risk must be weighed against the limited number of cases handled by the KPK overall). In the regions, the situation is reversed: the sums of money to be corrupted are smaller, but the risk of being caught is lower too.

The lower risk for corruptors in provinces and districts is significant because the circulation of corruption-prone funds at the local level is still immense. Although dispersed across 34 provinces and about 520 districts and cities, roughly 40 per cent of the Indonesian state budget is spent through local government institutions (Nasution 2016, p. 8). Accordingly, the gain to be had from corruption is high, while the potential costs are even more bearable than at the national level. While the Jakarta-based KPK indictments maintain a 100 per cent conviction rate in the anti-corruption courts, the rate of acquittal in the regions is around 30 per cent – and increasing (Lestari 2016). In 2013, local courts acquitted 16 corruption defendants; in 2014, that number increased to 28, and it shot up to 68 in 2015. It declined slightly to 56 in 2016, but was still much higher than in the pre-2015 years. Moreover, as mentioned above, those who received convictions mostly got off lightly.

Apart from local cases being handled by the police and the attorney general's office, what explains this lower risk for corruptors at the local level? One important factor is the weaker media scrutiny. While judges in Jakarta-based KPK cases are under great pressure by national television stations and print media, local cases receive much less attention. The more remote an area, the less likely it is to have its own newspaper or other media outlets. If they do exist, they are often owned by businesspeople closely linked to political elites, that is, the main actors of corruption (Lim 2012). The same dynamic is in place in regards to critical civil society groups. The further away a district is from urban centres, the fewer non-governmental organizations (NGOs) are active that could monitor and advocate against corruption. The lack of both media scrutiny and NGO activism in rural areas combine to give these territories low levels of transparency and accountability in government affairs – something often identified as a main cause of corruption (Brunetti and Weder 2003).

One specific aspect of the geography of corruption in Indonesia is the high intensity of corruption in peripheral areas with a history of separatism. In Aceh, for example, the government reached a peace agreement with former rebels in 2005, allowing them to take power in the province and many districts too. Since then, many ex-rebel leaders have used their public office to enrich themselves and distribute patronage to their former troops (Aspinall 2009). In that situation, Jakarta has been reluctant to lay corruption charges on former separatists in public office, fearing that doing so might make them abandon the peace accord. Similarly, a special 2001 autonomy deal for Papua led to native Papuans taking over all key political leadership positions at the provincial, city and district levels (although this deal was rejected by armed rebels and other radical groups). As in Aceh, the central government has been hesitant to move against highly corrupt Papuan officials, anxious that that might encourage them to switch to the pro-independence camp. While occasionally arresting the most blatant of the corruptors, the KPK and local law enforcement agencies have left the majority of the 42 district heads and mayors in Papua untouched. Thus, the political sensitivities of dealing with old and would-be rebels have added another element to the geography of corruption in Indonesia, providing those actors with further protection from the already weak anti-corruption campaign at the local level.

CONCLUSION: DEMOCRACY AND CORRUPTION IN INDONESIA

This chapter has argued that (a) democratization has not brought a significant decline in corruption in Indonesia, and that (b) this was due to a combination of factors, most notably the high cost of political competition. The first part of the argument initially seems counterintuitive. The CPI has recorded a gradual improvement for Indonesia, and the political opening, increased media freedom and intensified NGO activity would normally be seen as causing a decline in corruption. But the CPI and many other writings on corruption in Indonesia have mostly focused on the vastly expanded anti-corruption campaign in Indonesia – and indeed, there is no doubt that this campaign is incomparably more serious than at any other point in Indonesian history. Yet this alone does not constitute evidence for an actual decline in corruption itself. While corruption is inherently difficult to measure – hence the CPI's resort to perceptions – there are very few indications that corruption in political institutions and the judiciary is lower than under the Suharto regime. It has certainly changed its patterns, but the continued bluntness of politicians, state officials and judges in abusing their office for fundraising purposes does not speak to a fundamental alteration in the country's centuries-old patronage structures. With Klitgaard, I argue that the key reason for this phenomenon is that the potential gains of corruption still exceed the expected costs.

The circumstance that the possible gains of corruption remain high is a consequence of the Indonesian state's continued dominant role in managing and regulating the economy. In fact, under President Joko Widodo (since 2014), the role of state-owned enterprises in the national economy has expanded substantially, with new holding companies planned to absorb the contractually enforced divestment of foreign resource companies. As a result, holding public office – whether in the executive, legislature or judiciary – remains an immensely lucrative proposition. At the same time, while the costs of gaining such public positions have increased – both in terms of the investment needed to win it and the risk of punishment if found guilty of misusing it – they still don't exceed the benefits. Hence, unless the Indonesian state moves to (a) establish a credible party and campaign financing mechanism that could reduce the personal financial burden involving candidacies for public office; (b) significantly increase the risk of corruption by turning the KPK from a small-scale, Jakarta-based agency into a broad-based, nationwide operation; (c) decrease the role of the state in directing and regulating money flows; and (d) fight public perceptions in society that graft is acceptable as long as some benefits flow to everyone, the level of corruption in Indonesia is set to stay high, regardless of the ebbs and flows of its democratization process.

The Indonesian experience teaches us that there is no causal link, let alone an automatic one, between democratization and the prospect of reducing corruption. In a sense, this very expectation – seemingly confirmed by rising CPI scores – has even served to conceal the continued embeddedness of the problem. Policymakers crafting anti-corruption approaches in new democracies need to understand that democratic opening per se, more transparency and an anti-corruption agency – even one as aggressive as the KPK – are no guarantees for reduced corruption. Rather, successful corruption reduction strategies require work on the details: a functioning political funding regime; an anti-corruption body that not only produces high-profile arrests of a few but delivers a systematic deterrence against corruption at all levels; and an economic regulation network that reduces manipulation opportunities for public office holders. It is these areas that reformers need to turn to if they want democracy to lower the levels of corruption; the establishment of formal democracy in itself is not going to do the job for them.

REFERENCES

Angelino, K. 1931. *Colonial Policy, Volume II: The Dutch East Indies*. The Hague: Martinus Nijhoff.
Aspinall, E. 2009. Combatants to contractors: the political economy of peace in Aceh. *Indonesia* **87**, 1–34.
Aspinall, E. and M. Sukmajati (eds) 2016. *Electoral Dynamics in Indonesia: Money Politics, Patronage and Clientelism at the Grassroots*. Singapore: NUS Press.
Bolongaita, E. 2010. An exception to the rule? Why Indonesia's Anti-Corruption Commission succeeds where others don't – a comparison with the Philippines' Ombudsman. Chr. Michelsen Institute, Bergen.

Brunetti, A. and B. Weder. 2003. A free press is bad news for corruption. *Journal of Public Economics* **87**(7–8), 1801–24.

Butt, S. 2011. Anti-corruption reform in Indonesia: an obituary? *Bulletin of Indonesian Economic Studies* **47**(3), 381–94.

Butt, S. and T. Lindsey. 2011. Judicial mafia: the courts and state illegality in Indonesia. In E. Aspinall and G. van Klinken (eds), *The State and Illegality in Indonesia*. Leiden: KITLV, pp. 189–216.

Choi, J.-W. 2011. Measuring the performance of an anticorruption agency: the case of the KPK in Indonesia. *International Review of Public Administration* **16**(3), 45–63.

Cribb, R. 2000. *Historical Atlas of Indonesia*. London: Curzon.

Fealy, G. 1998. Ulama and politics in Indonesia: a history of Nahdlatul Ulama, 1952–1967. PhD Dissertation, Monash University.

Feith, H. 1967. The dynamics of Guided Democracy. In R. McVey (ed.), *Indonesia*. New Haven, CT: HRAF.

Hadiz, V. 2010. *Localising Power in Post-Authoritarian Indonesia: A Southeast Asia Perspective*. Stanford, CA: Stanford University Press.

Human Rights Watch. 2013. *"Letting the Big Fish Swim": Failures to Prosecute High-level Corruption in Uganda*. New York: Human Rights Watch.

ICW (Indonesian Corruption Watch) and Polling Center. 2017. Survei nasional anti-korupsi, April–Juni 2017. ICW and Polling Center, Jakarta (in Indonesian).

Klitgaard, R. 1998. *Controlling Corruption*. Berkeley, CA: University of California Press.

Lestari, S. 2016. ICW: Vonis bebas terdakwa kasus korupsi meningkat. *CNN Indonesia* (7 February) (in Indonesian).

Lev, D. 2000. *Legal Evolution and Political Authority in Indonesia: Selected Essays*. The Hague: Kluwer Law International.

Lim, M. 2012. *The League of Thirteen: Media Concentration in Indonesia*. Phoenix, AZ: Arizona State University and Ford Foundation.

McLeod, R.H. 2011. Institutionalized public sector corruption: a legacy of the Suharto franchise. In E. Aspinall and G. van Klinken (eds), *The State and Illegality in Indonesia*, pp. 45–64. Leiden: KITLV.

McLeod, R. 2014. A better bureaucracy? *New Mandala* (19 June).

Mietzner, M. 2015. Dysfunction by design: political finance and corruption in Indonesia. *Critical Asian Studies* **47**(4), 587–610.

Mietzner, M. and L. Misol. 2012. Military businesses in post-Suharto Indonesia: decline, reform and persistence. In J. Rueland, M.G. Manea, and H. Born (eds), *The Politics of Military Reform: Experiences from Indonesia and Nigeria*, pp. 101–22. Heidelberg: Springer.

Moreno, A. 2002. Corruption and democracy: a cultural assessment. *Comparative Sociology* **1**(3/4), 495–507.

Nasution, A. 2016. Government decentralization program in Indonesia. ADBI Working Paper 601. Asian Development Bank Institute, Tokyo.

Olken, B. 2007. Monitoring corruption: evidence from a field experiment in Indonesia. *Journal of Political Economy* **115**(2), 200–49.

Ricklefs, M. 1981. *A History of Modern Indonesia*. London and Basingstoke: Macmillan.

Rocamora, J. 1975. *Nationalism in Search of Ideology: The Indonesian Nationalist Party, 1946–1965*. Quezon City: University of the Philippines.

Schütte, S. 2012. Against the odds: anti-corruption reform in Indonesia. *Public Administration and Development* **32**(1), 38–48.

Sherlock, S. 2010. The parliament in Indonesia's decade of democracy: people's forum or chamber of cronies? In E. Aspinall and M. Mietzner (eds), *Problems of Democratisation in Indonesia: Elections, Institutions and Society*, pp. 160–80. Singapore: ISEAS.

Sitompul, M. 2017. Direktur penyidik KPK halangi penetapan tersangka Novanto. *Jurnas* (30 August) (in Indonesian).

Sohuturon, M. 2017. ICW: Koruptor rata-rata divonis 26 bulan penjara. *CNN Indonesia* (4 March) (in Indonesian).

Topsfield, J. 2017. Scandal-prone Indonesian speaker Setya Novanto named as embezzlement suspect. *Sydney Morning Herald* (18 July).
Triyono, A. 2015. Butuh Rp 75 miliar jadi bupati. *Kontan* (22 October) (in Indonesian).
Warf, B. 2016. Global geographies of corruption. *Geojournal* **81**(5), 657–69.

Index